Bacterial Genomes and Infectious Diseases

Bacterial Genomes and Infectious Diseases

Edited by

Voon L. Chan, PhD

Department of Medical Genetics and Microbiology
University of Toronto, Toronto, Ontario, Canada

Philip M. Sherman, MD, FRCPC

Department of Paediatrics
Department of Laboratory Medicine and Pathobiology
Hospital for Sick Children
University of Toronto, Toronto, Ontario, Canada

Billy Bourke, MD, FRCPI

Children's Research Center, Our Lady's Hospital for Sick Children,
School of Medicine and Medical Science,
Conway Institute for Biomolecular and Biomedical Research,
University College, Dublin, Ireland

HUMANA PRESS ✳ TOTOWA, NEW JERSEY

This publication is printed on acid-free paper. ∞
ANSI Z39.48-1984 (American Standards Institute)
Permanence of Paper for Printed Library Materials.

Production Editor: Jennifer Hackworth

Cover design by Patricia F. Cleary

For additional copies, pricing for bulk purchases, and/or information about other Humana titles, contact Humana at the above address or at any of the following numbers: Tel.: 973-256-1699; Fax: 973-256-8341; E-mail: orders@humanapr.com; or visit our Website: www.humanapress.com

Printed in the United States of America. 10 9 8 7 6 5 4 3 2 1
1-59745-152-5 (e-book)

Library of Congress Cataloging in Publication Data

Bacterial genomes and infectious diseases / edited by Voon L. Chan,
Philip M. Sherman, Billy Bourke.
     p. cm.
  Includes bibliographical references and index.
  ISBN 1-58829-496-X (alk. paper)
  1. Bacterial genomes. 2. Communicable diseases--Pathogenesis. I.
Chan, Voon L. II. Sherman, Philip M. III. Bourke, Billy.
    QH434.B334 2006
    616.9'201--dc22
                                                                    2005034347

Preface

The first bacterial genome, *Haemophilus influenzae*, was completely sequenced, annotated, and published in 1995. Today, more than 200 prokaryotic (archaeal and bacterial) genomes have been completed and over 500 prokaryotic genomes are in various stages of completion. Seventeen eukaryotic genomes plus four eukaryotic chromosomes have been completed. The concept of achieving better understanding of an organism through knowledge of the complete genomic sequence was first demonstrated in 1978 when the first bacteriophage genome, ΦX174, was sequenced. Complete genomic sequences of prokaryotes have led to a better understanding of the biology and evolution of the microbes, and, for pathogens, facilitated identification of new vaccine candidates, putative virulence genes, targets for antibiotics, new strategy for rapid diagnosis, and investigation of bacteria–host interactions and disease mechanisms.

Recent increased interest in microbial pathogens and infectious diseases is largely attributed to the re-emergence of infectious diseases like tuberculosis, emergence of new infectious diseases like AIDS and severe acute respiratory syndrome, the problem of an increasing rate of emergence of antibiotic-resistant variants of pathogens, and the fear of bioterrorism. Microbes are highly diverse and abundant in the biosphere. Less than 1% of these morphologically identified microbes can be cultured in vitro using standard techniques and conditions. With such abundance of microbes in nature, we can expect to see new variants and new species evolve and a small number will emerge as pathogens to humans.

In the first section of *Bacterial Genomes and Infectious Diseases*, some major general findings about bacterial genomes and their impact on strategy and approach for investigating mechanisms of pathogenesis of infectious diseases are discussed. Later chapters focus on the value and power of genomics, proteomics, glycomics, and bioinformatics as applied to selected specific bacterial pathogens.

Bacterial Genomes and Infectious Diseases is designed to provide valuable reading for senior microbiology, pathobiology and genetics undergraduate and graduate students, medical students, clinician scientists, infectious diseases clinicians, and medical microbiologists.

Voon L. Chan, PhD
Philip M. Sherman, MD, FRCPC
Billy Bourke, MD, FRCPI

Contents

Contributors

BEN ADLER • *Australian Bacterial Pathogenesis Program and Australian Research Council Centre of Excellence in Structural and Functional Microbial Genomics, Department of Microbiology, Monash University Clayton, Victoria, Australia*

DAVID C. ALEXANDER • *Montreal General Hospital Research Institute, McGill University Health Center, Montreal, Quebec, Canada*

JAMES B. BLISKA • *Department of Molecular Genetics and Microbiology, Center for Infectious Diseases, State University of New York at Stony Brook, Stony Brook, NY*

E. FIDELMA BOYD • *Department of Microbiology, University College Cork, National University of Ireland, Cork, Ireland*

BILLY BOURKE •*Children's Research Center, Our Lady's Hospital for Sick Children, School of Medicine and Medical Science, Conway Institute for Biomolecular and Biomedical Research, University College Dublin, Ireland*

JEAN-ROBERT BRISSON • *Institute for Biological Sciences, National Research Council, Ottawa, Ontario, Canada*

DIETER BULACH • *Australian Bacterial Pathogenesis Program, Department of Microbiology, Monash University Clayton, Victoria, Australia*

VOON L. (RICKY) CHAN • *Department of Medical Genetics and Microbiology, University of Toronto, Toronto, Ontario, Canada*

ZHONGMING GE • *Division of Comparative Medicine, Massachusetts Institute of Technology, Cambridge, MA*

JENS P. GRABENSTEIN • *School of Medicine, New York University, New York, NY*

HUEY-LAN HUANG • *The Campbell Family Institute for Breast Cancer Research, Ontario Cancer Institute, University Health Network and Department of Medical Biophysics, University of Toronto, Ontario, Canada*

KEITH IRETON • *Department of Molecular Biology and Microbiology, University of Central Florida, Orlando, FL*

HAROLD C. JARRELL • *Institute for Biological Sciences, National Research Council, Ottawa, Ontario, Canada*

JOHN KELLY • *Institute for Biological Sciences, National Research Council, Ottawa, Ontario, Canada*

JUN LIU • *Department of Medical Genetics and Microbiology, University of Toronto, Toronto, Ontario, Canada*

MARTIN J. MCGAVIN • *Division of Microbiology, Sunnybrook and Women's College Health Science Centre, Department of Laboratory Medicine and Pathobiology, University of Toronto, Toronto, Ontario, Canada*

YVONNE A. O'SHEA • *Department of Microbiology, University College Cork, National University of Ireland, Cork, Ireland*

ANNE MARIE QUIRKE • *Department of Microbiology, University College Cork, National University of Ireland, Cork, Ireland*

MICHELLE B. RYNDAK • *TB Center, Public Health Research Institute, Newark, NJ*

YUKO SASAKI • *Department of Bacterial Pathogenesis and Infection Control, National Institute of Infectious Diseases, Musashimurayama, Tokyo, Japan*

DAVID B. SCHAUER • *Division of Comparative Medicine and Biological Engineering Division, Massachusetts Institute of Technology, Cambridge, MA*

TORSTEN SEEMANN • *Victorian Bioinformatics Consortium, Clayton School of Information Technology, Monash University, Clayton, Victoria, Australia*

PHILIP M. SHERMAN • *Department of Paediatrics, Department of Laboratory Medicine and Pathobiology, Hospital for Sick Children, University of Toronto, Toronto, Canada*

CHRISTINE M. SZYMANSKI • *Institute for Biological Sciences, National Research Council, Ottawa, Ontario, Canada*

WEN-CHEN YEH • *The Cambell Family Institiute for Breast Cancer Research, Ontario Cancer Institute, University Health Network and Department of Medical Biophysics, University of Toronto, Toronto, Canada*

N. MARTIN YOUNG • *Institute for Biological Sciences, National Research Council, Ottawa, Ontario, Canada*

JERMYN S. WILLIAM • *Department of Microbiology, University College Cork, National University of Ireland, Cork, Ireland*

RICHARD L. ZUERNER • *Bacterial Diseases of Livestock Research Unit, National Animal Disease Center, USDA, Agricultural Research Service, Ames, IA*

Introduction

Billy Bourke

1. Genomic Data as a Cornerstone for Biomedical Science

A fundamental starting point for effective research in any scientific field is to first submit the constituents of that particular domain of science to a process of categorization, taxonomy, and systematics. Linnaeus, a compulsive cataloger, spent a lifetime classifying living species and Tycho Brahe laboriously mapped the position of the planets and stars of the heavens for 20 yr of his life (*1*). These enormous tasks embodied years of painstaking and repetitive observation and calculation. Although predicated on deeply unglamorous, day-to-day work, the resulting body of data formed the basis of all further scientific inquiry and discovery in the fields of botany and astronomy, thereafter.

The late 20th century will be remembered as a time of major discovery in many scientific fields. Undoubtedly, for the biological sciences, the major achievement of recent years has been the advent of genomics. Technology-driven, laborious, and, some might say, scientifically unexciting, genomic sequencing is reminiscent of the work of Linnaeus and Brahe. However, in also laying down a fundamental scientific cornerstone in biological sciences, the evolution of genomic sequencing represents an achievement equivalent to those established centuries ago by these great scientists.

2. The Birth of Bacterial Genome Sequencing

In the nonscientific community, the proposal to sequence complete genomes is usually linked with the "flag ship" Human Genome Project (*2*). However, the genomes of other organisms, in particular those of bacteria, in many respects form the vanguard of genome sequencing science. The first ever genome sequenced was that of the bacteriophage ΦX174, the 5500 base chromosome that was decoded by Sanger and his colleagues in 1978 (*3*). Around this time, interest arose in attempting to sequence the whole genome of *Escherichia coli* K-12, culminating in a more formal proposal from Fred Blattner in 1983 (*4*). However, efforts to complete the *E. coli* sequence became embroiled in controversy (*5*). Based on early predictions, Blattner and colleagues set a goal of sequencing 1 Mb per year. However, after nearly 4 yr of sequencing, only 1.4 Mb was completed (*6*). Projections for the *E. coli* sequencing project were undoubtedly overly optimistic and the whole process proved significantly more time consuming than expected. Blattner first had to finish the process of breaking the whole *E. coli* genome into 400 overlapping λ clones, then break each clone into random subclones before sequencing each clone and ordering them according to overlapping sequences (*6*).

Although the *E. coli* genome sequence project started with a healthy advantage compared with other bacterial sequencing efforts, in the early 1990s a dark horse appeared on the genomic horizon. By using rapid, fluorescence-based sequencing technology, a random ("shotgun") sequencing strategy, advanced computerization of data collection and processing, novel software for generating contigs, and efficient long polymerase chain reaction cloning to reduce the time needed to fill gaps in the sequence, The Institute of Genome Research (TIGR) successfully sequenced the whole *Haemophilus influenza* genome in about 1 yr *(7)*. The genome of *Mycoplasma genitalium*, the smallest genome known for a self-replicating organism (580 kb) (also from the TIGR group) followed soon after in the same year *(8)*. Using the novel dye-terminator fluorescence sequencing technology, Blattner and his colleagues went on to complete and publish the entire *E. coli* K12 genome sequence 2 yr later *(9)*.

The impact of this new technology and the interest in genome sequencing as a fundamental tool for understanding basic biological process, especially infectious diseases and their treatment, has led to an explosion in the numbers of organisms being sequenced. At the time of this writing, 179 prokaryotic genomes have been sequenced and approx 500 bacterial genomes are in the process of being sequenced. The rapid evolution of sequencing technology has been truly staggering and the accompanying costs have dropped substantially. The *M. genitalium* sequence was calculated to cost approx $0.30 per basepair. However, by 2002, random sequencing of genomes covering more than 99% of the whole sequence could be completed within a few days, at a cost of only $0.04 per basepair. Completion of a shotgun sequenced bacterial genome by closure of gaps and annotation can usually be accomplished within a few months and the cost of a completely annotated genome is less than $0.10 per basepair *(10)*. Therefore, with a cost of less than $100,000 for a small- to medium-sized organism, even the genomes of bacteria of lesser commercial interest can be sequenced. Such accessibility to whole genome sequencing, even for research groups without major financial resourcing, has led to a "democratization" of scientific exploration in microbiology worldwide *(11)*.

3. Pathogens and the Postgenomics Era

Although the generation and analysis of extensive volumes of sequence data is a major accomplishment, it is not an end in itself. Clearly, the motivation for sequencing genomes comes for a desire to understand the biology of living organisms. A major stimulus for research across all fields of biology is to understand and combat human diseases. It is not surprising then that many of the first organisms to be sequenced were important human pathogens. Indeed, given the small size and the paucity of intragenic DNA, the potential for the genome sequences of bacterial pathogens to yield biologically useful information of direct relevance to human disease outstrips that of the Human Genome Project, at least in the short to medium term.

The deluge of data generated by genome sequencing projects has forced a quantum leap in the application of computer science and bioinformatics to help analyze efficiently the information generated. This evolution of "biology *in silico*" is not the only interdisciplinary scientific alliance forged in the post genomic era. Microarray technology, proteomics, immunoinformatics, structural biology, and combinatorial chemistry, all of which are predicated on a knowledge of genome sequence, have opened the door to high-throughput technologies, increasing by orders of magnitude the efficiency with which novel virulence genes, potential drug targets, and vaccine strategies can be identified *(12–14)*.

Bacterial Genomes and Infectious Diseases focuses on how bacterial genomics has contributed to some of the major strides taken in understanding the basic biology of a variety of important human pathogens. Knowledge of the genetic content of individual pathogens has pointed toward novel virulence factors, provided unprecedented insights into pathogen evolution, uncovered key epidemiological relationships between different strains of the same organism, and helped forge important new links with other scientific disciplines for the exploration of infectious pathogenesis.

Bacterial genomics is at the cutting edge of a movement evolving in modern scientific research toward the integration of scientific subspecialities. The following chapters detail some of the initial fruits of this reductionist approach to understanding bacterial pathogens. Although much further information about individual organisms will come from analysis of existing data, the future challenge for microbiology and infectious disease in the postgenomic era lies in the integration of present knowledge with other areas of scientific research with a view to developing a better understanding of the biology of living organisms and the manner in which pathogens inflict disease.

References

1. Gribbin, J. (2002) Science: A History. Penguin, London, UK.
2. Venter, J. C., Adams, M. D., Myers, E. W., et al. (2001) The sequence of the human genome. *Science* **291**, 1304–1351.
3. Sanger, F., Air, G. M., Barrell, B. G., et al. (1977) Nucliotide sequence of bacteriophage phi X174 DNA. *Nature* **265**, 687–695.
4. Blattner, F. R. (1983) Biological frontiers. *Science* **222**, 719–720.
5. Danchin, A. (1995) Why sequence genomes? The Escherichia coli imbroglio. *Mol. Microbiol.* **18**, 371–376.
6. Nowak, R. (1995) Getting the bugs worked out. *Science* **267**, 172–174.
7. Fleischmann, R. D., Adams, M. D., White, O., et al. (1995) Whole-genome random sequencing and assembly of Haemophilus influenzae Rd. *Science* **269**, 496–512.
8. Fraser, C. M., Gocayne, J. D., White, O., et al. (1995) The minimal gene complement of Mycoplasma genitalium. *Science* **270**, 397–403.
9. Blattner, F. R., Plunkett, G., 3rd, Bloch, C. A., et al. (1997) The complete genome sequence of *Escherichia coli* K-12. *Science* **277**, 1453–1474.
10. Fraser, C. M., Eisen, J. A., Nelson, K. E., Paulsen, I. T., and Salzberg, S. L. (2002). The value of complete microbial genome sequencing (you get what you pay for). *J. Bacteriol.* **184**, 6403–6405.
11. Anonymous (2002) A genome fest; 25 years of pathogen genome sequencing. *Wellcome News* 10–11.
12. Hughes, D. (2003) Exploiting genomics, genetics and chemistry to combat antibiotic resistance. *Nat. Rev. Genet.* **4**, 432–441.
13. Meinke, A., Henics, T., and Nagy, E. (2004) Bacterial genomes pave the way to novel vaccines. *Curr. Opin. Microbiol.* **7**, 314–320.
14. De Groot, A. S. and Rappuoli, R. (2004) Genome-derived vaccines. *Expert Rev. Vaccines* **3**, 59–76.

1

Microbial Genomes

Voon Loong Chan

Summary

With more than 200 bacterial and archaeal genomes completely sequenced, and more than 500 genomes at various stages of completion, we begin to appreciate the enormous diversity of prokaryotic genomes in terms of chromosomal structure, gene content and organization, and the abundance and fluidity of accessory and mobile genetic elements. The genome of a bacterial species is composed of conserved core genes and variable accessory genes. Mobile genetic elements, such as plasmids, transposons, insertion sequences, integrons, prophages, genomic islands, and pathogenicity islands, are part of the accessory genes, which can have a significant influence on the phenotype and biology of the organism. These mobile elements facilitate interspecies and intraspecies genetic exchange. They play an important role in the pathogenicity of bacteria, and are a major contributor to species diversity. Further genomic analysis will likely uncover more interesting genetic elements like small (noncoding) RNA genes that can play a significant role in gene regulation.

Key Words: Genome diversity; plasmids; insertion elements; genomic islands; prophages; small RNA.

1. Introduction

The first bacterial genome sequenced to completion was *Haemophilus influenzae*. It was published in 1995 *(1)*. To date (Feb 1, 2006) 297 prokaryotic genomes have been sequenced, 272 bacterial and 25 archaeal. In addition, there are more than 500 prokaryotic genomes that are at various stages of completion. Interest in prokaryotic genomes is still growing at an escalating rate (Fig. 1). The major impetus for determining the complete genome sequence of an organism is to gain a better understanding of the biology and evolution of the microbes; and for pathogens, to identify new vaccine candidates, putative virulence genes, and targets for therapeutics, including antibiotics. Emergence of new antibiotic-resistant isolates of pathogens spurs the search for new vaccines, antimicrobials, and novel approaches in the prevention of infection.

The complete genome sequence of a bacterium provides valuable information on the genome size and topology. Many sequence analysis programs and algorithms have been developed, and their capacity and power have improved with the growth of the data base *(2)*.

With the aim of identifying the function of all the genes in a particular genome, the nucleotide and predicted amino acid sequences are then compared with sequences in the public database using Basic Local Alignment Search Tool (BLAST). Algorithms have also been developed to align the whole genome of two closely related bacteria. A

From: *Bacterial Genomes and Infectious Diseases*
Edited by: V. L. Chan, P. M. Sherman, and B. Bourke © Humana Press Inc., Totowa, NJ

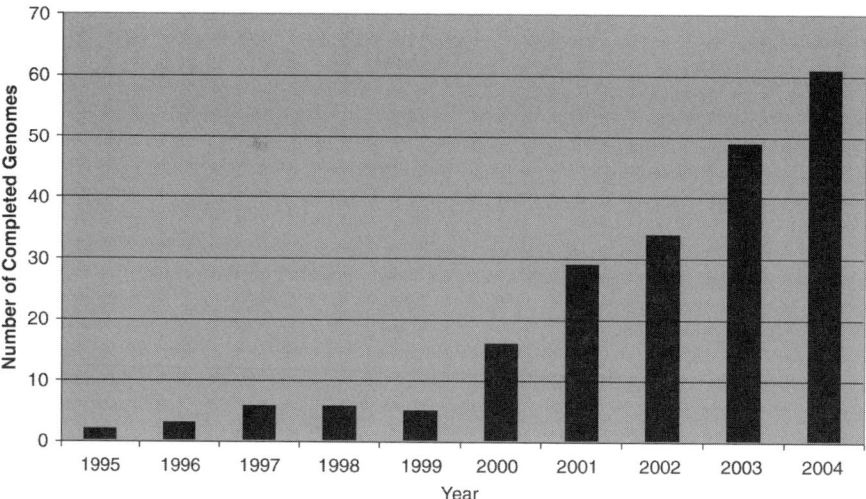

Fig. 1. Number of completed genomes per year from 1995 to 2004. The numbers for each year were derived from two web sites: http://www.ncbi.nlm.nih.gov/genomes/MICROBES/Complete.html and http://www.genomeonline.org.

high percentage of protein genes can be identified through similarity to known genes. However, for most of the genomes sequenced, 25% or more of identified open reading frames (ORFs) do not match any known genes. To achieve full understanding of the biology and pathogenic mechanisms of a bacterial pathogen, we need to have a good understanding of the function of all the gene products and how the genes are regulated.

2. Structure and Diversity of Microbial (Bacterial and Archaeal) Genomes

The structure of a microbial genome can be analyzed and compared with other genomes at various levels, including guanine and cytosine (G+C) content, overall size, topology, organization of genes, and the presence and abundance of accessory and mobile genetic elements. Limited information of the structure of bacterial genomes was available before the era of genomic sequencing. Most of the early structural information was derived from physical mapping using restriction enzyme cleavage and pulsed-field gel electrophoresis. This has been reviewed extensively (3–6). It is abundantly clear from physical mapping information and genomic sequence data that bacterial, as well as archaeal, genomes are highly diversified.

2.1. Chromosome Size

Chromosome sizes of sequenced bacteria range from 580 kb of *Mycoplasma genitalium (7)* to 9105 kb of *Bradyrhizobium japonicum (8)*. The distribution of the sizes of bacterial genomes is shown in Fig. 2. Ninety percent of the bacterial genomes sequenced are less than 5.5 Mb. There are clusters of genomes around 0.9–1.3, 1.7–2.5, 3.3–3.7, and 4.5–4.9 Mb. Variations in the size of bacterial genomes are expected because there is huge diversity among bacteria in morphology, metabolic capability, and the ability to survive and grow in various environments or hosts. For some bacterial species, multiple strains have been sequenced: *Bacillus anthracis, Bacillus cereus, Brucella melitensis, Chlamydophila pneumoniae, Escherichia coli, Leptospira interrogans, Myco-*

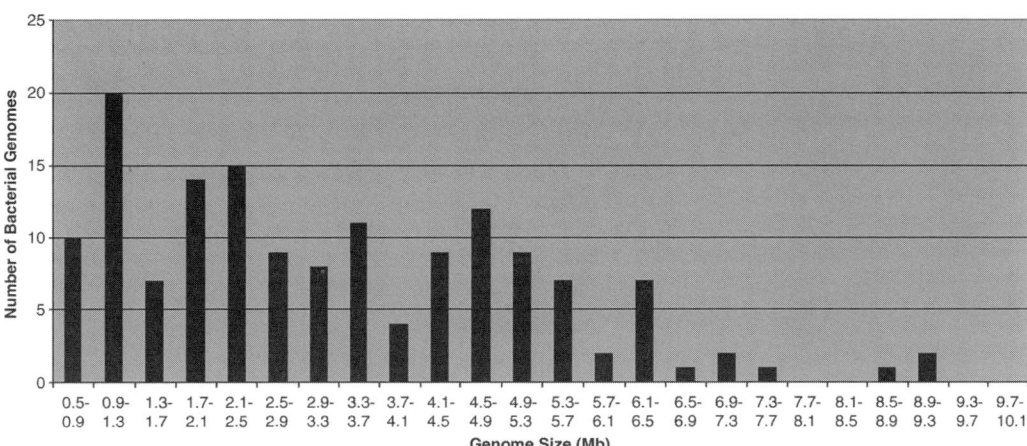

Fig. 2. Size distribution of bacterial genomes. The sizes of the bacterial genomes were from two web sites: http://www.ncbi.nlm.nih.gov/genomes/MICROBES/Complete.html and http://www.genomeonline.org.When multiple strains of the same species were sequenced the mean size of the different strains were used, provided the size variation was less than 10%. However, if the size variation exceeded 10%, they were entered as separate genomes.

bacterium tuberculosis, Neisseria meningitidis, Prochorococus marinus, Salmonella enterica Typhi, Shigella flexneri, Staphylococcus aureus, Streptococcus agalactiae, Streptococcus pyogenes, Tropheryma whipplei, Vibro vulnificus, Xylella fastidiosa, and *Yersinia pestis* (*see* http://www.genomeonline.org; http://www.ncbi.nlm.nih.gov/genomes/MICROBES/Complete.html). With the exception of *E. coli* and *P. marinus,* different strains of the same species have a very similar genome size. *E. coli K12*-MG1655 is 4.6 Mb, whereas the largest strain sequenced, *E. coli* O157:H7 (Sakai) is 5.6 Mb, 18% larger. Two ecotypes (strains of the same species occupying a different niche) of *P. marinus,* MED4 and MIT9313, were recently sequenced. The genome of both strains consists of a single circular chromosome. The chromosome of MED4, a high-light-adapted strain, is 1658 kb and that of MIT9313, a low-light-adapted strain, is 2411 kb. That is, the genome of ecotype MED4 is only 68.8% the size of MIT9313 and yet, on the basis of their ribosomal DNA sequences, they are classified as the same species.

There are fewer archaeal genomes sequenced than bacterial genomes. Even though chromosomes of archaeal genomes can also differ greatly in size, ranging from the smallest at 500 kb for *Nanoarchaeum equitans (9)* to 5751 kb for *Methanosarcina acetivorans (10),* the sizes of archaeal genomes are not as varied as those of bacteria. Fourteen of the 20 archaeal species sequenced to date have a genome size between 1.5 and 2.3 Mb (*see* Fig. 3).

2.2. Chromosome Topology and Number

Bacteria generally have a single, double-stranded circular DNA chromosome. However, a number of bacterial species have a linear chromosome *(5,11).* The first linear bacterial chromosome identified was that of *Borrelia burgdorferi (12).* A number of *Streptomyces* species, including *S. lividans* 66 *(13), S. coelicolor* A3 *(14),* and S. *avermitilis (15),* also have a linear chromosome. It is likely that the *Streptomyces* linear chromosome evolved from an ancestral circular chromosome *(11).*

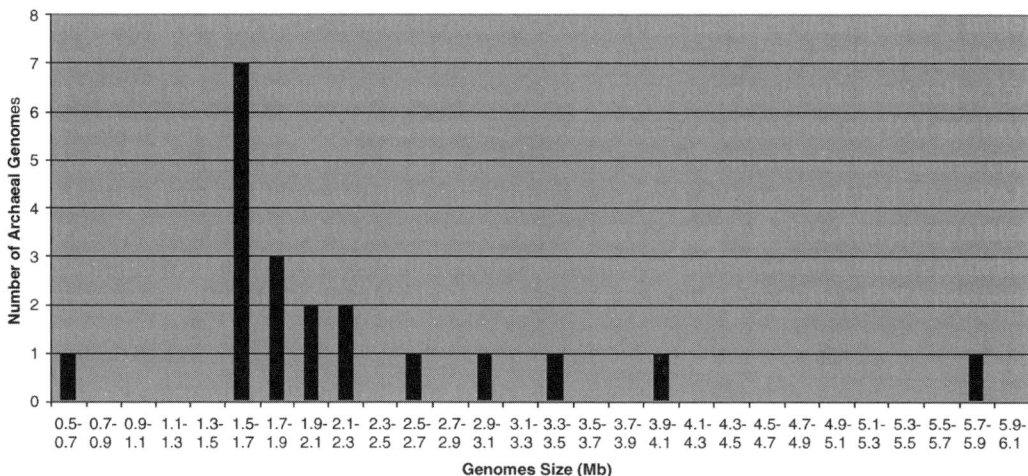

Fig. 3. Size distribution of archaeal genomes. The sizes of the archaeal genomes were from two web sites: http://www.ncbi.nlm.nih.gov/genomes/MICROBES/Complete.html and http://www.genomeonline.org.

Agrobacterium tumefaciens, an α *Proteobacteria,* has two chromosomes, one linear and one circular *(16,17)*. Other α *Proteobacteria, Brucella melitensis biovar suis* 1330 *(18)* and *B. melitensis* 16M *(18)*, also have two chromosomes. *B. melitensis,* which causes abortion in sheep and goats and undulant fever in humans, has three biovars (biotypes), whereas *Brucella suis*, which infects swine, has four biovars. *B. suis* biovar 1, biovar 2, and biovar 4 all have two circular chromosomes. The two chromosomes in biovar 2 and biovar 4 are of similar size, but are different from the two chromosomes in biovar 1 suggesting chromosome rearrangement. *B. suis* biovar 3 has only one large circular chromosome, which is about the sum of the two chromosomes found in the other biovar types *(19)*. So far, there are no other reports of different strains of the same species having different numbers of chromosomes, therefore it is probably a rare occurrence. Other α *Proteobacteria* that contain two circular chromosomes are *Rhizobium meliloti* and *Rhodobacter sphaeroides (5)*.

Leptospira interrogans, a causative agent of leptospirosis in humans, has a 4.33 Mb large circular chromosome and a small 359 kb circular chromosome *(20,21)*. *Leptospira borgpetersenii* serovar Hardjobovis, belonging to the same phylogenic pathogen group as *L. interrogans*, also has one large (3.61 Mb) circular and one small (317 kb) circular chromosomes (*see* Chapter 7). The genome structure of *Leptospira* spp. is thus different from that of two other sequenced spirochetes, *Treponema pallidum (22)* and *B. burgdorferi (23)*. The *T. pallidum* genome is a circular chromosome of 1138 kb, whereas *B. burgdorferi* has a linear chromosome of 911 kb.

Among the γ *Proteobacteria*, most of the genomes have a single circular chromosome. A well-known exception is the genus *Vibro*. Three *Vibro* species have been sequenced, *V. cholerae (24)*, *V. vulnificus (25)*, and *V. parahaemolyticus (26)* and all three genomes contain two circular chromosomes.

Burkholderia cepacia, a member of the β-subdivision of *Proteobacteria*, is found in soils and waters and is recognized as a serious respiratory pathogen for patients with cystic fibrosis. Although the genome is still being sequenced, pulsed-field gel electrophoresis analysis of *B. cepacia* isolates identified multiple circular chromosomes, between two and four chromosomes with overall genome sizes in the range of 5 to 9 Mb *(27,28)*. The complexity and plasticity of the *B. cepacia* genome is likely associated with the presence of multiple copies of an array of insertion sequences (IS) elements *(27)*.

Deinococcus radiodurans, a Gram-positive bacterium known for its extreme resistance to radiation and oxidative stress, has a genome with two circular chromosomes (2.65 and 0.412 Mb) and a megaplasmid of 0.178 Mb *(29)*. Part of the mechanisms for the resistance to radiation is owing to the polyploidy nature of the genome and the redundancy of DNA repair genes.

2.3. Molar G+C Content Variation

G+C content of bacterial genomes can range from as low as 24.9% for *Mycoplasma mobile (30)* to 74% for *Micrococcus luteus (5)*. Bacteria of the same species generally have a very similar G+C content with less than 1% deviation. Before the genomic era, G+C content played an important role in identifying and classifying bacteria. Close phylogenetically related species tend to have similar G+C content. However, distantly related bacteria also can have very similar G+C content. Therefore, G+C content cannot be employed alone as the basis for phylogenetic analysis, but it is an important parameter of a particular bacterial species that can be accurately obtained from the genome sequence. Programs used in the annotation of a sequenced genome identify G+C content of all the ORFs and RNA genes. The G+C content of an individual ORF is usually very close to that of the whole genome, unless the genes are either very small or have been acquired in recent evolutionary time from distantly related bacteria with different G+C content via a horizontal gene transfer (HGT) process. Genomic islands (GEIs) and pathogenicity islands (PAIs) are clusters of genes acquired through HGT, and generally have a G+C content significantly different from the average of the whole genome. Bacteria with different G+C content have different codon usage patterns (*see* **Subheading 2.5.**).

2.4. Variation in Genome Structure Among Bacteria of the Same Species and Closely Related Species

Comparative analyses of genomic sequences of closely related bacteria provide valuable knowledge on phylogenetic relationships and evolutionary mechanisms of the species. For example, analysis of the genomic sequences of four *Rickettsia* species, *R. typhi (31)*, *R. sibirica (32)*, *R. conorii (33)*, and *R. prowazekii (34)* demonstrates major gene loss during the evolution of *R. prowazeki* and prevalence of gene decay in all four obligate intracellular *Rickettsia* species. Genes acquired by HGT do not play a major role in the evolution of these species, unlike *E. coli, Salmonella typhimurium, Shigella*, and many other species.

Eight closely related species of *Mycoplasma* have been sequenced (*see* www.genomes online.org/). Like *Rickettsia*, evolution of the smaller genome (less than 1 Mb) species, *M. hyopneumoniae (35)*, *M. mobile (30)*, *M. gallisepticum (36)*, *M. pulmonis (37)*, *M. pneumoniae (38)*, and *M. genitalium (7)* involved major gene losses (*see* Chapter 10).

The value of obtaining a genome sequence of multiple strains of the same species was first demonstrated when the sequence of *E. coli* K12-MG1655 *(39)*, a nonpathogenic strain, and *E. coli* O157:H7 EDL933 *(40)*, an enterohemorrhagic strain, were compared and analyzed *(40)*. This analysis permitted the identification of putative virulence genes, and provided insight into the evolution of pathogenic strains from nonvirulent strains. Presently, five *E. coli* strains have been sequenced. Analysis of these genomes has shown that the genomes of *E. coli* are highly diverse and the acquisition of clusters of genes, as GEIs, has played a major role in the evolution of *E. coli*.

Multiple strains of *S. aureus* also have been sequenced. Comparative analysis of these strains has provided insight regarding the mechanisms of evolution of this species, and the development of antibiotic resistant strains *(41*; Chapter 11). Having multiple strains of many bacterial species sequenced, we now have a better appreciation of the speciation processes, and are in a better position to provide a molecular concept of what constitutes a bacterial species. Lan and Reeves *(42)* defined bacteria of the same species as sharing a common core of genes, which are present in 95% of the isolates. The genome also possesses accessory genes that are found in 1–95% of the strains. Comparative analysis of multiple strains of a few species clearly shows that the current bacterial species comprise many ecotypes, each of which has features commonly attributed to a species in the eukaryotic species concept *(43)*.

2.5. Codon Usage Diversity Among Microbes

The standard genetic code was established in the 1960s. It consists of 61 triplet nucleotides (codons) coding for the common 20 amino acids and 3 codons coding chain termination. The genetic code is degenerate as some amino acids are coded by up to six "synonymous" codons. The code was found to be identical for many organisms including tobacco mosaic virus, human beings, *E. coli*, and bacteriophages, and, thus, was considered universal.

The first deviation from the standard genetic code was observed in 1981 when it was found that mammalian mitochondria used AUA for methionine instead of isoleucine, and UGA for tryptophan instead of chain termination *(44)*. Many deviations have since been observed in other organisms, including both lower eukaryotes and bacteria *(45, 46)*. In bacteria, nonstandard genetic codes have been found in *Micrococcus*, *Mycoplasma*, and *Spiroplasma*. Similar to mitochondrial DNA, the stop codon UGA in AT-rich genomes of *Mycoplasma* and *Spiroplasma* is assigned to encode tryptophan. In *Mycoplasma capricolum*, CGG, an arginine codon, appears unassigned; that is, it does not code for any amino acid. In *Micrococcus luteus*, with a GC-rich genome (74%), the arginine AGA and the isoleucine AUA codons also appear to be unassigned *(47)*. As further genome sequences are completed and analyzed, it is likely that further non-standard genetic code usages will be identified.

Different genomes have their own characteristic pattern of synonymous codon usage *(48)*. This property has been used to identify pathogenicity islands and HGT clusters in bacterial genomes *(49–52)*. Synonymous codon usage patterns in prokaryotes are affected by two major factors, the overall G+C content of the genome and the optimal growth temperature *(53)*.

3. Orientation and Organization of Genes at the Genomic Level

3.1. Origins of Chromosome Replication

Initiation of chromosome replication involves several regulated steps: (1) binding of initiator protein (DnaA) to sites located within the origin of replication (*oriC*); (2)

local unwinding of the *oriC* region; and (3) loading the DNA helicase and other proteins that are required to form replication forks. In bacteria, chromosomal replication initiated at the *oriC* region proceeds bidirectionally until the replication forks reach the termination site, *terC* for a circular chromosome or chromosomal ends for a linear chromosome. During initiation of chromosomal replication, the initiator protein, DnaA, interacts with repetitive nonpalindromic nonamer sequences, the DnaA boxes, located within the *oriC* region. Sequences of bacterial *oriC* regions are conserved only among closely related species. Sizes of the bacterial origin of replication vary from approx 200 to 1000 bp and generally contain several DnaA boxes and an AT-rich region. The *oriC* region is always located within the intergenic region, frequently within the *rnpA-rmpH-dnaA-dnaN-recF-gyrB-rnpA* genes cluster, usually next to the *dnaA* gene *(54)*.

Three methods are used to identify the origin of chromosomal replication in a newly sequenced genome. The first, and most reliable, method is based on bias (asymmetry) in nucleotide composition observed in most bacterial chromosomes. The asymmetry changes its polarity at the origin and the terminus of chromosome replication *(55)*. This asymmetry is usually measured as the normalized difference in the content of complementary GC nucleotides, GC skew (G–C)/(G+C). The *oriC* region can be identified using the Oriloc computer program *(56)* that analyzes chromosome asymmetry. However, the origin of replication predicted by this program does not always correspond to the position of a functional replication origin, as shown in the genome of *H. pylori (57)*, *Synechococcus sp. (54)*, and *Mesorhizobium loti (54)*.

The second method used to identify the *oriC* region is based on locating a cluster of DnaA boxes (three or more). The best characterized DnaA box (5'-TTATCCACA-3') is that of *E. coli*. It appears that the sequence of the DnaA box alters with the change in the genomic G+C content *(54)*. The consensus DnaA box derived from 76 chromosomes is $T_{93.7}T_{94.9}A_{78.9}T_{98.7}C_{88.1}C_{90.6}A_{96.6}C_{96.0}A_{95.3}$ *(54)*.

The third method involves identifying the location of the *dnaA* gene. *dnaA* genes are found in all bacterial genomes sequenced to date, with the exception of *Wigglesworthia glossinidia (58)* and *Blochmannia floridanus (59)*. Each is a small genome γ-Proteobacterium. In addition, the chromosomes of these two endosymbionts do not contain any clusters of DnaA boxes. Interestingly, many of the sequenced genomes of Mollicutes (e.g., *Mycoplasma gallisepticum, M. penetrans, M. pneumoniae, Phytoplasma asteris,* and *Ureaplasma urelyticum*) also do not contain any clusters of DnaA boxes.

3.2. Gene Clustering: Origin and Conservation

Clusters of cotranscribed genes, operons, often composed of genes with similar function, are the basic organization of genes and the units of gene regulation in prokaryotes. Lawrence and Roth *(60)* proposed the Selfish Operon Model to explain the evolution of gene clusters. This model proposes that the gene cluster was initially beneficial to the genes themselves and not to the host organisms. Genes organized into clusters can be transmitted by HGT or vertical transmission. Comparative analysis of bacterial and archaeal genomes sequenced showed that only a few operons are conserved among distantly related species *(61–63)*. In general, gene order in prokaryotes is poorly conserved and undergoes frequent gene rearrangements. This feature is well illustrated by the lack of genome-wide colinearity of orthologous genes among four ε-proteobacteria, *Campylobacter jejuni, Helicobacter pylori, Helicobacter hepaticus,* and *Wolinella succinogenes,* all members of the order Campylobacteraceae *(64)*. The conserved gene clusters comprise genes and operons involved in translation, transcription, energy production, and membrane transport. The most conserved gene cluster observed is the ribosomal protein genes, whose gene order is largely preserved throughout the bacteria domain and in some archaeal species *(61–63)*.

4. Accessory and Mobile Genetic Elements

4.1. Plasmids: Structure, Size, Topology, and Evolutionary Role

Plasmids are extrachromosomal genetic elements of diverse sizes ranging from a few kilo basepairs to more than 400 kb. The most common structure of plasmids is a double-stranded circular DNA molecule. However, linear plasmids are found in all species of the *Borrelia* genus and in many species of *Streptomyces (65)*. Linear plasmids have also been isolated from *Rhodococcus fascians (66)*, *Nocardia opaca (67)*, and *Clavibacter michiganensis* subsp. Sepedonicus *(68)*.

Many bacterial pathogen genomes were found to possess one or more plasmids that encode a diverse array of virulence factors, including toxins, adherence pili, type IV pili, hemolysins, capsules, and type III secretion systems *(69)*. Most virulence plasmids are large and have modular structures, with clusters of replication genes, virulence genes, and DNA transfer-related genes *(70)*. Virulence plasmids often contain IS elements, transposons, and cryptic phages. IS elements may facilitate recombination between the host chromosome and megaplasmid, as observed in *Deinococcus radiodurans* R1 *(29)*. Some plasmids have limited host ranges, although others are able to replicate in distantly related bacterial hosts and, thus, play a major role in the horizontal transfer of genes and in the evolution of microbial genomes.

4.2. Insertion Sequence and Transposons

An IS is a discrete segment of DNA, commonly between 1 and 1.5 kb, that can transpose directly to new sites on the genome. An IS element usually has a short inverted repeat (IR) sequence at both ends (i.e., IRL and IRR). The sequences of IRL and IRR of each IS element are very similar, but not identical. The IR sequences, which can vary in length from 9 to 41 bp, are characteristic for the particular IS and are required for transposition. Transposition of an IS to a new site leads to a duplication of a small sequence of the target DNA. The size of the duplication is characteristic of the IS element, ranging from 2 to 13 bp. ISs encode only the protein(s) that is required for transposition. Most of the ISs encode a transposase (Tpase) that has a conserved triplet of amino acids, Asp (D), Asp, and Glu (E) *(71)*. The Tpase, which binds and processes IR sequences during transposition, is encoded by a single or sometimes two *orf* that correspond to the entire length of the element. Some IS elements encode enzymes related to the serine and tyrosine site-specific recombinases or the rolling circle replicase *(72)*. At least 19 different families of IS elements have been identified in prokaryotic genomes *(71)*.

ISs are found on the chromosomes of many different sequenced bacterial and archaeal genomes *(71)*. No IS element was identified in a few sequenced genomes, notably, *Aeropyrum pernix* K1, *Aquifex aeolicus* VF5, *Bacillus subtillis* 168, *Buchnera* sp. strain APS, *Chlamydia muridarum, Chlamydia trachomatis* D/UW-3/CX, *C. trachomatis*

MOPN, *Chlamydophila pneumoniae* AR39, *C. pneumoniae* CWL029, *C. pneumoniae* J138, *Methanobacterium thermautotrophicum* DH, *Mycoplasma genitalium* G37, *Mycoplasma pneumoniae* M129, *Pasteurella multocida* PM70, *Pyrococcus horikkoshii* OT3, *Pyrococcus abyssi, Rickettsia prowazekii, Synechocystis sp.* PCC6803, and *Treponema pallidum* subsp. *pallidum*. However, subsequent analysis by other workers have uncovered IS in *R. prowazekii, P. abyssi, Synechocystis, A. pernix K1, C. muridarum*, and *C. trachomatis (71)*.

In some bacterial lineages, many IS elements are found in high copy numbers. The genomes of *Bordetella pertussis* and *Bordetella parapertussis* have 261 and 112 copies of different IS elements, respectively *(73)*. These elements play an important role in chromosomal rearrangements and generation of pseudogenes in *B. pertussis* and *B. parapertussis (74)*. The genome of *Burkholderia mallei* with two chromosomes possesses 171 complete and partial IS elements accounting for about 3.1% of the genome. These elements likely have mediated extensive genome-wide insertions, deletions, and inversion mutations in *B. mallei (75)*. *Yersinia pestis* is another genome with abundant IS elements. A total of 140 complete and partial IS elements amounting to 3.7% of the genome were identified *(76)*. Of the 149 pseudogenes found in *Y. pestis* 51 are caused by IS-insertion mutations.

The complete genome sequence of two strains of *Shigella flexneri* serotype 2a, strain 2457T *(77)* and strain 301 *(78)* were determined and analyzed. Strain 2457T contains a total of 284 IS elements, which account for 6.7% (309.4 kb) of the chromosome. The 301 strain contains 247 complete and 67 partial IS elements. These elements are believed to be the major cause of the dynamic nature of the *S. flexneri* chromosome *(77,78)*.

The genome of *Leptospira borgpetersenii* serovar Hardjobovis also has a diverse and high number of IS elements, amounting to more than 6% of its genome. This is significantly higher than that of *Leptospira interrogans* serovar Lai and serovar Copenhageni, both around 2% of the genome (*see* Chapter 7). The IS elements in *L. borgpetersenii* play a major role in the generation of large number (161) of pseudogenes in the genome.

A transposon, unlike an IS element, is a more complex type of mobile genetic element. Transposons contain genes that encode proteins required for transposition and proteins that have other functions including, for example, resistance to antibiotics or heavy metals. An important group of transposons is the conjugative transposon (CTn), which is able to mobilize from one bacterial cell to another of the same, or different, species by a conjugation-like process that requires cell-to-cell contact *(79)*. Conjugative transposons are found in many bacterial genera, and are particularly common among the Gram-positive streptococci and enterococci. A number of well-characterized CTns are able to transfer to commensal bacteria of the human gastrointestinal tract in anaerobic filter matings *(80)*. A major concern is that CTns with multiple antibiotic resistance genes transmitted to commensal bacteria could potentially then transfer the resistance genes to pathogens in the gastrointestinal tract.

Some of the conjugative transposons (e.g., Tn916 from *Enterococcus faecalis* and CTnDot from *Bacteroides thetaiotaomicron*) have been classified as integrative and conjugative elements (ICEs). ICEs are a heterogenous group of self-transmissible mobile genetic elements that can mobilize like a conjugative plasmid and integrate into a host chromosome like a lysogenic phage. Many ICEs contain antibiotic and metal resistance genes like transposons. They may also contain DNA repair genes and genes encoding virulence factors *(81)*. ICEs promote genome plasticity and provide a major contribution to lateral gene flow in prokaryotes *(81)*.

4.3. Integrons and Mobile Gene Cassettes

Integrons are genetic elements that contain the determinants of the components of a site-specific recombination system, which capture and promote expression of mobile gene cassettes. A basic integron consists of an integrase gene (*int*) of the tyrosine recombinase family *(82)* and a proximal recombination site, called the *attl* site. A gene cassette is a discrete mobile genetic unit that contains one ORF (or more) followed by a recombination site (*attC*) belonging to a family of sites known as 59-base elements *(83)*. A gene cassette often contains an antibiotic resistance gene and can be integrated into an integron by a site-specific recombination reaction between the *attl* site and the *attC* site catalyzed by the integrase. Cassettes have been shown to exist as free covalent circular DNA molecules generated by excision through an Intl-mediated reaction from an existing integron with an integrated cassette. The covalent circular DNA cassettes are not able to replicate, but are able to integrate into another integron or secondary sites on the chromosome. Integrons containing up to eight resistance cassettes have been identified in multiple antibiotic-resistant clinical isolates *(84)*. These multi-antibiotic resistance integrons are found in plasmids and transposons and, thus, can be transmitted horizontally to diverse bacterial species allowing rapid evolution of antibiotic-resistant bacterial species. A new form of integron, a super-integron (SI), first identified in *Vibro cholera*, is defined as a chromosomal array of a large number of gene cassettes, mobilized by an integron-like integrase *(85)*. SIs are widely distributed among ε-proteobacteria, including many species of *Vibrio, Shewanella,* and *Xanthomonas* and *Pseudomonas* *(86,87;* Chapter 13). These chromosomal SIs have been proposed as ancestors for multiresistant integrons and play an important role in the evolution of more adaptive strains *(86)*.

4.4. Introns: Structure, Distribution, and Evolution

An intron is a segment of DNA that is transcribed as part of a premature RNA, but is ultimately spliced out by ligating the exon sequences that flank it. In higher eukaryotic genomes, intron sequences (spliceosomal introns) are widely distributed. In prokaryotes, introns (group I and group II) are rare mobile genetic elements. Group I and II introns are predicted to form distinct secondary structures *(88)*. The first bacterial intron identified was the group I self-splicing intron found in the thymidylate synthase (*td*) gene of T4 bacteriophage *(89)*. Group I introns are found in many bacterial species, mainly in tRNA genes *(90)*. However, a group I intron was identified recently in the *recA* gene of *Bacillus anthracis* *(91)*. Sizes of group I introns range from 200 to 3000 nt. Larger introns often have ORFs, which are either freestanding or in-frame with the upstream exon that encodes endonuclease and maturase *(90)*. Group I introns are able to spread through a population of homologous (cognate) intron-less alleles. This homing process is dependent on the intron-encoded DNA endonuclease.

Group II introns were first discovered in organelles of plants, fungi, and other lower eukaryotes, but have since been found in bacterial genomes of diverse lineages *(92)*. Like group I introns, group II introns may contain an ORF and the sizes varies from approx 600 to 3000 nt. Group II intron RNAs are predicted to form a conserved structure with six domains (I to VI), and are able to self-splice using a similar mechanism as that of higher eukaryotic spliceosomal introns. Domains V and VI share secondary structural features with U6 small nuclear RNA (snRNA) and U2 snRNA, respectively of the spliceosomal complex. Because of these similarities, group II introns are believed to have been predecessors of spliceosomal introns *(93,94)*. Like group I introns, group II intron RNAs can reverse splice into cognate intron-less DNA sites (retrohoming) but, in addition, they can also transpose to nonallelic sites (retrotransposition) *(95)*. Recently, *Lactococcus lactis* group II intron, LI.LtrB, was shown to be mobilized by conjugation to *Enterococcus facalis* and the integrated intron was able to retrotranspose to multiple chromosomal sites of the recipient cell *(96)*.

4.5. Prophages: Distribution, Influence on Chromosome Structure, and Virulence

Prophages are latent phases of temperate bacteriophages and are commonly chromosomal genetic elements. However, some prophages (e.g., P1 and N15) are extrachromosomal elements like plasmids. Prophages can constitute as much as 20% of a bacterium's genome and are major contributors to species diversity *(97,98)*. For example, the *E. coli* O157 strain Sakai genome has 18 prophages (including cryptic phages), whereas that of *E. coli* K-12 only has 11 *(97)*. The sequenced geonome of *S. enterica* serovar Typhi CT18 has seven distinct prophage-like elements collectively encompassing more than 180 kb, representing about 3.8% of the genome.

Some of the phage genes cause major phenotypic changes of the host cell, including antigenic variation, resistance to infection by related phages, and increased virulence. The first phage-encoded virulence factor identified was the diphtheria toxin coded by the β-phage of *Corynebacterium diphtheriae* *(99)*. Numerous phage-encoded virulence factors have since been identified *(69,100)*.These include many extracellular toxins (e.g., Shiga toxins by H-19B phage in *E. coli*, neurotoxin by β-phage in *Clostridium botulinum*, cholera toxin by CTX in *V. cholerae*), type III secretion effector proteins (SopE, SseI, and SspH1 by SopE, GIFSY-2, and GIFSY-3 phages, respectively in *S. enterica*), as well as proteins that alter antigenicity, various hydrolytic enzymes, and adhesins. Many of these virulence factors are encoded by genes of lysogenic phages that are not essential for phage replication and growth, but enhance the fitness of the bacteria. These prophage genes are called morons *(101)*.

The integration site for many prophages has been characterized. Many prophages are integrated into tRNA genes and often the attP site of the phage reconstitutes the gene to produce a normal tRNA molecule *(98,102,103)*. Many of the PAIs and GEIs are localized at tRNA genes *(104)*, thus providing support that the origin of these elements is associated with prophages.

About half of the sequenced genomes of prokaryotes have defined prophages or phage remnants. Archaea and intracellular eubacterial pathogens with a small genome (<1 Mbp) have no detectible integrated prophages *(97)*. Presumably, these intracellular bacteria have undergone major genome reductions over time in which all nonessential genes for the intracellular niche have been deleted.

4.6. Pathogenicity and Genomic Islands: Structure, Distribution, and Function

PAIs were first discovered in the genomes of pathogenic *E. coli* and subsequently identified in many other pathogens *(104)*. PAIs are associated with pathogenicity of bacteria

(104). The size of PAIs ranges from about 10 to more than 100 kb. In sequenced genomes, PAIs are recognized as clusters of genes with a G+C content and codon usage that is different from that of the whole genome. PAIs contain virulence genes and often contain phage- and/or plasmid-derived sequences, including transfer genes, integrases, and IS elements. PAIs often are integrated into tRNA genes and flanked by direct repeats. Some PAIs show instability that is likely because of excision catalyzed by phage-encoded enzymes *(105).*

Genetic elements similar to PAIs, but with no known virulence genes, are found in many of the sequenced bacterial genomes. These genetic elements and islands are given a general name–GEIs. GEIs are diverse and may contain genes for environmental adaptation (ecological islands), for symbiosis survival (symbiosis islands), and for parasitic life PAIs *(105,106).* Smaller islands (<10 kb) are known as genomic islets *(105).* Genes located in GEIs and genomic islets are part of the flexible gene pool of a bacterial genome and are evolutionarily recent genes acquired through HGT. These islands can greatly increase the fitness of bacteria, and thus, are major driving forces of bacterial evolution.

5. Small (Noncoding) RNA (sRNA) Genes: Distribution and Role in Virulence

Prior to completion of the *E. coli* genomic sequence only 10 sRNA genes were known. In the past 3 yr, using a variety of new approaches over 50 new sRNA- encoding genes in *E. coli (107)* have been identified. These systematic approaches were based on the presence of a σ-70 promoter and ρ-independent transcription terminator within a sequence of less than 400 nt, conservation of the genes within closely related species, conservation of secondary structure, microarrays, and cDNA cloning of the transcriptome *(108–113).*

The sRNA genes, in general, are not well conserved among distantly related bacterial species. For example only 16 of the 55 *E. coli* sRNA examined were conserved in *Y. pestis* and only one gene was conserved beyond *Y. pestis.* Fifty-five percent of the sRNA genes are between 50 and 250 nt long *(107).* Only a small number of the new sRNAs have been characterized. For those that have been characterized, expression and functional studies indicate that many have multiple targets. Some of the target transcripts encode global regulators, supporting a role in global control *(114–116).* Global regulator sRNAs generally act in *trans* by base pairing with the target mRNAs to increase or decrease their expression through either changes in the stability of mRNAs or in the efficiency of translation. In *E. coli,* sRNAs are involved in regulating a variety of cellular processes, including carbon storage and utilization *(117),* response to iron limitation *(114),* response to oxidative stress *(118),* transition to stationary phase *(119),* response to acid conditions *(120),* and expression of outer membrane proteins (Omp) *(121).*

The prevalence of sRNA genes in bacteria other than *E. coli* and closely related species is not known, largely because there is no reliable algorithm that can be used to screen diverse genome sequences for these genes with such heterogenous sequences and structures. A recent study that used a bioinformatics approach based on detecting small RNA genes with a σ-54 promoter sequence and ρ-independent terminator within intergenic regions of the *V. cholerae* genome identified four sRNA (*Qrr1-4*) genes that control quorum sensing *(122).* Quorum-sensing controls diverse genes, including many that encode virulence factors *(123).*

A bioinformatics approach was also used to search the *Pseudomonas aeruginosa* genome for functional homologs of *ryhB*, a sRNA gene identified in *E. coli* to be repressed by the ferric regulatory protein (Fur) in iron-replete conditions. Under iron-starvation conditions, sRNA RyhB is made and negatively regulates the expression of genes encoding iron-binding proteins (e.g., *sdh* and *bfrB*) and superoxide dismutase (*sodB*) by binding and destabilizing the target mRNAs *(124)*. The search led to the identification of two RyhB functional homologs, PrrF1 and PrrF2, in *P. aeruginosa (125)*. PrrF1 and PrrF2 are orthologs with over 95% identical nucleotide sequences. Two *prrF* sequence homologs were found in *Pseudomonas putida, Pseudomonas fluorescens*, and *Pseudomonas syringae (125)*. A recently sequenced genome of *Rickettsia typhi* was annotated to have 39 noncoding RNA genes *(126)*. Their identification was facilitated by information gained from the sRNA genes of *E. coli*. It is likely that many more sRNA genes will be identified in future sequenced bacterial genomes and many will be shown to have a regulatory role in virulence and other cellular processes.

6. Conclusions

With more than 200 prokaryotic genomes sequenced and analyzed, we have gathered an enormous amount of data and gained a significant amount of knowledge on bacterial and archaeal genomes. Bacterial genomes are highly diversified in terms of chromosomal size, copy number, topology, and G+C content. Most bacterial chromosomes are actively undergoing rearrangement, deletion, duplication, and acquisition of DNA through HGT.

Most genomes contain accessory GEIs and mobile genetic elements. These islands and elements contain species- and strain-specific genes, including determinants for fitness, virulence, and antibiotic resistance. In order to gain a better understanding of the evolution and emergence of pathogenic strains and species, future studies need to focus more on genes located in these sites and elements.

Acknowledgments

The author is grateful for the insightful and thoughtful comments on the chapter by Drs. Philip Sherman, Billy Bourke, Penny Chan, and Gavin Clark. Assistance on preparation of the figures by Dave Ng is much appreciated.

References

1. Fleischmann, R. D., Adams, M. D., White, O., et al. (1995) Whole-genome random sequencing and assembly of *Haemophilus influenzae* Rd. *Science* **269,** 496–512.
2. Sakzberg, S. L. and Delcher, A. L. (2004) Tools for gene finding and whole genome composition, in *Microbial Genomes* (Fraser, C. M., Read, T. D., and Nelson, K. E., eds.). Humana Press, Totowa, NJ, pp. 19–31.
3. Cole, S. T. and Saint Girons, I. (1994) Bacterial genomics. *FEMS Microbiol. Rev.* **14,** 139–160.
4. Fonstein, M. and Haselkorn, R. (1995) Physical mapping of bacterial genomes. *J. Bacteriol.* **177,** 3361–3369.
5. Casjens, S. (1998) The diverse and dynamic structure of bacterial genomes. *Ann. Rev. Genet.* **32,** 339–377.
6. Danchin, A., Guerdoux-Jamet, P., Moszer, I., and Nitschke, P. (2000) Mapping the bacterial cell architecture into the chromosome. *Philos. Trans. R. Soc. Lond. B. Biol. Sci.* **355,** 179–190.

7. Fraser, C. M., Gocayne, J. D., White, O., et al. (1995) The minimal gene complement of *Mycoplasma genitalium. Science* **270,** 397–403.

8. Kaneko, T., Nakamura, Y., Sato, S., et al. (2002) Complete genomic sequence of nitrogen-fixing symbiotic bacterium *Bradyrhizobium japonicum* USDA110. *DNA Res.* **9,** 189–197.

9. Waters, E., Hohn, M. J., Ahel, I., et al. (2003) The genome of *Nanoarchaeum equitans*: insights into early archaeal evolution and derived parasitism. *Proc. Natl. Acad. Sci. USA* **100,** 12,984–12,988.

10. Galagan, J. E., Nusbaum, C., Roy, A., et al. (2002) The genome of *M. acetivorans* reveals extensive metabolic and physiological diversity. *Genome Res.* **12,** 532–542.

11. Volff, J.-N. and Altenbuchner, J. (2000) A new begininng with new ends:linearisation of circular chromosomes during bacterial evolution. *FEMS Microbiol. Letts.* **186,** 143–150.

12. Ferdows, M. S. and Barbour, A. G. (1989) Megabase-sized linear DNA in the bacterium *Borrelia burgdorferi,* the lyme disease agent. *Proc. Natl. Acad. Sci. USA* **86,** 5969–5973.

13. Lin, Y.-L., Kieser, H. M., Hopwood, D. A., and Chen, C. W. (1993) The chromosomal DNA of *Streptomyces lividans* 66 is linear. *Mol. Microbiol.* **10,** 923–933.

14. Bentley, S. D., Chater, K. F., Cerdeno-Tarraga, A. M., et al. (2002) Complete genome sequence of the model actinomycete *Streptomyces coelicolor* A3(2). *Nature* **417,** 141–147.

15. Ikeda, H., Ishikawa, J., Hanamoto, A., et al. (2003) Complete genome sequence and comparative analysis of the industrial microorganism *Streptomyces avermitilis. Nat. Biotechnol.* **21,** 526–531.

16. Wood, D. W., Setubal, J. C., Kaul, R., et al. (2001) The genome of the natural genetic engineer *Agrobacterium tumefaciens* C58. *Science* **294,** 2317–2323.

17. Goodner, B., Hinkle, G., Gattung, S., et al. (2001) Genome sequence of the plant pathogen and biotechnology agent *Agrobacterium tumefaciens* C58. *Science* **294,** 2323–2328.

18. Paulsen, I. T., Seshadri, R., Nelson, K. E., et al. (2002) The *Brucella suis* genome reveals fundamental similarities between animal and plant pathogens and symbionts. *Proc. Natl. Acad. Sci. USA* **99,** 13,148–13,153.

19. Jumas-Bailak, E., Michaux-Charachon, S., Bourg, G., O'Callaghan, D., and Ramuz, M. (1998) Differences in chromosome number and genome rearrangements in the genus *Brucella. Mol. Microbiol.* **27,** 99–106.

20. Ren, S. X., Fu, G., Jiang, X. G., et al. (2003) Unique physiological and pathogenic features of *Leptospira interrogans* revealed by whole-genome sequencing. *Nature* **422,** 888–893.

21. Nascimento, A. L., Ko, A. I., Martins, E. A., et al. (2004) Comparative genomics of two *Leptospira interrogans* serovars reveals novel insights into physiology and pathogenesis. *J. Bacteriol.* **186,** 2164–2172.

22. Fraser, C. M., Norris, S. J., Weinstock, G. M., et al. (1998) Complete genome sequence of *Treponema pallidum*, the syphilis spirochete. *Science* **281,** 375–388.

23. Fraser, C. M., Casjens, S., Huang, W. M., et al. (1997) Genomic sequence of a Lyme disease spirochaete, *Borrelia burgdorferi. Nature* **390,** 580–586.

24. Heidelberg, J. F., Eisen, J. A., Nelson, W. C., et al. (2000) DNA sequence of both chromosomes of the cholera pathogen *Vibrio cholerae. Nature* **406,** 477–483.

25. Chen, C. Y., Wu, K. M., Chang, Y. C., et al. (2003) Comparative genome analysis of *Vibrio vulnificus*, a marine pathogen. *Genome Res.* **13,** 2577–2587.

26. Makino, K., Oshima, K., Kurokawa, K., et al. (2003) Genome sequence of *Vibrio parahaemolyticus*: a pathogenic mechanism distinct from that of *V. cholerae. Lancet* **361,** 743–749.

27. Lessie, T. G., Hendrickson, W., Manning, B. D., and Devereux, R. (1996) Genomic complexity and plasticity of *Burkholderia cepacia. FEMS Microbiol. Lett.* **144,** 117–128.

28. Wigley, P. and Burton, N. F. (2000) Multiple chromosomes in *Burkholderia cepacia* and *B. gladioli* and their distribution in clinical and environmental strains of *B. cepacia. J. Appl. Microbiol.* **88,** 914–918.

29. White, O., Eisen, J. A., Heidelberg, J. F., et al. (1999) Genome sequence of the radioresistant bacterium *Deinococcus radiodurans* R1. *Science* **286,** 1571–1577.

30. Jaffe, J. D., Stange-Thomann, N., Smith, C., et al. (2004) The complete genome and proteome of *Mycoplasma mobile. Genome Res.* **14,** 1447–1461.

31. McLeod, M. P., Qin, X., Karpathy, S. E., et al. *(*2004) Complete genome sequence of *Rickettsia typhi* and comparison with sequences of other rickettsiae. *J. Bacteriol.* **18,** 5842–5855.

32. Malek, J. A., Wierzbowski, J. M., Tao, W., et al. (2004) Protein interaction mapping on a functional shotgun sequence of *Rickettsia sibirica. Nucleic Acids Res.* **32,** 1059–1064.

33. Ogata, H., Audic, S., Renesto-Audiffren, P., et al. (2001) Mechanisms of evolution in *Rickettsia conorii* and *R. prowazekii. Science* **293,** 2093–2098.

34. Andersson, S. G., Zomorodipour, A., Andersson, J. O., et al. (1998) The genome sequence of *Rickettsia prowazekii* and the origin of mitochondria. *Nature* **396,** 133–140.

35. Minion, F. C., Lefkowitz, E. J., Madsen, M. L., Cleary, B. J., Swartzell, S. M., and Mahairas, G. G. (2004) The genome sequence of *Mycoplasma hyopneumoniae* strain 232, the agent of swine mycoplasmosis. *J. Bacteriol.* **186,** 7123–7133.

36. Papazisi, L., Gorton, T. S., Kutish, G., et al. (2003) The complete genome sequence of the avian pathogen *Mycoplasma gallisepticum* strain R(low). *Microbiology* **149,** 2307–2316.

37. Chambaud, I., Heilig, R., Ferris, S., et al. (2001) The complete genome sequence of the murine respiratory pathogen *Mycoplasma pulmonis. Nucleic Acids Res.* **29,** 2145–2153.

38. Himmelreich, R., Hilbert, H., Plagens, H., Pirkl, E., Li, B. C., and Herrmann, R. (1996) Complete sequence analysis of the genome of the bacterium *Mycoplasma pneumoniae. Nucleic Acids Res.* **24,** 4420–4449.

39. Blattner, F. R., Plunkett, G. 3rd, Bloch, C. A., et al. (1997) The complete genome sequence of *Escherichia coli* K-12. *Science* **277,** 1453–1474.

40. Perna, N. T., Plunkett, G. 3rd, Burland, V., et al. (2001) Genome sequence of enterohaemorrhagic *Escherichia coli* O157:H7. *Nature* **409,** 529–533.

41. Lindsay, J. A. and Holden, M. T. (2004) *Staphylococcus aureus*: superbug, super genome? *Trends Microbiol.* **12,** 378–385.

42. Lan, R. and Reeves, P. R. (2000) Intraspecies variation in bacterial genomes: the need for a species genome concept. *Trends Microbiol.* **8,** 396–401.

43. Cohan, F. M. (2002) What are bacterial species? *Annu. Rev. Microbiol.* **56,** 457–487.

44. Anderson, S., Bankier, A. T., Barrell, B. G., et al. (1981) Sequence and organization of the human mitochondrial genome. *Nature* **290,** 457–465.

45. Osawa, S., Jukes, T. H., Watanabe, K., and Muto, A. (1992) Recent evidence for evolution of the genetic code. *Microbiol. Rev.* **56,** 229–264.

46. Santos, M. A., Ueda, T., Watanabe, K., and Tuite, M. F. (1997) The non-standard genetic code of *Candida* spp.: an evolving genetic code of a novel mechanism for adaptation? *Mol. Microbiol.* **26,** 423–431.

47. Kano, A., Ohama, T., Abe, R., and Osawa, S. (1993) Unassigned or nonsense codons in *Micrococcus luteus. J. Mol. Biol.* **230,** 51–56.

48. Grantham, R., Gautier, C., and Gouy, M. (1980) Codon frequencies in 119 individual genes confirm consistent choices of degenerate bases according to genome type. *Nucleic Acids Res.* **8,** 1893–1912.

49. Ochman, H., Lawrence, J. G., and Grolsman, E. A. (2000) Lateral gene transfer and the nature of bacterial innovation. *Nature* **405,** 299–304.

50. Kanaya, S., Kinouchi, M., Abe, T., et al. (2001) Analysis of codon usage diversity of bacterial genes with a self-organizing map (SOM): characterization of horizontally transferred genes with emphasis on the *E. coli* O157 genome. *Gene* **276,** 89–99.

51. Wang, H. C., Badger, J., Kearney, P., and Li, M. (2001) Analysis of codon usage patterns of bacterial genomes using the self-organizing map. *Mol. Biol. Evol.* **18,** 792–800.

52. Chen, L. L. and Zhang, C. T. (2003) Seven GC-rich microbial genomes adopt similar codon usage patterns regardless of their phylogenetic lineages. *Biochem. Biophys. Res. Commun.* **306,** 310–317.

53. Lynn, D. J., Singer, G. A., and Hickey, D. A. (2002) Synonymous codon usage is subject to selection in thermophilic bacteria. *Nucleic Acids Res.* **30,** 4272–4277.

54. Mackiewicz, P., Zakrzewska-Czerwinska, J., Zawilak, A., Dudek, M. R., and Cebrat, S. (2004) Where does bacterial replication start? Rules for predicting the *oriC* region. *Nucleic Acids Res.* **32,** 3781–3791.

55. Lobry, J. R. (1996) Origin of replication of *Mycoplasma genitalium*. *Science* **272,** 745–746.

56. Frank, A. C. and Lobry, J. R. (2000) Oriloc: prediction of replication boundaries in unannotated bacterial chromosomes. *Bioinformatics* **16,** 560–561.

57. Zawilak, A., Cebrat, S., Mackiewicz, P., et al. (2001) Identification of a putative chromosomal replication origin from *Helicobacter pylori* and its interaction with the initiator protein DnaA. *Nucleic Acids Res.* **29,** 2251–2259.

58. Akman, L., Yamashita, A., Watanabe, H., et al.(2002) Genome sequence of the endocellular obligate symbiont of tsetse flies, *Wigglesworthia glossinidia*. *Nat. Genet.* **32,** 402–407.

59. Gil, R., Silva, F. J., Zientz, E., et al. (2003) The genome sequence of *Blochmannia floridanus*: comparative analysis of reduced genomes. *Proc. Natl. Acad. Sci. USA* **100,** 9388–9393.

60. Lawrence, J. G. and Roth, J. R. (1996) Selfish operons: horizontal transfer may drive the evolution of gene clusters. *Genetics* **143,** 1843–1860.

61. Watanabe, H., Mori, H., Itoh, T., and Gojobori, T. (1997) Genome plasticity as a paradigm of eubacteria evolution. *J. Mol. Evol.* **44,** S57–S64.

62. Siefert, J. L., Martin, K. A., Abdi, F., Widger, W. R., and Fox, G. E. (1997) Conserved gene clusters in bacterial genomes provide further support for the primacy of RNA. *J. Mol. Evol.* **45,** 467–472.

63. Wolf, Y. I., Rogozin, I. B., Kondrashov, A. S., and Koonin, E. V. (2001) Genome alignment, evolution of prokaryotic genome organization, and prediction of gene function using genomic context. *Genome Res.* **11,** 356–372.

64. Eppinger, M., Baar, C., Raddatz, G., Huson, D. H., and Schuster, S. C. (2004) Comparative analysis of four Campylobacterales. *Nat. Rev. Microbiol.* **2,** 872–885.

65. Hinnebusch, J. and Tilly, K. (1993) Linear plasmids and chromosomes in bacteria. *Mol. Microbiol.* **10,** 917–922.

66. Crespi, M., Messens, E., Caplan, A. B., van Montagu, M., and Desomer, J. (1992) Fasciation induction by the phytopathogen *Rhodococcus fascians* depends upon a linear plasmid encoding a cytokinin synthase gene. *EMBO J.* **11,** 795–804.

67. Kalkus, J., Reh, M., and Schlegel, H. G. (1990) Hydrogen autotrophy of *Nocardia opaca* strains is encoded by linear megaplasmids. *J. Gen. Microbiol.* **136,** 1145–1151.

68. Brown, S. E., Knudson, D. L., and Ishimaru, C. A. (2002) Linear plasmid in the genome of *Clavibacter michiganensis* subsp. sepedonicus. *J. Bacteriol.* **184,** 2841–2844.

69. Davis, B. M. and Waldor, M. K. (2002) Mobile genetic elements and bacterial pathogenesis, in *Mobile DNA II* (Craig, N. L., Gellert, M., and Lambowitz, A. M., eds.). ASM Press, Washington, DC, pp. 1040–1059.

70. Thomas, C. M. (2000) Paradigms of plasmid organization. *Mol. Microbiol.* **37,** 485–491.

71. Chandler, M. and Mahillon, J. (2002) Insertion sequences revisited, in *Mobile DNA II* (Craig, N. L., Gellert, M., and Lambowitz, A. M., eds.). ASM Press, Washington, DC, pp. 305–366.

72. Curcio, M. J. and Derbyshire, K. M. (2003) The outs and ins of transposition: from mu to kangaroo. *Nat. Rev. Mol. Cell. Biol.* **4**, 865–877.

73. Parkhill, J., Sebaihia, M., Preston, A., et al. (2003) Comparative analysis of the genome sequences of *Bordetella pertussis, Bordetella parapertussis* and *Bordetella bronchiseptica*. *Nat. Genet.* **35**, 32–40.

74. Preston, A., Parkhill, J., and Maskell, D. J. (2004) The bordetellae: lessons from genomics. *Nat. Rev. Microbiol.* **2**, 379–390.

75. Nierman, W. C., DeShazer, D., Kim, H. S., et al. (2004) Structural flexibility in the *Burkholderia mallei* genome. *Proc. Natl. Acad. Sci. USA* **101**, 14,246–14,251.

76. Parkhill, J., Wren, B. W., Thomson, N. R., et al. (2001) Genome sequence of *Yersinia pestis*, the causative agent of plague. *Nature* **413**, 523–527.

77. Wei, J., Goldberg, M. B., Burland, V., et al. (2003) Complete genome sequence and comparative genomics of *Shigella flexneri* serotype 2a strain 2457T. *Infect. Immun.* **71**, 2775–2786.

78. Jin, Q., Yuan, Z., Xu, J., et al. (2002) Genome sequence of *Shigella flexneri* 2a: insights into pathogenicity through comparison with genomes of *Escherichia coli* K12 and O157. *Nucleic Acids Res.* **30**, 4432–4441.

79. Scott, K. P., Melville, C. M., Barbosa, T. M., and Flint, H. J. (2000) Occurrence of the new tetracycline resistance gene *tet(W)* in bacteria from the human gut. *Antimicrob. Agents Chemother.* **44**, 775–777.

80. Scott, K. P. (2002) The role of conjugative transposons in spreading antibiotic resistance between bacteria that inhabit the gastrointestinal tract. *Cell. Mol. Life Sci.* **59**, 2071–2082.

81. Burrus, V. and Waldor, M. K. (2004) Shaping bacterial genomes with integrative and conjugative elements. *Res. Microbiol.* **155**, 376–386.

82. Nunes-Duby, S. E., Kwon, H. J., Tirumalai, R. S., Ellenberger, T., and Landy, A. (1998) Similarities and differences among 105 members of the Int family of site-specific recombinases. *Nucleic Acids Res.* **26**, 391–406.

83. Hall, R. M., Brookes, D. E., and Stokes, H. W. (1991) Site-specific insertion of genes into integrons: role of the 59-base element and determination of the recombination crossover point. *Mol. Microbiol.* **5**, 1941–1959.

84. Naas, T., Mikami, Y., Imai, T., Poirel, L., and Nordmann, P. (2001) Characterization of In53, a class 1 plasmid- and composite transposon-located integron of *Escherichia coli* which carries an unusual array of gene cassettes. *J. Bacteriol.* **183**, 235–249.

85. Mazel, D., Dychinco, B., Webb, V. A., and Davies, J. (1998) A distinctive class of integron in the *Vibrio cholerae* genome. *Science* **280**, 605–608.

86. Rowe-Magnus, D. A., Guerout, A. M., Ploncard, P., Dychinco, B., Davies, J., and Mazel, D. (2001) The evolutionary history of chromosomal super-integrons provides an ancestry for multiresistant integrons. *Proc. Natl. Acad. Sci. USA* **98**, 652–657.

87. Vaisvila, R., Morgan, R. D., Posfai, J., and Raleigh, E. A. (2001) Discovery and distribution of super-integrons among pseudomonads. *Mol. Microbiol.* **42**, 587–601.

88. Belfort, M., Reaban, M. E., Coetzee, T., and Dalgaard, J. Z. (1995) Prokaryotic introns and inteins: a panoply of form and function. *J. Bacteriol.* **177**, 3897–3903.

89. Chu, F. K., Maley, G. F., Maley, F., and Belfort, M. (1984) Intervening sequence in the thymidylate synthase gene of bacteriophage T4. *Proc. Natl. Acad. Sci. USA* **81**, 3049–3053.

90. Edgell, D. R., Belfort, M., and Shub, D. A. (2000) Barriers to intron promiscuity in bacteria. *J. Bacteriol.* **182,** 5281–5289.

91. Ko, M., Choi, H., and Park, C. (2002). Group I sel-splicing intron in the *recA* gene of *Bacillus anthracis. J. Bacteriol.* **184,** 3917–3922.

92. Dai, L. and Zimmerly, S. (2002) Compilation and analysis of group II intron insertions in bacterial genomes: evidence for retroelement behavior. *Nucleic Acids Res.* **30,** 1091–1102.

93. Michel, F. and Ferat, J. L. (1995) Structure and activities of group II introns. *Annu. Rev. Biochem.* **64,** 435–461.

94. Qin, P. Z. and Pyle, A. M. (1998) The architectural organization and mechanistic function of group II intron structural elements. *Curr. Opin. Struct. Biol.* **8,** 301–308.

95. Cousineau, B., Lawrence, S., Smith, D., and Belfort, M. (2000) Retrotransposition of bacterial group II intron. *Nature* **404,** 1018–1021.

96. Belhocine, K., Plante, I., and Cousineau, B. (2004) Conjugation mediates transfer of the Ll.LtrB group II intron between different bacterial species. *Mol. Microbiol.* **51,** 1459–1469.

97. Casjens, S. (2003) Prophages and bacterial genomics: what have we learned so far? *Mol. Microbiol.* **49,** 277–300.

98. Canchaya, C., Fournous, G., and Brussow, H. (2004) The impact of prophages on bacterial chromosomes. *Mol. Microbiol.* **53,** 9–18.

99. Freeman, V. J. (1951) Studies on the virulence of bacteriophage-infected strains of *Corynebacterium diphtheriae. J. Bacteriol.* **61,** 675–688.

100. Brussow, H., Canchaya, C., and Hardt, W. D. (2004) Phages and the evolution of bacterial pathogens: from genomic rearrangements to lysogenic conversion. *Microbiol. Mol. Biol. Rev.* **68,** 560–602.

101. Hendrix, R. W., Lawrence, J. G., Hatfull, G. F., and Casjens, S. (2000) The origins and ongoing evolution of viruses. *Trends Microbiol.* **8,** 504–508.

102. Campbell, A. M. (1992) Chromosomal insertion sites for phages and plasmids. *J. Bacteriol.* **174,** 7495–7499.

103. Campbell, A. (2002) Eubacterial genomes, in *Mobile DNA II* (Craig, N. L., Gellert, M., and Lambowitz, A. M., eds.). ASM Press, Washington, DC, pp. 1024–1039.

104. Blum, G., Ott, M., Lischewski, A., et al. (1994) Excision of large DNA regions termed pathogenicity islands from tRNA-specific loci in the chromosome of an *Escherichia coli* wild-type pathogen. *Infect. Immun.* **62,** 606–614.

105. Hacker, J. and Carniel, E. (2001) Ecological fitness, genomic islands and bacterial pathogenicity. A Darwinian view of the evolution of microbes. *EMBO Rep.* **2,** 376–381.

106. Dobrindt, U., Hochhut, B., Hentschel, U., and Hacker, J. (2004) Genomic islands in pathogenic and environmental microorganisms. *Nat. Rev. Microbiol.* **2,** 414–424.

107. Hershberg, R., Altuvia, S., and Margalit, H. (2003) A survey of small RNA-encoding genes in *Escherichia coli. Nucleic Acids Res.* **31,** 1813–1820.

108. Argaman, L., Hershberg, R., Vogel, J., et al. (2001) Novel small RNA-encoding genes in the intergenic regions of *Escherichia coli. Curr. Biol.* **11,** 941–950.

109. Wassarman, K. M., Repoila, F., Rosenow, C., Storz, G., and Gottesman, S. (2001) Identification of novel small RNAs using comparative genomics and microarrays. *Genes Dev.* **15,** 1637–1651.

110. Rivas, E., Klein, R. J., Jones, T. A., and Eddy, S. R. (2001) Computational identification of noncoding RNAs in *E. coli* by comparative genomics. *Curr. Biol.* **11,** 1369–1373.

111. Chen, S., Lesnik, E. A., Hall, T. A., et al. (2002) A bioinformatics based approach to discover small RNA genes in the *Escherichia coli* genome. *Biosystems* **65,** 157–177.

112. Tjaden, B., Saxena, R. M., Stolyar, S., Haynor, D. R., Kolker, E., and Rosenow, C. (2002) Transcriptome analysis of *Escherichia coli* using high-density oligonucleotide probe arrays. *Nucleic Acids Res.* **30,** 3732–3738.

113. Vogel, J., Bartels, V., Tang, T. H., et al. (2003) RNomics in *Escherichia coli* detects new sRNA species and indicates parallel transcriptional output in bacteria. *Nucleic Acids Res.* **31,** 6435–6443.

114. Masse, E., Majdalani, N., and Gottesman, S. (2003) Regulatory roles for small RNAs in bacteria. *Curr. Opin. Microbiol.* **6,** 120–124.

115. Storz, G., Opdyke, J. A., and Zhang, A. (2004) Controlling mRNA stability and translation with small, noncoding RNAs. *Curr. Opin. Microbiol.* **7,** 140–144.

116. Altuvia, S. (2004) Regulatory small RNAs: the key to coordinating global regulatory circuits. *J. Bacteriol.* **186,** 6679–6680.

117. Romeo, T. (1998) Global regulation by the small RNA-binding protein CsrA and the noncoding RNA molecule CsrB. *Mol. Microbiol.* **29,** 1321–1330.

118. Zhang, A., Altuvia, S., Tiwari, A., Argaman, L., Hengge-Aronis, R., and Storz, G. (1998) The OxyS regulatory RNA represses *rpoS* translation and binds the Hfq (HF-I) protein. *EMBO J.* **17,** 6061–6068.

119. Repoila, F., Majdalani, N., and Gottesman, S. (2003) Small non-coding RNAs, co-ordinators of adaptation processes in *Escherichia coli*: the RpoS paradigm. *Mol. Microbiol.* **48,** 855–861.

120. Opdyke, J. A., Kang, J. G., and Storz, G. (2004) GadY, a small-RNA regulator of acid response genes in *Escherichia coli*. *J. Bacteriol.* **186,** 6698–6705.

121. Chen, S., Zhang, A., Blyn, L. B., and Storz, G. (2004) MicC, a second small-RNA regulator of Omp protein expression in *Escherichia coli*. *J. Bacteriol.* **186,** 6689–6697.

122. Lenz, D. H., Mok, K. C., Lilley, B. N., Kulkarni, R. V., Wingreen, N. S., and Bassler, B. L. (2004) The small RNA chaperone Hfq and multiple small RNAs control quorum sensing in *Vibrio harveyi* and *Vibrio cholerae*. *Cell* **118,** 69–82.

123. Henke, J. M. and Bassler, B. L. (2004) Three parallel quorum-sensing systems regulate gene expression in *Vibrio harveyi*. *J. Bacteriol.* **186,** 6902–6914.

124. Masse, E., Escorcia, F. E., and Gottesman, S. (2003) Coupled degradation of a small regulatory RNA and its mRNA targets in *Escherichia coli*. *Genes and Dev.* **17,** 2374–2383.

125. Wilderman, P. J., Sowa, N. A., FitzGerald, D. J., et al. (2004) Identification of tandem duplicate regulatory small RNAs in *Pseudomonas aeruginosa* involved in iron homeostasis. *Proc. Natl. Acad. Sci. USA* **101,** 9792–9797.

126. McLeod, M. P., Qin, X., Karpathy, S. E., et al. (2004) Complete genome sequence of *Rickettsia typhi* and comparison with sequences of other rickettsiae. *J. Bacteriol.* **186,** 5842–5855.

2

Evolution and Origin of Virulence Isolates

Voon Loong Chan, Philip M. Sherman, and Billy Bourke

Summary

Perhaps the most significant benefit of microbial genomic sequences is the knowledge gained on the molecular process of genome evolution in microbes derived from comparative genomic analysis. Genetic variations are the driving forces of evolution. These are generated not only through base substitutions, small deletions and insertions, major DNA rearrangements and deletions, but also through DNA acquisition by horizontal gene transfer. Pathogens are evolved from diverse bacterial species. The molecular mechanisms involved are diverse, and likely affected by the conditions of the microenvironment inhabited by the evolving bacterial species.

Key Words: Evolution; virulence; mutations; horizontal gene transfer.

1. Introduction

As a result of the ever increasing number of genomes being sequenced, we have a much better understanding of the molecular processes of evolution of prokaryotes, especially bacteria. Comparative genomic analysis between virulent and nonvirulent strains of the same species provide valuable insight regarding mechanisms of evolution of virulence, whereas comparative analysis of genomes of closely related species provide a much better understanding of how new species have evolved.

Genetic variations (mutations) by generating a large pool of diverse variants for natural selection are the driving forces of evolution. Such genetic variations can be generated through local base changes, deletions, and acquisition of new DNA through horizontal transfer and DNA rearrangements. With constant change in environmental conditions, some new variant strains may prove to be more fit for a new microenvironment and, thus, survive, replicate, and become the predominant clone.

In this chapter, the key elements of biological evolution are discussed and some critical information gained from comparative analyses of sequenced bacterial genomes on the evolution of virulent strains and species are presented.

2. Genetic Variation

A significant source of mutant genes and mutants is small mutations like base substitutions, addition, or deletion of a small number of bases (resulting in frameshift mutations). Regions of the genome with homopolymeric tracts are "hot spots" for frameshift mutations. These mutations could occur during DNA replication or repair. The rate of infidelity of DNA polymerases during DNA replication or repair is usually low, but it could be greatly enhanced by the presence of mutagens, generated either endogenously

From: *Bacterial Genomes and Infectious Diseases*
Edited by: V. L. Chan, P. M. Sherman, and B. Bourke © Humana Press Inc., Totowa, NJ

(e.g., free radicals generated through oxidative respiration), or from the environment. This source of mutations is paramount for those bacteria that have little opportunity to acquire new genes through horizontal gene transfer (HGT).

Mutations can also be generated through DNA rearrangement caused by intrachromosomal homologous recombination between repeated sequences, e.g., insertion sequence (IS) elements and *rrn* operons. An inversion results when the repeated sequences involved in the recombination are in reverse orientation on the chromosome. Deletions and tandem duplications are generated when the repeated sequences are in the same, or tandem, orientation. DNA rearrangement can generate a new gene through fusion of distantly linked domains. Some bacterial genomes have a great abundance of IS elements, while others have few or none (*see* Chapter 1). For example, *Bordetella pertussis* and *Bordetella parapertussis* genomes have 261 and 112 copies of different IS elements respectively, whereas *Bordetella bronchiseptica* RB50 strain has none *(1)*. *B. pertussis* and *B. parapertussis* are believed to have evolved from different *B. bronchiseptica*-like ancestors. The acquisition of IS elements and their proliferation in *B. pertussis* and *B. parapertussis* is a major contributor to the generation of a large number of pseudogenes in these two bacterial genomes. This process likely has played an important role in the evolution of these two species.

Perhaps the most important strategy for evolving new or variant genes is the acquisition of genetic material from distantly or closely related species by HGT. These acquired genes, which have evolved in the donor organisms, are likely functional and may provide enhanced ability for the recipient cells to survive and grow in new niches (e.g., in the presence of antibiotics, heavy metals, or iron-limitation). Certain bacteria are naturally competent and can acquire these DNA fragments through transformation. HGT can also be mediated by a virus (transduction) or a conjugative plasmid or conjugative transposon (conjugation). Analyses of bacterial genomes sequenced to date indicate a huge diversity in the amount of HGT genes that are accumulated ranging from 0% for some small genomes (*Rickettsia prowazekii*, *Borrelia burgdorferi* and *Mycoplasma genitalium*) to 17% in *Synechocystis* PCC6803, and *Escherichia coli* K12 with 12.8% *(2)*.

3. Evolution Genes

Arber *(3)* postulated the existence of "evolution" genes. The products of these genes function to promote genetic variation or moderate the frequency of genetic variation for the benefit of the evolution of the micro-organism population. Genes involved in site-specific recombination are examples of evolution genes. Site-specific recombination systems are widespread in bacteria. In these systems, recombinases catalyze the inversion of DNA fragments that are flanked by consensus sequences. Genomes containing secondary consensus sequences at diverse sites provide the opportunity for a low level of aberrant novel inversion events, thereby leading to a pool of genetic variants. Transposase genes of IS elements are another example of evolution genes that promote genetic variation (*see* Chapter 1). Mismatched base pairs generated during DNA replication or DNA repair would also result in mutations if not corrected before cell division. The newly inserted noncomplementary nucleotide is generally removed by the methylation-dependent mismatch DNA repair enzymes *(4)*. Other DNA lesions can be repaired by base or nucleotide excision repair enzymes. These DNA repair genes are evolution genes that modulate the frequency of genetic variation.

Restriction-modification (RM) systems are composed of genes that encode a DNA restriction endonuclease and a modification methylase. A type II restriction enzyme recognizes a specific DNA sequence and cleaves it if the sequence is not methylated. The associated DNA methylase recognizes the same sequence and methylates a key residue within the specific sequence, thereby preventing cleavage of host DNA by its own restriction enzyme. Thus, RM systems are believed to play a role in reducing the uptake of foreign DNA to a tolerable level. However, DNA fragments which have been cleaved by restriction enzymes often have cohesive ends, and these DNA fragments are recombinogenic, thus facilitating recombination with host DNA. In this way RM systems can have a dual role in the evolution of prokaryotic genomes.

Analyses of sequenced prokaryotic genomes show that RM systems are widespread and some genomes have an abundance of RM genes. For example, 23 DNA methyltransferase genes were identified in the sequenced genome of *H. pylori* 26695 with the RM genes amounting to more than 1% of the genome *(5)*. Typically, genes encoding restriction enzymes are clustered with their cognate methyltransferase genes *(6)*. DNA methyltransferase genes are much easier to identify compared with genes for restriction enzymes because the former contain coding sequences for many characteristic protein sequence motifs *(7)*. Some RM gene complexes were shown to behave as selfish gene entities *(8)*. Close linkage of the restriction and modification genes of a RM genetic element favors mobility by HGT independent of the host chromosome. Furthermore, host cells that have lost the RM genetic element are killed because some of the target sites on the chromosome are no longer modified by the labile methylase and thus subjected to degradation by the stable restriction enzyme. There is abundant evidence that RM genetic elements are extensively transferred horizontally between distantly related genomes *(9)*. These genes are often linked with mobile genetic elements, including plasmids, prophages, transposons, and integrons.

Arber *(3)* suggested that bacterial viruses and plasmids also should be considered primarily as genetic elements with evolutionary functions because they are natural gene vectors and are involved in DNA acquisition. These elements promote genetic variation of host cells within a population and, therefore, enhance evolution.

4. Origin of Virulence Strains and Species

The first bacteria on this planet probably evolved about 3.8 billion yr ago. Virulence factors started to evolve about 1 billion years ago during the coevolution of bacteria and unicellular eukaryotic organisms *(10)*. Virulence factors are commonly defined as gene products that facilitate the bacterial interaction, subversion, and destruction of host cells, or the neutralization of host defense mechanisms. Bacterial toxins that are specific for highly conserved proteins like heterotrimeric G proteins, small G proteins, and actin are likely to have evolved in early time. However, some virulence factors evolved recently (since the arrival of higher eukaryotes). Other virulence factors of human-restricted bacterial pathogens like *Streptococcus pyogenes, Shigella* spp., and *Salmonella enterica* likely evolved in the past 1 million yr (the time frame of human evolution) *(10)*.

Virulent species have evolved from numerous divergent clades of bacteria. Prevalence of pathogenic bacterial species has been estimated from various environmental sources, including the biota from animal surfaces and digestive tracts. These studies showed that pathogens represent only a small percentage of microbial species *(11–14)*. Pathogenic strains and species are thought to be derived or evolved from either nonpathogenic siblings or closely related species.

A bacterial pathogen is defined as a microbe capable of causing host damage that results from either direct bacterial actions or from the host immune response to infection. There are many properties of a bacterium that can determine whether or not it will be a pathogen. As well as the production of a virulence factor, bacteria may need to infect the host or at least establish close contact and avoid host defenses. A pathogen may evolve from a commensal bacterium or a nonpathogen through various genetic mechanisms; such as the acquisition of virulence genes by HGT, major genomic deletions, genomic rearrangements, or point mutations.

The species of *E. coli* comprises not only nonpathogenic and commensal strains, but also different intestinal (intestinal pathogenic *E. coli* [IPEC]) and extraintestinal pathogenic (extraintestinal pathogenic *E. coli* [ExPEC]) strains that cause diseases in humans and some animals. To date, complete genome sequences have been published for five strains of *E. coli*, including nonpathogenic K12 strains MG1655 *(15)* and W3110 (http://ecoli/aist-nara.ac.jp/), two enterohemorrhagic *E. coli* (EHEC) O157:H7 strains *(16, 17)*, and one urosepsis (ExPEC) strain *(18)*. In addition, there are four pathogenic strains whose genomic sequences are at various stages of completion (*see* http://www.genome online.org). Comparison of the genome sequences of pathogenic and nonpathogenic *E. coli* strains identified a common core sequence of 4.1 Mb representing the backbone of the chromosome. This is interspersed with variable genomic islands (GEIs) containing strain-specific DNA sequences, some of which may contribute to virulence. Several of these GEIs in pathogenic *E. coli* strains represent pathogenicity islands (PAIs) or other mobile genetic elements that contain virulence genes *(16–19)*.

Genome analysis of the nonpathogenic probiotic *E. coli* strain Nissle 1917 identified four major GEIs (I–IV). The GEI II DNA region contains the *iuc, sat,* and *iha* genes that encode the aerobactin siderophore system, the serine protease Sat (autotransporter), and the putative adherence-conferring protein Iha, respectively *(20)*. These three genes are found and similarly organized in the *pheV*-associated GEI of the pathogenic (ExPEC) *E. coli* CFT073 strain, but the island of the pathogenic strain also contains the complete *hly* and *pap* gene clusters coding for the important virulence factors α-hemolysin and P fimbria, respectively. Interestingly, the *pap* operon in the GEI II of the probiotic strain is disrupted and partially deleted probably because of insertion of IS10 elements, and consecutive recombination events. These events, and the loss of the *hly* gene, were important steps in the evolution of the nonpathogenic Nissle 1917 strain *(20)*. Presumably the reverse could also occur, that is, a nonpathogenic *E. coli* strain, like Nissle 1917, on acquiring a wild type *hly* gene and a functional *pap* operon, could be transformed to a virulent strain.

Shigella species evolved from commensal *E. coli* strains in relatively recent evolutionary history. *Shigella* species and enteroinvasive *E. coli* (EIEC) both cause dysentery and have very similar phenotypic properties. Phylogenetic analysis indicates that *Shigella* and EIEC should be considered as a single pathovar of *E. coli (21)*. Evolution of pathogenic *Shigella* involved acquisition of a virulence plasmid, two PAIs (SHI-1 and SHI-2) *(2)* and the deletion of two virulence suppressor genes, *ompT* and *cadA (22,23)*.

Yersinia pestis, the causative agent of plague, also evolved recently in evolutionary time (1500–20,000 yr ago). It evolved from *Yersinia pseudotuberculosis (24)*. Two fully virulent *Y. pestis* strains, CO92 (Orientalis strain) *(25)* and KIM (Mediaevalis strain), *(26)* were sequenced in 2001 and 2002, respectively, followed by a nonvirulent strain 91001 *(27)* in 2004. The genome of *Y. pseudotuberculosis* IP32953, an entero-pathogen of humans and animals, was recently sequenced *(28)*. All three sequenced *Y. pestis* strains and *Y. pseudotuberculosis* possess a virulence plasmid pYV (70.3 kb) (pCD1 in *Y. pseudotuberculosis*) that encodes a type III secretion system, which is respon-sible for injecting a number of cytotoxins and effectors into host cells *(29*; Chapter 12*)*. *Y. pestis* possesses two additional plasmids, pPCP1 (9.6 kb) and pMT1 (102 kb), which encode plasminogen activator, murine toxin, and capsule-like antigen, respectively. The chromosome of *Y. pestis* CO92 (4.65 Mb) and *Y. pestis* KIM (4.60 Mb) are scat-tered with 21 GEIs including a cluster encoding a type-III secretion system that shows similar gene content and order to the Spi2 type III secretion system of *S. enterica* sero-var Typhimurium *(30)*. Comparative analysis of the genomes of *Y. pseudotuberculosis* IP32953 and *Y. pestis* CO92 and KIM, identified 112 genes that are found in the two *Y. pestis* strains, but not in IP32953 *(28)*. Further analysis of 19 *Yersinia* strains showed that 32 of the 112 genes, located in six clusters, are unique to *Y. pestis (28)*. These six clusters include many genes that encode phage-related proteins, membrane proteins, and proteins with no known function. None of these are known virulence factors, but their role in pathogenesis deserves to be explored.

Evolution of *Y. pestis* from *Y. pseudotuberculosis* also involved major accumulation of pseudogenes, many of which result from insertion of IS elements. At least nine of the pseudogenes are mutated regulatory genes, which likely have contributed to the pheno-type of *Y. pestis,* including its virulence *(28)*.

Vibrio cholerae is the etiological agent of cholera, a severe diarrheal disease ende-mic in many areas of Southern Asia and the Indian subcontinent. The emergence of pathogenic strains from nonpathogenic environmental strains involved the acquisition of at least three essential elements: (1) the PAI VPI-1, which encodes the type IV toxin coregulated pilus, an essential colonization factor and the receptor for phage CTXφ, (2) CTXφ that contains the *ctx AB* genes that encode cholera toxin (CT) *(31)*, and (3) transcriptional regulator genes (*toxRS*) and (*toxT* and *tcpPH*) encoded on the core chro-mosome and on the VPI-1, respectively *(32,33*; *see* Chapter 13*)*. Determinants for resis-tance to trimethoprim and aminoglycosides are encoded by super integrons, and integra-tive and conjugative elements encode resistance to sulfamethoxazole, trimethoprim, chloramphenicol, and streptomycin *(34)*.

Unlike the evolution of *E. coli* pathogens, the evolution of obligate intracellular pathogens of both mammals and arthropods, the Rickettsiales (*Rickettsia, Anaplasma,* and *Wolbachia*), involves major genome reduction. It is estimated that a few thousand genes were lost at the early stage of evolution prior to the divergence of *Rickettsia* and *Wolbachia* spp. *(35)*. *R. prowazekii*, the causative agent of typhus, has the smallest genome of the group with 1.1 Mb. Genomes of *Rickettsia* spp. have characteristic low levels of coding content and a high number of pseudogenes *(36)*. *R. prowazekii* is the extreme with 24% of the genome as noncoding DNA *(37)*. Noncoding DNA and pseudo-genes are intermediate stages of genome reduction. It appears that elimination of genes was most pronounced at an early stage of the transition to the intracellular environment *(36)*, but the process is ongoing in the modern species *(38)*.

Mycoplasma and *Ureaplasma* are two genera members of the taxonomic class of Mollicutes. *Mycoplasmas*, like all mollicutes, are bacteria that do not have a cell wall. *Mycoplasma* spp are all obligate parasites of various hosts and have a small genome, generally between 0.6 and 1.4 Mb. At least five species of *Mycoplasma* (*M. pneumoniae, M. genitalium, M. hominis, M. fermentans, M. penetrans*) and one *Ureaplasma* spp. (*U. urealyticum*) are human pathogens (*see* Chapter 10). Their small genomes are thought to be the result of reductive evolution from a low guanine + cytosine Grampositive bacterial ancestor that is common with *Clostridium* spp. and *Bacillus* spp *(39)*. Comparative analysis of the sequenced genomes of mycoplasma and ureaplasma species showed high levels of divergence and little conservation of gene order, except between *M. genitalium* and *M. pneumoniae* (*40*; Chapter 10). *Mycoplasmas* evolved different mechanisms of varying the antigenic structure of lipoproteins or adhesions, which they encode, to evade the host immune system (*41–45*; Chapter 10).

Three closely related *Bordetella* species, *B. pertussis, B. parapertussis*, and *B. bronchiseptica*, were recently sequenced *(46)*. These Gram-negative β-proteobacteria colonize the respiratory tract of mammals. *B. pertussis* is a strict human pathogen and is the causative agent of whooping cough. *B. parapertussis* infects both humans and sheep, and also can cause whooping cough in human infants. *B. bronchiseptica* has a broader host range causing respiratory infections in a wide range of animals and only occasionally humans. Comparative analysis of the three genomes indicates that *B. pertussis* and *B. parapertussis* evolved independently from a *B. bronchiseptica*-like ancestor *(46)*. Loss of genes and gene function play a major role in the evolution process *(1)*. *B. pertussis* Tohama I (4.1 Mb) and *B. parapertussis* strain 12822 (4.8 Mb) are not only significantly smaller than *B. bronchiseptica* RB50 (5.3 Mb), but also have a higher percentage of pseudogenes present in their genomes, 9.4 and 5.0 vs 0.4%, respectively.

Staphylococcus aureus, the causative agent of a wide range of human diseases, including carbuncles, food poisoning, bacteremia, necrotizing pneumonia, toxic shock syndrome, and endocarditis, is an important nosocomial and community-acquired pathogen. *S. aureus* encodes a large number of virulence factors that promote adhesion, colonization, cell–cell interactions, immune-system evasion, and tissue damage (*see* Chapter 11). Multilocus sequence typing of a large population of clinical isolates showed that the population structure of *S. aureus* is highly clonal *(47)*. Five *S. aureus* strains have been sequenced, including hospital-acquired methicillin-resistant strains (MRSAs), methilicillin-sensitive strains (MSSA), and vancomycin intermediate susceptible strains (*48–50*). The main differences between the strains are in accessory genetic elements, including Staphylococcal chromosome cassette (SCC) elements, PAIs, GEIs, transposons, prophages, plasmids, and insertion elements *(50)*. Virulence factors are encoded by the PAIs and lysogenic phages; therefore, it would appear as though evolution of virulence involved acquiring these mobile genetic elements (*50,51*; Chapter 11).

5. Conclusions

Comparative analyses of genomic sequences of virulent and nonvirulent strains and species provide great insight regarding the mechanisms of evolution of bacterial pathogens. Many bacterial lineages, especially those that occupy microenvironments that are

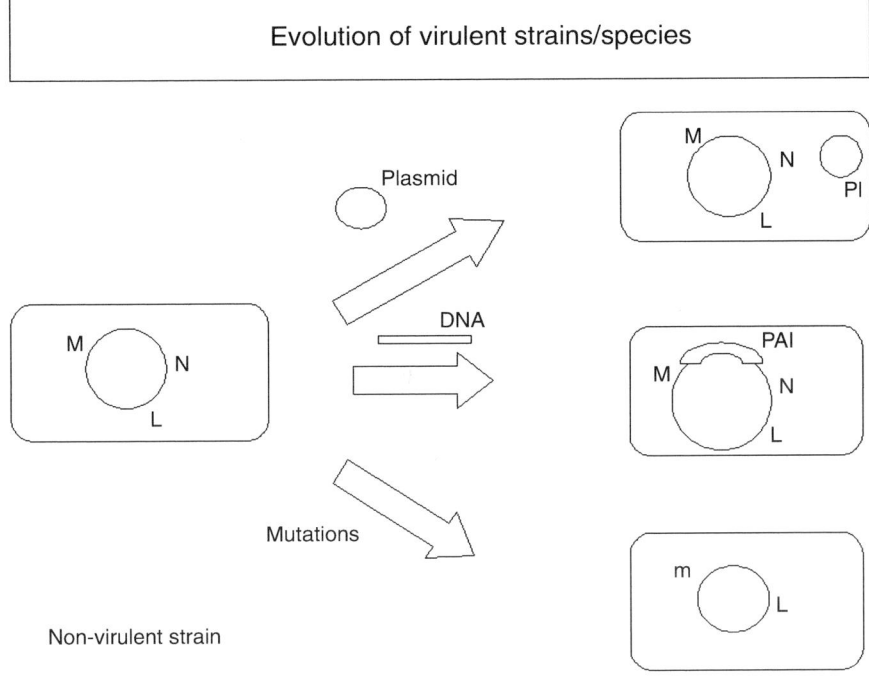

Evolution of virulent strains/species

Fig. 1. Mechanisms of evolution of virulent strains and species. Mechanisms are not exclusive. Evolution of virulent strains from a nonvirulent strain may include one or more mechanisms, whereas evolution of a virulent species generally would involve multiple mechanisms. The letters M, N, and L are different genes, and m is a mutant gene of M. PAI, pathogenicity island; pl, plasmid.

rich in micro-organisms and phages, acquired their virulence genes from another species through HGT. However, intracellular nonpathogenic bacterial lineages evolved to virulent strains or species through endogenous mutations, including deletions, insertions, gene duplications, gene fusions, and rearrangements (Fig. 1).

Acknowledgments

The authors are grateful for the insightful and thoughtful comments on the chapter by Drs. Penny Chan and Gavin Clark. Assistance on preparation of the figure by Chee Chan is much appreciated.

References

1. Preston, A., Parkhill, J., and Maskell, D. J. (2004) The bordetellae: lessons from genomics. *Nat. Rev. Microbiol.* **2,** 379–390.
2. Ochman, H., Lawrence, J. G., and Grolsman, E. A. (2000) Lateral gene transfer and the nature of bacterial innovation. *Nature* **405,** 299–304.
3. Arber, W. (2000) Genetic variation: mechanisms and impact on microbial evolution. *FEMS Microbiol. Rev.* **24,** 1–7.
4. Schofield, M. J. and Hsieh, P. (2003). DNA mismatch repair: molecular mechanisms and biological function. *Annu. Rev. Microbiol.* **57,** 579–608.

5. Tomb, J. F., White, O., Kerlavage, A. R., et al. (1997) The complete genome sequence of the gastric pathogen *Helicobacter pylori*. *Nature* **388,** 539–547.

6. Bickle, T. A. and Kruger, D. H. (1993) Biology of DNA restriction. *Microbiol. Review* **57,** 434–450.

7. Roberts, R. J. (1998) Bioinformatics: a new world of restriction and modification enzymes. *NEB Transcript* **9,** 1–4.

8. Nakayama, Y. and Kobayashi, I. (1998) Restriction-modification gene complexes as selfish gene entities: Roles of regulatory system in their estabishment, maintenance, and apoptotic mutual exclusion. *Proc. Natl. Acad. Sci. USA* **95,** 6442–6447.

9. Kobayashi, I. (2001) Behavior of restriction-modification systems as selfish mobile elements and their impact on genome evolution. *Nucleic Acids Res.* **29,** 3742–3756.

10. Brussow, H., Canchaya, C., and Hardt, W. D. (2004) Phages and the evolution of bacterial pathogens: from genomic rearrangements to lysogenic conversion. *Microbiol. Mol. Biol. Rev.* **68,** 560–602.

11. Kroes, I., Lepp, P. W., and Relman, D. A. (1999) Bacterial diversity within the human subgingival crevice. *Proc. Natl. Acad. Sci. USA* **96,** 14,547–14,552.

12. DeLong, E. F. (2002) Microbial population genomics and ecology. *Curr. Opin. Microbiol.* **5,** 520–524.

13. Hugenholtz, P. (2002) Exploring prokaryotic diversity in the genomic era. *Genome Biol.* **3,** REVIEWS0003.

14. Torsvik, V. and Ovreas, L. (2002) Microbial diversity and function in soil: from genes to ecosystems. *Curr. Opin. Microbiol.* **5,** 240–245.

15. Blattner, F. R., Plunkett, G. 3rd, Bloch, C. A., et al. (1997) The complete genome sequence of *Escherichia coli* K-12. *Science* **277,** 1453–1474.

16. Perna, N. T., Plunkett, G. 3rd, Burland, V., et al. (2001) Genome sequence of enterohaemorrhagic *Escherichia coli* O157:H7. *Nature* **409,** 529–533.

17. Hayashi, T., Makino, K., Ohnishi, M., et al. (2001) Complete genome sequence of enterohemorrhagic *Escherichia coli* O157:H7 and genomic comparison with a laboratory strain K-12. *DNA Res.* **8,** 11–22.

18. Welch, R. A., Burland, V., Plunkett, G. 3rd, et al. (2002) Extensive mosaic structure revealed by the complete genome sequence of uropathogenic *Escherichia coli*. *Proc. Natl. Acad. Sci. USA* **99,** 17,020–17,024.

19. Dobrindt, U., Hochhut, B., Hentschel, U., and Hacker, J. (2004) Genomic islands in pathogenic and environmental microorganisms. *Nat. Rev. Microbiol.* **2,** 414–424.

20. Grozdanov, L., Raasch, C., Schulze, J., et al. (2004) Analysis of the genome structure of the nonpathogenic probiotic *Escherichia coli* strain Nissle 1917. *J. Bacteriol.* **186,** 5432–5441.

21. Lan, R., Alles, M. C., Donohoe, K., Martinez, M. B., and Reeves, P. R. (2004) Molecular evolutionary relationships of enteroinvasive *Escherichia coli* and *Shigella* spp. *Infect. Immun.* **72,** 5080–5088.

22. Nakata, N., Tobe, T., Fukuda, I., et al. (1993) The absence of a surface protease, OmpT, determines the intercellular spreading ability of *Shigella*: the relationship between the *ompT* and *kcpA* loci. *Mol. Microbiol.* **9,** 459–468.

23. Maurelli, A. T., Fernandez, R. E., Bloch, C. A., Rode, C. K., and Fasano, A. (1998) "Black holes" and bacterial pathogenicity: a large genomic deletion that enhances the virulence of *Shigella* spp. and enteroinvasive *Escherichia coli*. *Proc. Natl. Acad. Sci. USA* **95,** 3943–3948.

24. Achtman, M., Zurth, K., Morelli, G., Torrea, G., Guiyoule, A., and Carniel, E. (1999) *Yersinia pestis*, the cause of plague, is a recently emerged clone of *Yersinia pseudotuberculosis*. *Proc. Natl. Acad. Sci. USA* **96,** 14,043–14,048.

25. Parkhill, J., Wren, B. W., Thomson, N. R., et al. (2001) Genome sequence of *Yersinia pestis*, the causative agent of plague. *Nature* **413,** 523–527.

26. Deng, W., Burland, V., Plunkett, G. 3rd, et al. (2002) Genome sequence of *Yersinia pestis* KIM *J. Bacteriol.* **184,** 4601–4611.

27. Song, Y., Tong, Z., Wang, J., et al. (2004) Complete genome sequence of *Yersinia pestis* strain 91001, an isolate avirulent to humans. *DNA Res.* **11,** 179–197.

28. Chain, P. S., Carniel, E., Larimer, F. W., et al. (2004) Insights into the evolution of *Yersinia pestis* through whole-genome comparison with *Yersinia pseudotuberculosis. Proc. Natl. Acad. Sci. USA* **101,** 13,826–13,831.

29. Cornelis, G. R. (2002) *Yersinia* type III secretion: send in the effectors. *J. Cell Biol.* **158,** 401–408.

30. Wren, B. W. (2003) The yersiniae—a model genus to study the rapid evolution of bacterial pathogens. *Nat. Rev. Microbiol.* **1,** 55–64.

31. Waldor, M. K. and Mekalanos, J. J. (1996) Lysogenic conversion by a filamentous phage encoding cholera toxin. *Science* **272,** 1910–1914.

32. Faruque, S. M., Asadulghani, Saha, M. N., et al. (1998) Analysis of clinical and environmental strains of nontoxigenic *Vibrio cholerae* for susceptibility to CTXPhi: molecular basis for origination of new strains with epidemic potential. *Infect Immun.* **66,** 5819–5825.

33. Dziejman, M., Balon, E., Boyd, D., Fraser, C. M., Heidelberg, J. F., and Mekalanos, J. J. (2002) Comparative genomic analysis of *Vibrio cholerae*: genes that correlate with cholera endemic and pandemic disease. *Proc. Natl. Acad. Sci. USA* **99,** 1556–1561.

34. Hochhut, B. and Waldor, M. K. (1999) Site-specific integration of the conjugal *Vibrio cholerae* SXT element into *prfC. Mol. Microbiol.* **32,** 99–110.

35. Batut, J., Andersson, S. G., and O'Callaghan, D. (2004) The evolution of chronic infection strategies in the alpha-proteobacteria. *Nat. Rev. Microbiol.* **2,** 933–945.

36. Boussau, B., Karlberg, E. O., Frank, A. C., Legault, B. A., and Andersson, S. G. (2004) Computational inference of scenarios for alpha-proteobacterial genome evolution. *Proc. Natl. Acad. Sci. USA* **101,** 9722–9727.

37. Andersson, S. G., Zomorodipour, A., Andersson, J. O., et al. (1998) The genome sequence of *Rickettsia prowazekii* and the origin of mitochondria. *Nature* **396,** 133–140.

38. Amiri, H., Davids, W., and Andersson, S. G. (2003) Birth and death of orphan genes in *Rickettsia. Mol. Biol. Evol.* **20,** 1575–1587.

39. Maniloff, J. (2002) Phylogeny and evolution, in *Molecular Biology and Pathogenicity of Mycoplasmas* (Rasin, S. and Herrmann, R., eds.). Kluwer Academic/Plenum Publishers, New York, pp. 31–44.

40. Sasaki, Y., Ishikawa, J., Yamashita, A., et al. (2002) The complete genomic sequence of *Mycoplasma penetrans*, an intracellular bacterial pathogen in humans. *Nucleic Acids Res.* **30,** 5293–5300.

41. Neyrolles, O., Chambaud, I., Ferris, S., et al. (1999) Phase variations of the *Mycoplasma penetrans* main surface lipoprotein increase antigenic diversity. *Infect. Immun.* **67,** 1569–1578.

42. Glew, M. D., Papazisi, L., Poumarat, F., Bergonier, D., Rosengarten, R., and Citti, C. (2000) Characterization of a multigene family undergoing high-frequency DNA rearrangements and coding for abundant variable surface proteins in *Mycoplasma agalactiae. Infect. Immun.* **68,** 4539–4548.

43. Roske, K., Blanchard, A., Chambaud, I., et al. (2001) Phase variation among major surface antigens of *Mycoplasma penetrans. Infect. Immun.* **69,** 7642–7651.

44. Liu, L., Panangala, V. S., and Dybvig, K. (2002) Trinucleotide GAA repeats dictate *pMGA* gene expression in *Mycoplasma gallisepticum* by affecting spacing between flanking regions. *J. Bacteriol.* **184,** 1335–1339.

45. Winner, F., Markova, I., Much, P., et al. (2003) Phenotypic switching in *Mycoplasma gallisepticum* hemadsorption is governed by a high-frequency, reversible point mutation. *Infect. Immun.* **71,** 1265–1273.
46. Parkhill, J., Sebaihia, M., Preston, A., et al. (2003) Comparative analysis of the genome sequences of *Bordetella pertussis, Bordetella parapertussis* and *Bordetella bronchiseptica. Nat. Genet.* **35,** 32–40.
47. Feil, E. J., Cooper, J. E., Grundmann, H., et al. (2003) How clonal is *Staphylococcus aureus*? *J. Bacteriol.* **185,** 3307–3316.
48. Kuroda, M., Ohta, T., Uchiyama, I., et al. (2001) Whole genome sequencing of meticillin-resistant *Staphylococcus aureus. Lancet* **357,** 1225–1240.
49. Baba, T., Takeuchi, F., Kuroda, M., et al. (2002) Genome and virulence determinants of high virulence community-acquired MRSA. *Lancet* **359,** 1819–1827.
50. Holden, M. T., Feil, E. J., Lindsay, J. A., et al. (2004) Complete genomes of two clinical *Staphylococcus aureus* strains: evidence for the rapid evolution of virulence and drug resistance. *Proc. Natl. Acad. Sci. USA* **101,** 9786–9791.
51. Lindsay, J. A. and Holden, M.T. (2004) *Staphylococcus aureus*: superbug, super genome? *Trends Microbiol.* **12,** 378–385.

3

Genomic Approach to Understanding Infectious Disease Mechanisms

Voon Loong Chan, Philip M. Sherman, and Billy Bourke

Summary

Close to 100 genomes of bacterial pathogens have been sequenced, and yet most of the genomes sequenced have approx 25% of their open reading frames annotated as proteins with no known function. When genomic sequences of virulent and nonvirulent strains of a particular species are available, comparative genomic analysis is a powerful tool to identify putative virulence genes. Variation in virulence between strains of the same species is a common phenomenon. Availability of the genome sequence of a pathogen permits the application of DNA microarrays to investigate the genetic basis of this variation. DNA microarray technology has facilitated the identification of putative virulence determinants, host specificity genes, and bacterial and host genes that are activated or repressed during an infection. Isogenic mutants and suitable virulence assays are critical in verifying the role of the putative virulence genes identified.

Key Words: Comparative genomic analysis; microarray; virulence genes; bacteria–host interaction.

1. Introduction

As of late 2004, the genomes of 95 bacteria that infect humans were sequenced and annotated. We have gained a huge amount of information on the diverse structure of bacterial genomes and the genes they encode, and have a fair understanding of their evolutionary history. Unfortunately, for most of the pathogenic genomes sequenced, 25% or more of the open reading frames (ORFs) identified do not match any known genes and, thus, their functions are unknown (so called *FUN* genes). These *FUN* genes are often species- or strain-specific. To gain full understanding of the mechanisms of pathogenesis, we need to focus research and mount a consolidated effort using multi-disciplinary approaches to probe the structure and function of these *FUN* genes.

In this chapter, genomic approaches to gaining understanding of infectious disease mechanisms of different pathogens will be discussed. Two innovative approaches, DNA microarrays and whole genome polymerase chain reaction (PCR) scanning (WGPScanning), have been developed to analyze and compare the gene contents of the sequenced bacterial strains with that of other isolates of the same or closely related species. This chapter will illustrate how DNA microarrays and WGPScanning are invaluable in defining intraspecies diversity and identifying putative virulence genes. In the final section, the value of generating isogenic mutants of a putative virulence gene, and the necessity of biochemical and virulence assays, are illustrated.

From: *Bacterial Genomes and Infectious Diseases*
Edited by: V. L. Chan, P. M. Sherman, and B. Bourke © Humana Press Inc., Totowa, NJ

2. Comparative Genomic Analysis of Virulent and Nonvirulent Strains

Yersinia pestis, the causative agent of bubonic and pneumonic plagues that killed 25 million people in Europe between 1346 and 1350, is an example where genomic sequence is available for virulent and nonvirulent strains. Two fully virulent strains, CO92 (Orientalis strain) *(1)* and KIM (Mediaevalis strain) *(2)* were the first sequenced. Recently, the genome of a nonvirulent strain 91001 was determined *(3)*. Comparative genomic analysis identified a 33-kb prophage-like fragment that is shared by CO92 and KIM but is absent in the nonvirulent 91001 strain. Song et al. *(3)* proposed that, because CO92 and KIM are virulent to humans and strain 91001 is only virulent to mice, the prophage-like fragment might contribute to pathogenicity in humans. This hypothesis needs to be verified by constructing mutants with mutations in genes located in this fragment.

Shigella is a major pathogen, but its pathogenicity is restricted to humans. *Shigella* species evolved from a commensal *Escherichia coli* strain in relatively recent evolutionary history (*see* Chapter 2). Recent phylogenetic analyses indicate that *Shigella* should be considered as a single pathovar of *E. coli (4)*. The complete genome sequence of two strains of *Shigella flexneri* serotype 2a, strain 2457T *(5)* and strain 301 *(6)*, were determined and compared with that of the enterohemorrhagic *E. coli* O157:H7 and non-pathogenic *E. coli* K12. Analyses of the genome sequences of *S. flexneri* 2457T and 301 identified many new candidate genes for involvement in different infectious stages of bacterial invasion during translocation through the colonic mucosa *(5,6)*. These new candidate virulence genes are all located on genomic islands (GEIs) and some encode proteins that have similarity to known virulence factors of other pathogens. Mutants need to be constructed with mutations in these putative virulence genes and then tested in an appropriate model system. Unfortunately, there is not an appropriate animal model system available for *S. flexneri* strains (for instance, they have not been shown to cause intestinal diseases in mice). Wei et al. *(5)* proposed constructing a transgenic mouse expressing a human receptor for *S. flexneri* adhesion. However, this can not be done until the human-specific adhesin and its cognate receptor have been identified.

3. DNA Microarray Analysis of Bacterial Pathogens

A DNA microarray is a high-density array of nucleic acid targets immobilized on a glass slide or a silicon chip. The nucleic acids are either denatured complementary DNA or genomic products amplified by PCR for spotted DNA arrays. For oligonucleotides arrays, oligonucleotides complementary to specific gene sequences are generally synthesized *in situ*. DNA microarray was first used in parallel detection and the analysis of the expression patterns of thousands of genes in plant tissues *(7)* and human cells exposed to different conditions *(8)*. Since complete bacterial genome sequences have been available, DNA microarrays have been used to analyze genome-wide gene expression of many bacteria and also to compare the whole genetic content of the sequenced bacterial strains with those of closely related species or other isolates of the same species *(9,10)*. As discussed in **Subheadings 3.** and **4.**, microarray has facilitated the identification of putative virulence determinants and genes that determine host specificity.

The genus *Brucella* consists of six species, *B. melitensis* (primarily in goats and sheep), *B. abortus* (cattle), *B. suis* (pig), *B. canis* (dogs), *B. ovis* (sheep), and *B. neotomae* (wood rats). *B. abortus, B. melitensis*, and *B. suis* are also pathogenic to humans; all causing similar serious disease consequences. The complete genome sequence of *B. melitensis*

16M was determined and a whole-genome high-density oligonucleotide DNA microarray was developed and used to compare genomes of other *Brucella* species *(11)*. Hybridization results showed that the great majority of the ORFs in *B. melitensis* 16M are present in all five *Brucella* genomes examined, supporting the suggestion that *Brucella* is a monospecific genus with limited genetic diversity *(12)*. Many of the putative deleted genes are clustered into nine GEIs. Comparison of *B. melitensis* 16M with two *Brucella* species that are nonpathogenic for humans, *B. ovis* REO198 and *B. neotomae* 5K33, is of great interest because it may identify genes that are responsible for pathogenicity in humans. Microarray hybridization, followed by PCR cloning and sequencing, identified 84 ORFs that are found in the 16M genome, but are absent or partially deleted in the genome of *B. ovis* REO198. Eighty of the 84 ORFs were clustered in five regions corresponding to GEIs (GI-1, GI-2, GI-5, GI-7, and GI-9) of the 16M genome *(11)*. The genome of *B. neotomae* 5K33 is very similar to that of 16M; only 17 ORFs were identified as deleted or altered, including a region containing 10 contiguous ORFs corresponding to about 7.5 kb in the GI-6 of 16M. The GEIs (GI-1, GI-2, GI-5, GI-7, and GI-9) missing in *B. ovis*, are present in *Brucella* species that are pathogenic to humans. However, *B. neotomae*, a species not pathogenic to humans, also has these islands. It is possible that some of these genes in *B. neotomae* are expressed at a much lower level or are inactivated, because microarray hybridization cannot identify minor base changes and thus will not detect most pseudogenes. Further studies involving proteomic analysis and mutagenesis of specific genes are needed to understand the role of these GEIs in host adaptation and virulence of the *Brucella* species.

Porphyromonas gingivalis is a Gram-negative oral anarobe associated with periodontal disease in humans. The genome of a virulent strain of *P. gingivalis* W50 was recently sequenced *(13)*. A DNA microarray was prepared from PCR amplicons derived from the predicted ORFs of this virulent strain. This microarray was used as a reference to compare the genome of W50 with that of a nonvirulent strain, ATCC33277, with the hope of identifying genes that are required for virulence. Hybridization results showed that 154 ORFs (7% of the total) that are present in W50 are highly divergent or absent in the nonvirulent strain ATCC33277. Interestingly, many of these divergent genes are clustered into three regions of the W50 genome with a lower guanine and cytosine content than the rest of the genome, corresponding to GEIs which are clusters of genes that were likely acquired through horizontal gene transfer. These regions contain putative virulence genes, including genes for capsular polysaccharide synthesis, a lipoprotein gene (*rag B*), and many species-specific hypothetical genes. These GEIs should be the focus for a future pathogenesis study of this pathogen.

Chlamydia trachomatis is an obligate intracellular pathogen of humans. Isolates are differentiated into biovars based on their in vitro infection properties and type of disease caused. They are grouped into 15 serovars based on antigenic variation of the major outer membrane protein, OmpA. Serovars A to C are the etiological agents of trachoma; serovars D to K and L1 to L3 cause cervicitis and urethritis or lymphogranuloma venereum , respectively *(14)*. The genomes of three *Chlamydia* species, *C. trachomatis*, *C. pneumoniae*, and *C. psittaci* have been sequenced *(15–17)*. Comparative analysis of these genomes (1.04–1.3 Mb) showed a high degree of conservation in terms of gene content and gene order with the exception of one polymorphic region named the plasticity zone (PZ), which showed a significant amount of variation among the different species. A DNA microarray was developed based on the genomic sequence of *C. trach-omatis* serovar D and used in analyzing the genomic diversity of the 15 serovars. The hybridization results showed that the genomes of the 15 serovars are highly conserved (>99%). In contrast, similar studies done with isolates of *Helicobacter pylori (18)* and *Staphylococcus aureus (19)* identified major intraspecies diversity. Variation in the *OmpA* gene was observed as expected, and was confirmed by gene cloning and sequencing. Serovar B showed the greatest diversity with a maximum of eight deleted genes. Deletions were also observed with serovars A, Ba, C, L1, L2, and L3. No gene deletion was observed in serovars E, F, G, H, I, J, and K. Without exception, all the deleted genes observed localized to the PZ region *(20)*. To precisely define the PZ region of *C. trachomatis*, PCR primers were designed to amplify the region as four overlapping fragments and sequenced. PCR amplification results showed that most of the ocular and genital serovars have a large internal deletion compared with the intact cytotoxin gene of *C. muridarum,* a mouse-adapted strain. Deletions in the ocular serovars (A, Ba, and C) are smaller than those observed in the genital serovars (D, E, F, G, I, and K).

The DNA sequences obtained from all the serovars were aligned with the mouse-adapted *C. muridarum* cytotoxin *(TC0438)* gene. Analysis of the sequence showed that each serovar possessed a unique combination of DNA sequences and ORFs, but none encoded a full-length cytotoxin. It appears that the genital serovars have a large ORF that could encode both the uridine diphosphate (UDP)-glucose binding and the glycosyltransferase domains, while the ORF of the ocular serovars could encode only the UDP-glucose binding domain. Deletion in the genome of the lymphogranuloma venereum serovars removed the encoding sequence of both domains. These results suggest the UDP-glucose binding and the glycosyltransferase domains of the chlamydia cytotoxin, a homolog of the clostridial toxin, may have an important role in urogenital infection *(20)*.

Microarrays have also been used to analyze intraspecies diversity and to identify strain-specific genes of many bacterial pathogens, including *H. pylori*, *Streptococcus pneumoniae*, *Salmonella enterica* serovar Typhimurium, *Campylobacter jejuni (21,22)*, *Vibrio cholerae (23)*, *E. coli*, and many others. Microarray analysis of 15 *H. pylori* strains showed great genetic diversity, 362 genes (22% of all *H. pylori* genes) are not conserved among all 15 strains *(9)*.

An Affymetrix high-density oligonucleotide array based on the genomic sequence of *S. pneumoniae* serotype 4 strain was used to analyze genetic diversity of 20 clinical *S. pneumoniae* isolates and 9 oral streptococcal isolates. Most of the *S. pneumoniae* strains differed from the sequenced strain by 8–11%. In contrast, the nine oral streptococci strongly diverged from the sequenced strain, as only 15–61% of their DNA hybridized with the reference array. This study identified 470 *S. pneumoniae* strain-specific genes, which were not detected in at least one of the strains examined *(24)*. A desired future goal is to be able to associate a specific gene(s) lost to a specific disease phenotype.

S. pneumoniae infection can lead to bacteremia and meningitis. Invasive diseases are the result of spread of the bacteria from the nasopharynx, the site of colonization, to the lung and bloodstream with possible sequelae of septicemia or meningitis. With the aim of identifying virulence determinants of *S. pneumoniae*, a whole-genome microarray was used to analyze the bacterial genes that are expressed in vivo. The study involved isolating total RNA from *S. pneumoniae* isolated from infected blood, infected cerebrospinal fluid (CSF), and bacteria attached to a pharyngeal epithelial cell (ECC) line in vitro. Gene expression levels at these three sites were compared with levels when *S. pneumoniae* was grown in semisynthetic casein liquid medium. Such in vivo studies are limited by the difficulties of obtaining sufficient quantities of pure and relatively intact bacterial RNA from infected tissues. Interestingly, the majority of the genes (92% in the blood, 85% in CSF, and 90% after ECC) were expressed in a similar fashion as growth in culture medium. However, distinct patterns of gene expression in each anatomical site can be identified. Amazingly, only eight genes showed similar alterations in gene expression during bacterial growth in blood, in CSF, or during ECC *(25)*. Two of the eight genes encode pspA *(26)* and prtA *(27)* previously characterized as virulence factors, three involved in manganese acquisition and transport (*psa* operon), two in energy metabolism and one transporter. Orihuela et al. *(25)* postulated that these eight genes may be part of the core set of genes required for virulence and, therefore, deserve further investigation.

DNA microarrays also were successfully used in analyzing whole-genome gene expression (transcriptome) of uropathogenic *E. coli* strain CFT073 during urinary tract infection of CBA/J mice *(28)*. Total RNA was isolated from CFT073 bacteria obtained directly from the urine of infected mice. The in vivo transcription profiles were compared with those of CFT073 grown statically to exponential phase in rich medium. Overall, transcription of 313 genes was found elevated, whereas that of 207 genes was reduced. Of the 313 CFT073 genes that were to be elevated, only 45 genes were unique to the uropathogenic strain and not found in nonpathogenic *E. coli* K12. The author proposed that these 45 are candidate virulence genes for urinary tract infection. Twenty-five of these genes have previously been implicated in virulence. These include genes involved in iron acquisition, capsule synthesis, and synthesis of microcin secretion proteins. Thirteen new candidate virulence genes encoding hypothetical proteins were also identified.

4. DNA Microarray Analysis of Host–Pathogen Interaction

Understanding the molecular basis of the host response to bacterial infections is critical for understanding disease mechanisms involved, and in preventing disease and tissue damage resulting from the host response. The global transcription effects on host cells by various bacterial pathogens including *Listeria monocytogenes*, *Salmonella typhimurium*, *Pseudomonas aeruginosa*, *Bordetella pertussis*, *Mycobacterium tuberculosis*, *H. pylori*, and *Chlamydia trachomatis* have been analyzed by using microarray technology *(29,30)*. Rosenberger et al. *(31)* identified novel macrophage genes whose level of expression are altered in *S. typhimurium* infection or when treated with lipopolysaccharide. Similarly, Cohen et al. *(32)* identified 74 upregulated RNAs and 23 downregulated host RNAs in *L. monocytogenes*-infected human promyelocytic THP1 cells. Infection of human bronchial epithelial cells by *B. pertussis* results in an increase in transcriptional levels of 33 genes and decrease in transcriptional levels of 65 genes *(33)*. Many of the upregulated genes encode proinflammatory cytokines (e.g., interleukin-8, interleukin-6, and growth-related oncogene-1) and many of the downregulated genes encode transcriptional factors and cellular adhesion molecules.

H. pylori is a human gastric pathogen that causes gastritis, gastric ulcer, and duodenal ulcer, and is implicated to enhance the incidence of mucosa-associated lymphoid tissue lymphoma and gastric cancer. The genomes of *H. pylori* strains sequenced contain a pathogenicity island (PAI) denoted *cag* (*cag*PAI), which has been shown to be a major virulence factor encoding about 30 proteins *(34;* Chapter 6). The *cag*PAI encodes a type IV secretion system that facilitates transfer of virulence factors to the host cells *(35).* Nagasako et al. *(36)* used cDNA microarrays to compare the effects of *cag*PAI-positive and *cag*PAI-negative *H. pylori* in altering the gene expression of infected gastric epithelial cells. This study showed that a host protein, Smad5, was upregulated in *cag*PAI-positive *H. pylori*-infected cells, but not the *cag*PAI-negative cells. The upregulated expression of Smad5 mRNA is required for induction of apoptosis of gastric epithelial cells because inhibition of Smad5 mRNA expression by Smad5-specific siRNA prevents apoptosis *(36).*

5. Whole-Genome PCR Scanning Analysis

Another genomic analysis strategy, WGPScanning, was developed by Ohnishi et al. *(37).* The principle of the WGPScanning method is to design a set of PCR primer pairs so that amplified segments overlapp with adjacent segments at both ends and cover the whole genome of a reference sequenced strain. By comparing the amplified fragments with those of the reference genome, one can determine whether the target regions are arranged in the same order, and whether the segments between target regions have undergone any major structural changes, including deletions and insertions. Ohnishi et al. *(37)* developed a WGPScanning primer pair set based on the genomic sequence of *E. coli* O157 Sakai to analyze eight O157 strains in order to determine the genomic diversity of this lineage. The results indicate that significant genomic diversity exists among O157 strains. Variation of prophages located in GEIs is a major contributor of genomic diversity of the O157 lineage.

WGPScanning was used in analyzing 19 serotype M3 group A *Streptococcus* strains isolated from patients in Ontario, Canada *(38).* This scanning method permitted the authors to detect genetic diversity in the *sclb* gene that encodes SclB, a collagen-like surface protein, which has been implicated in host–pathogen interaction. This minor sequence diversity was not detected by DNA microarray. Thus, WGPScanning is useful in complementing the microarray method.

6. Construction and Characterization
of Isogenic Mutants of Putative Virulence Genes

Bioinformatics and various innovative genomic tools discussed in **Subheadings 2.–5.** dramatically speed up the process of identification of putative virulence genes of various bacterial pathogens. To verify and define the role of these genes in pathogenesis, isogenic mutants need to be constructed and analyzed using different virulence assays or disease animal models. Our laboratory recently reported the cloning and characterization of a lipoprotein gene, *jlpA,* of *C. jejuni (39).* The *jlpA* gene is located in a *C. jejuni*-specific region of the genome *(40,41).* An isogenic *C. jejuni* mutant containing an insertion in *jlpA* exhibited a reduced level of adhesion to HEp2 cells ($20.2 \pm 2.1\%$ relative to wild-type), implicating possible involvement in cell adhesion. A deletion *jlpA C. jejuni* mutant has also been constructed and shows a similar defect in adhesion to HEp2 cells. Adhesion of wild-type *C. jejuni* to HEp2 cells is inhibited by anti-JlpA antibodies and by the presence of excess purified GST–JlpA fusion protein further implicating a role for JlpA in adhesion *(39).* JlpA binds to surface-localized Hsp90α of HEp2 cells, resulting in activation of nuclear factor κB (NF-κB) *(42)*. The activation occurred in a dose-dependent manner. Geldanamycin, an antibiotic that binds to the N-terminal of Hsp90, inhibits the induction of NF-κB activation by JlpA *(42)*. Induction of NF-κB activation in HEp-2 cells by JlpA was blocked by pretreatment of GST-JlpA (10 μg/mL) with anti-GST-JlpA antibodies, and GST-JlpA was unable to induce the activation of NF-κB in anti-Hsp90α antibody-blocked (5 μg/mL) HEp-2 cells *(42)*. These results further indicate that interaction between JlpA and Hsp90α is critical in triggering the signaling pathway leading to NF-κB activation. To further define the affects of JlpA on the host cell signaling pathways, knockout mice with defects in different signaling pathways would be invaluable (*see* Chapter 4).

7. Conclusions

The value of determining the complete genome sequence of pathogens is clear, as it permits the identification of some virulence genes by comparing the predicted ORFs to the data base. GEIs and mobile genetic elements can be identified. Comparative genomic analysis with nonpathogenic strains will further identify putative virulence determinants. Microarray and WGPScanning are powerful tools that can be used to identify diversity and plasticity of the genome of the sequenced species, and also strain-specific genes and genes that determine host range. Microarray analysis is also an efficient tool for examining the effects of bacterial infection on the host global transcription profile, which could lead to a better understanding of disease mechanisms. Mutants of the putative virulent genes need to be constructed and analyzed in different virulence assays, including animal models, before their role(s) in pathogenesis can be confirmed.

Acknowledgments

The authors are grateful for the insightful and thoughtful comments on the chapter by Drs. Penny Chan and Gavin Clark. The *Campylobacter* work described was funded by the Crohn's and Colitis Foundation of Canada.

References

1. Parkhill, J., Wren, B. W., Thomson, N. R., et al. (2001) Genome sequence of *Yersinia pestis*, the causative agent of plague. *Nature* **413,** 523–527.
2. Deng, W., Burland, V., Plunkett, G. III., et al. (2002) Genome sequence of *Yersinia pestis* KIM. *J. Bacteriol.* **184,** 4601–4611.
3. Song, Y., Tong, Z., Wang, J., et al. (2004) Complete genome sequence of *Yersinia pestis* strain 91001, an isolate avirulent to humans. *DNA Res.* **11,** 179–197.
4. Lan, R., Alles, M. C., Donohoe, K., Martinez, M. B., and Reeves, P. R. (2004) Molecular evolutionary relationships of enteroinvasive *Escherichia coli* and *Shigella* spp. *Infect. Immun.* **72,** 5080–5088.
5. Wei, J., Goldberg, M. B., Burland, V., et al. (2003) Complete genome sequence and comparative genomics of *Shigella flexneri* serotype 2a strain 2457T. *Infect. Immun.* **71,** 2775–2786.
6. Jin, Q., Yuan, Z., Xu, J., et al. (2002) Genome sequence of *Shigella flexneri* 2a: insights into pathogenicity through comparison with genomes of *Escherichia coli* K12 and O157. *Nucleic Acids Res.* **30,** 4432–4441.
7. Schena, M., Shalon, D., Davis, R. W., and Brown, P. O. (1995) Quantitative monitoring of gene expression patterns with a complementary DNA microarray. *Science* **270,** 467–470.

8. Schena, M., Shalon, D., Heller, R., Chai, A., Brown, P. O., and Davis, R. W. (1996) Parallel human genome analysis: microarray-based expression monitoring of 1000 genes. *Proc. Natl. Acad. Sci. USA* **93,** 10,614–10,619.

9. Joyce, E. A., Chan, K., Salama, N. R., and Falkow, S. (2002) Redefining bacterial populations: a post-genomic reformation. *Nat. Rev. Genet.* **3,** 462–473.

10. Conway, T. and Schoolnik, G. K. (2003) Microarray expression profiling: capturing a genome-wide portrait of the transcriptome. *Mol. Microbiol.* **47,** 879–889.

11. Rajashekara, G., Glasner, J. D., Glover, D. A., and Splitter, G. A. (2004) Comparative whole-genome hybridization reveals genomic islands in *Brucella* species. *J. Bacteriol.* **186,** 5040–5051.

12. Gandara, B., Merino, A. L., Rogel, M. A., and Martinez-Romero, E. (2001) Limited genetic diversity of *Brucella* spp. *J. Clin. Microbiol.* **39,** 235–240.

13. Nelson, K. E., Fleischmann, R. D., DeBoy, R. T., et al. (2003) Complete genome sequence of the oral pathogenic Bacterium *Porphyromonas gingivalis* strain W83. *J. Bacteriol.* **185,** 5591–5601.

14. Schachter, J. (1978) Chlamydial infections (first of three parts). *N. Engl. J. Med.* **298,** 428–435.

15. Stephens, R. S., Kalman, S., Lammel, C., et al. (1998) Genome sequence of an obligate intracellular pathogen of humans: *Chlamydia trachomatis*. *Science* **282,** 754–759.

16. Read, T. D., Brunham, R. C., Shen, C., et al. (2000) Genome sequences of *Chlamydia trachomatis* MoPn and *Chlamydia pneumoniae* AR39. *Nucleic Acids Res.* **28,** 1397–1406.

17. Shirai, M., Hirakawa, H., Kimoto, M., et al. (2000) Comparison of whole genome sequences of *Chlamydia pneumoniae* J138 from Japan and CWL029 from USA. *Nucleic Acids Res.* **28,** 2311–2314.

18. Salama, N., Guillemin, K., McDaniel, T. K., Sherlock, G., Tompkins, L., and Falkow, S. (2000) A whole-genome microarray reveals genetic diversity among *Helicobacter pylori* strains. *Proc. Natl. Acad. Sci. USA* **97,** 14,668–14,673.

19. Fitzgerald, J. R., Sturdevant, D. E., Mackie, S. M., Gill, S. R., and Musser, J. M. (2001) Evolutionary genomics of *Staphylococcus aureus*: insights into the origin of methicillin-resistant strains and the toxic shock syndrome epidemic. *Proc. Natl. Acad. Sci. USA* **98,** 8821–8826.

20. Carlson, J. H., Hughes, S., Hogan, D., et al. (2004) Polymorphisms in the *Chlamydia trachomatis* cytotoxin locus associated with ocular and genital isolates. *Infect. Immun.* **72,** 7063–7072.

21. Leonard, E. E. II, Tompkins, L. S., Falkow, S., and Nachamkin, I. (2004) Comparison of *Campylobacter jejuni* isolates implicated in Guillain-Barre syndrome and strains that cause enteritis by a DNA microarray. *Infect. Immun.* **72,** 1199–1203.

22. Taboada, E. N., Acedillo, R. R., Carrillo, C. D., et al. (2004) Large-scale comparative genomics meta-analysis of *Campylobacter jejuni* isolates reveals low level of genome plasticity. *J. Clin. Microbiol.* **42,** 4566–4576.

23. Dziejman, M., Balon, E., Boyd, D., Fraser, C. M., Heidelberg, J. F., and Mekalanos, J. J. (2002) Comparative genomic analysis of *Vibrio cholerae*: genes that correlate with cholera endemic and pandemic disease. *Proc. Natl. Acad. Sci. USA* **99,** 1556–1561.

24. Hakenbeck, R., Balmelle, N., Weber, B., Gardes, C., Keck, W., and de Saizieu, A. (2001) Mosaic genes and mosaic chromosomes: intra- and interspecies genomic variation of *Streptococcus pneumoniae*. *Infect. Immun.* **69,** 2477–2486.

25. Orihuela, C. J., Radin, J. N., Sublett, J. E., Gao, G., Kaushal, D., and Tuomanen, E. I. (2004) Microarray analysis of pneumococcal gene expression during invasive disease. *Infect. Immun.* **72,** 5582–5596.

26. Tu, A. H., Fulgham, R. L., McCrory, M. A., Briles, D. E., and Szalai, A. J. (1999) Pneumo-coccal surface protein A inhibits complement activation by *Streptococcus pneumoniae*. *Infect. Immun.* **67,** 4720–4724.

27. Bethe, G., Nau, R., Wellmer, A., et al. (2001) The cell wall-associated serine protease PrtA: a highly conserved virulence factor of *Streptococcus pneumoniae*. *FEMS Microbiol. Lett.* **205,** 99–104.

28. Snyder, J. A., Haugen, B. J., Buckles, E. L., et al. (2004) Transcriptome of uropathogenic *Escherichia coli* during urinary tract infection. *Infect. Immun.* **72,** 6373–6381.

29. Rappuoli, R. (2000) Pushing the limits of cellular microbiology: microarrays to study bac-teria-host cell intimate contacts. *Proc. Natl. Acad. Sci. USA* **97,** 13,467–13,469.

30. Bryant, P. A, Venter, D., Robins-Browne, R., and Curtis, N. (2004) Chips with everything: DNA microarrays in infectious diseases. *Lancet Infect. Dis.* **4,** 100–111.

31. Rosenberger, C. M., Scott, M. G., Gold, M. R., Hancock, R. E., and Finlay, B. B. (2000) *Salmonella typhimurium* infection and lipopolysaccharide stimulation induce similar changes in macrophage gene expression. *J. Immunol.* **164,** 5894–5904.

32. Cohen, P., Bouaboula, M., Bellis, M., et al. (2000) Monitoring cellular responses to *Lis-teria monocytogenes* with oligonucleotide arrays. *J. Biol. Chem.* **275,** 11,181–11,190.

33. Belcher, C. E., Drenkow, J., Kehoe, B., et al. (2000) From the cover: the transcriptional responses of respiratory epithelial cells to *Bordetella pertussis* reveal host defensive and pathogen counter-defensive strategies. *Proc. Natl. Acad. Sci. USA* **97,** 13,847–13,852.

34. Alm, R. A., Ling, L. S., Moir, D. T., et al. (1999) Genomic-sequence comparison of two unrelated isolates of the human gastric pathogen *Helicobacter pylori*. *Nature* **397,** 176–180.

35. Covacci, A., Telford, J. L., Del Giudice, G., Parsonnet, J., and Rappuoli, R. (1999) *Helico-bacter pylori* virulence and genetic geography. *Science* **284,** 1328–1333.

36. Nagasako, T., Sugiyama, T., Mizushima, T., Miura, Y., Kato, M., and Asaka, M. (2003) Up-regulated Smad5 mediates apoptosis of gastric epithelial cells induced by *Helicobacter pylori* infection. *J. Biol. Chem.* **278,** 4821–4825.

37. Ohnishi, M., Terajima, J., Kurokawa, K., et al. (2002) Genomic diversity of enterohemorr-hagic *Escherichia coli* O157 revealed by whole genome PCR scanning. *Proc. Natl. Acad. Sci. USA* **99,** 17,043–17,048.

38. Beres, S. B., Sylva, G. L., Sturdevant, D. E., et al. (2004) Genome-wide molecular dissec-tion of serotype M3 group A *Streptococcus* strains causing two epidemics of invasive infec-tions. *Proc. Natl. Acad. Sci. USA* **101,** 11,833–11,838.

39. Jin, S., Joe, A., Lynett, J., Hani, E. K., Sherman, P. M., and Chan, V. L. (2001). JlpA, a novel surface-exposed lipoprotein specific to *Campylobacter jejuni,* mediates adherence to host epithelial cells. *Mol. Microbiol*. **39,** 1225–1236.

40. Chan, V. L., Hani, E. K., Joe, A., Lynett, J., Ng, D., and Steele, M. (2000). The hippuricase gene and other unique genes of *Campylobacter jejuni,* in *Campylobacter* (Blaser, M. J. and Nachamkin, I., eds.). ASM Press, Washington, DC, pp. 455–463.

41. Parkhill, J., Wren, B. W., Mungall, K., et al. (2000) The genome sequence of the food-borne pathogen *Campylobacter jejuni* reveals hypervariable sequences. *Nature* **403,** 665–668.

42. Jin, S., Song, Y. C., Emili, A., Sherman, P. M., and Chan, V. L. (2003) JlpA of *Camp-ylobacter jejuni* interacts with surface-exposed heat shock protein 90α and triggers signal-ing pathways leading to the activation of NF-κB and p38 MAP kinase in epithelial cells. *Cell. Microbiol.* **5,** 165–174.

4

Knockout and Disease Models in Toll-Like Receptor-Mediated Immunity

Huey-Lan Huang and Wen-Chen Yeh

Summary

In recent years, innate immunity has been one of the most intensively studied areas in immunology. It secures the first line of host defense against various microorganisms, including bacteria and viruses. Innate immune responses are highly conserved throughout evolution, and one important family of sensors responsible for initiating innate immune responses in mammals, the Toll-like receptors (TLRs), has recently been discovered. In this chapter, we will introduce the importance of innate immunity and TLR signals. We will then systematically discuss individual TLRs, their ligands, and the key downstream signaling molecules and pathways. Most of what we currently know is based on multidisciplinary approaches and supported by the phenotypes of specific knockout mice. Finally, we will briefly discuss the infectious diseases in humans that are caused by the mutations of TLR signals.

Key Words: Innate immunity; Toll-like receptor (TLR); MyD88; IRAK; TRAF6; TRIF.

1. Introduction

When the host is infected with bacteria or other micropathogens, the innate immune system is immediately turned on before the onset of medical symptoms. In fact, the majority of initial symptoms exhibited by the host, such as fever, mucosal secretions, and coughing, are manifestations of the innate immune system initiated to curtail the propagation of foreign pathogens. Following these initial defense and inflammatory responses, more specific lymphocyte-mediated immune responses against pathogens are activated. Activation of these adaptive immune responses, primarily because of antigen-specific antibodies and cytotoxic T cells, also depends on the systematic action of innate immunity. Despite its critical function, the innate immune system had not garnered much attention until 6 or 7 yr ago, when one of the major pathogen-sensing receptor systems was discovered in mammals.

Innate immunity is conserved through evolution. In fact, it is the major immune system for nonvertebrate organisms. Many pioneering studies on the components and the signaling pathways of innate immunity were carried out in fruit flies (*Drosophila*) and plants (*Arabidopsis*) (Fig. 1) *(1–8)*. In contrast, vertebrates utilize both innate immunity and the more sophisticated adaptive immune response to combat pathogens. Although the latter response has been well characterized, particularly in mammals, our understanding of innate immunity in various organisms is still at its infancy stage.

From: *Bacterial Genomes and Infectious Diseases*
Edited by: V. L. Chan, P. M. Sherman, and B. Bourke © Humana Press Inc., Totowa, NJ

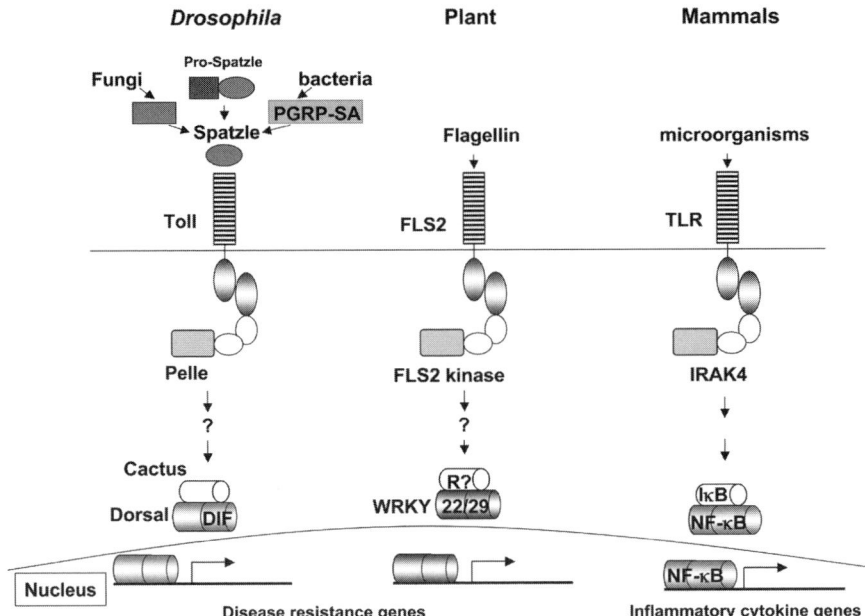

Fig. 1. The parallel signaling pathways for innate immunity in *Drosophila*, the plant *Arabidopsis*, and mammals. All have pathogen-recognizing Toll-like receptors (TLR) which contain leucine-rich repeat motifs present in their extracellular domains. The TLR in plants is called flagellin-sensitive 2 (FLS2). After binding to components of pathogens, all of these receptors recruit adaptors and activate Pelle-like kinases (interleukin-1 receptor associated kinase [IRAK4] in mammals and FLS2 kinase in plants) leading to activation of nuclear factor κB (NF-κB)-like transcription factors through the degradation of NF-κB inhibitors. This results in transcription of mRNA for disease resistance genes and inflammatory cytokines. Furthermore these signaling pathways also trigger MAP kinase activation in *Drosophila*, *Arabidopsis*, as well as in mammals.

The innate immune defense begins by taking advantage of natural physical and chemical barriers on the surface of hosts' bodies. Some examples are tight junctions between the epithelial cells, cilia movement in the respiratory tract, fatty acids on the skin, and enzymes or low pH in the gastrointestinal tract *(9)*. If pathogenic microorganisms are successful in breaking through these barriers, cellular components of the innate immune system, including neutrophils, macrophages/monocytes, and dendritic cells (DCs), are recruited. These cells recognize pathogens mainly by their general molecular patterns, and then respond by releasing inflammatory cytokines and phagocytosing invading microorganisms. Inflammatory cytokines such as tumor necrosis factor (TNF)-α, interleukin (IL)-6, IL-1, IL-12, and IL-8 are produced and they signal through various cytokine receptors to trigger inflammatory responses and to attract additional monocytes/macrophages, DCs, and neutrophils to the pathogen-invaded tissue (Fig. 2). Further-more, professional antigen-presenting DC undergo maturation and activation, and trigger the adaptive immune response by interacting with naïve T cells and presenting processed antigens derived from the pathogens (Fig. 3).

Table 1 summarizes some major differences between innate and adaptive immunity *(9)*. The latter is characterized by its specificity and diversity, as the adaptive immune response requires recognition of many specific antigens, and relies on a diverse reservoir

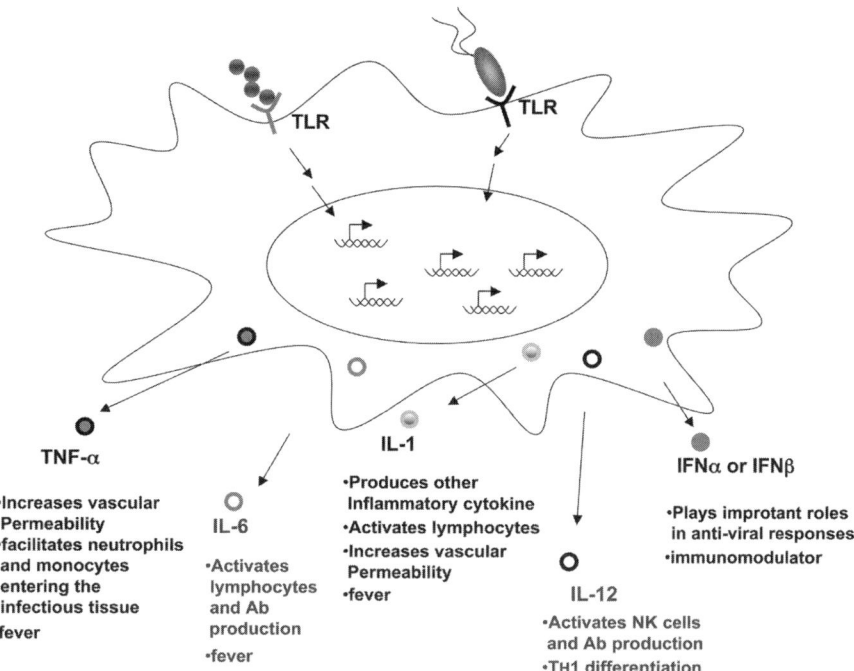

Fig 2. The inflammatory cytokines and their functions. After pathogen recognition, Toll-like receptors signaling pathways trigger inflammatory cytokine production. These cytokines can activate vascular endothelial cells and lymphocytes, facilitate the penetration of neutrophils and monocytes into the infected tissue, act as chemokines, increase the fluid volume in local tissue, and may cause systemic fever (including swelling, redness, and pain, which are typical inflammatory responses).

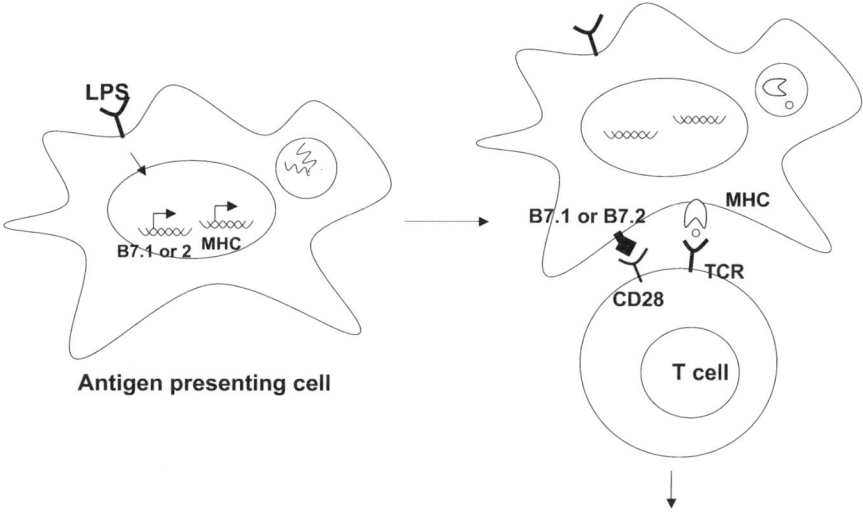

Fig. 3. Maturation of antigen presenting cells and activation of naïve T cells. TLR signaling pathways not only trigger inflammatory cytokine production, but also upregulate major histocompatibility complex (MHC) B7.1 and B7.2 costimulatory molecules on the surface of dendritic cells (DCs) or macrophages. TLR activation also leads to DC maturation. The processed pathogen, acting as antigen, combines with MHC molecules on DCs.

Table 1
Comparison of Factors Involved in Innate Immunity and Adaptive Immunity

	Innate Immunity	Adaptive immunity
TLRs as major sensor	★	
Evolutionary conserved from invertebrate (e.g., Drosophila) to vertebrate (e.g., Human)	★	
Trigger immediate immune responses (e.g., inflammation) against pathogens	★	
Activate signals for the initiation of adaptive immunity	★	
TLRs expressed by the particular cells (e.g., macrophages, monocytes, DC)	★	
Using antibodies or cytotoxic T cells against pathogens		★
Only in vertebrates		★
Require gene rearrangement		★
Need several days for clonal expansion		★

Innate immunity uses Toll-like receptors (TLRs) as sensors, and triggers immediate immune responses via secreted inflammatory cytokines in contrast to adaptive immunity which uses antibody (Ab) and T-cell receptors (TCR) to recognize pathogens. Furthermore, pathogen-specific Ab and TCR requires activation by innate immunity, therefore the adaptive response is generated several days later than the innate immune response. TLRs are directly transcribed and translated from the genome, but Ab and TCR transcription occurs from rearranged genes so that they can recognize thousands of different epitopes. Innate immunity is evolutionary conserved from invertebrates to vertebrates, but adaptive immunity is only present in vertebrates.

of T-cell receptors and antibody molecules generated through somatic permutations. In contrast, innate immunity requires a much smaller set of gross sensors that identify molecular patterns derived from micropathogens, and initiates production of inflammatory cytokines swiftly and efficiently. This set of pathogen pattern sensors is represented by Toll-like receptors (TLR). These germline-encoded receptors contain variable leucine-rich repeats in their extracellular domain, which are required for pathogen recognition. The intracellular portion of TLRs contains a critically conserved domain, the Toll/IL-1 receptor (TIR) domain (further explained in **Subheading 2.**), which mediates downstream signaling.

2. Toll-Like Receptors and Their Ligands

The first line of immune response in mammals against microbial pathogens is the recognition of pathogen-associated molecular patterns by TLR family members. As the cornerstone for initiating innate immune responses, the TLR pathogen recognition system is evolutionarily conserved. The Toll protein was originally identified in *Drosophila* as a receptor playing a key role in development, as well as in antifungal responses. In mammals, at least eleven TLRs have been reported and most of these receptors have been linked to recognition of specific pathogen-associated ligands. Based on the cytoplasmic domains of *Drosophila* Toll and some mammalian TLRs, they are also homologous to members of the IL-1 receptor (IL-1R) family. Furthermore, the intracellular domains of **TLRs** and **IL-1R** are homologous to a region in the plant **R** gene product, thus designating these regions as "TIR" domains *(6,10)*.

Signaling mechanisms mediated by TIR domains are remarkably similar amongst different organisms and various TLR/IL-1R family proteins. The apical signaling adap-

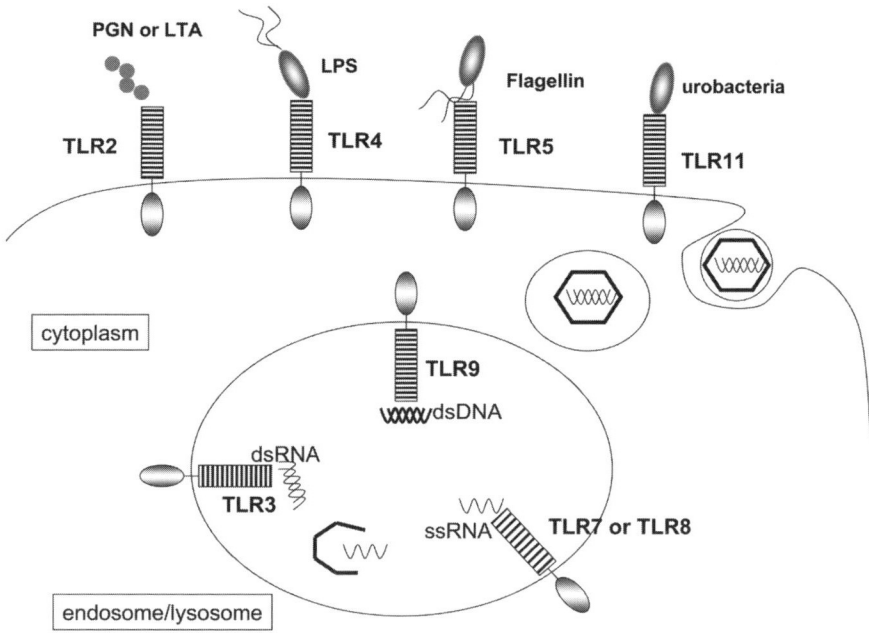

Fig. 4. TLRs and their ligands. *See* **Subheading 2.** for more details.

tors that are directly recruited to *Drosophila* Toll or mammalian TLR/IL-1R also contain TIR domains that interact with the receptors through their TIR counterparts. Here, we will first describe various TLRs and their specific ligands (Fig. 4) *(6,11)* and functions based on biochemical and genetic studies. In the next section, we will focus on TIR-containing adaptors and other key signaling molecules.

2.1. TLR4

In 1997, a paper published in *Nature* by Medzhitov et al. *(12)* jumpstarted the study of human TLR. These researchers identified the first mammalian TLR, a homolog of the *Drosophila* Toll protein, which was later named TLR4. They found that overexpression of a constitutively active TLR4 mutant induces nuclear factor κB (NF-κB) activation, and production of inflammatory cytokines is dependent on NF-κB. This publication initiated a new era of immunological research into the puzzle of innate immune receptors.

In parallel to the landmark study previously mentioned, groups led by Beutler and Malo discovered that TLR4 is a potential sensor for lipopolysaccharide (LPS) *(13,14)*. LPS is a component of Gram-negative bacterial cell walls, and is responsible for the sepsis syndrome induced by overreacting innate immune responses. Sepsis syndrome can lead to septic shock, which causes multiple organ failure and is a severe health threat accounting for the death of many in-patients. They discovered that LPS-hyposensitive mice, C3H/HeJ and C57BL/10ScCr, possess mutated an *lps* allele ($lps^{d/d}$) that causes the animals to respond poorly upon LPS challenge. Using high-resolution genetic methods, these researchers identified the *lps* allele as the *tlr4* gene *(13)*. C3H/HeJ and C57BL/10ScCr mice contain a point mutation (a substitution of a highly conserved proline by histidine) and a deletion of the *tlr4* gene, respectively. In addition, TLR4-knockout mice, generated by Akira's group, are also resistant to LPS-induced shock, corroborating the phenotypes identified in naturally mutated mice *(15)*. TLR4-deficient cells derived from knockout mice also showed impaired production of TNF-α, proliferation, MHC class II expression, and NF-κB activation induced by LPS. Taken together, TLR4 is a critical receptor necessary for LPS signal initiation.

Further studies suggested that MD-2 and CD14 can cooperate with TLR4 for LPS response *(16,17)*. MD-2 was found as a molecule physically associated with TLR4 on the cell surface and in the endoplasmic reticulum/*cis* Golgi by coimmunoprecipitation, flow cytometry, and fractionation techniques. Cotransfection of TLR4 and MD-2 enhances NF-κB activation, a key TLR downstream signaling event, in the absence or presence of LPS. Additional evidence supporting a critical role for MD-2 in TLR4/LPS response was provided by studies of knockout mice *(18)*. In MD-2-deficient embryonic fibroblasts, TLR4 cannot be secreted to the plasma membrane and predominantly resides in the Golgi. Furthermore, TNF-α, IL-6, and IL-12 production induced by LPS are severely impaired in MD-2 knockout mice, similar to TLR4-deficient mice. MD-2 knockout mice are also more susceptible to infection by *Salmonella typhimurium* than wild-type mice, suggesting an essential role for MD-2 in immune responses against bacteria.

CD14 is another protein with a role in LPS-induced TLR4 signaling. Using resonance energy transfer technique, Petty's group showed that LPS triggers a physical association between CD14 and TLR4 *(19)*. Using cross-linking combined with radioimmunoprecipitation, this group also showed that LPS directly binds to CD14, which then associates with the TLR4/MD-2 complex in a CD-14-dependent manner.

In addition to LPS, the following molecules were also reported to be TLR4 ligands: the fusion protein from respiratory syncytial virus, a human respiratory pathogen *(20)*; Chlamydial heat shock protein 60 from bacteria *Chlamydia pneumonine (21)*; envelope protein from mouse mammary tumor virus *(22)*; antitumor agent Taxol derived from the yew tree *(23)*; heat shock protein 70 *(24,25)*, hyaluronic acid *(26)*, heparan sulfate *(27)*, and fibrinogen *(28)* from the host.

2.2. TLR2, TLR6, and TLR1

TLR2 recognizes compounds from many microorganisms, including peptidoglycan (PGN) and lipoteichoic acid from Gram-positive bacteria (e.g., *Staphylococcus aureus* or *Bacillus subtilis*), and lipoproteins from Gram-negative bacteria (e.g., *Borrelia burgdorferi*, *Treponema pallidum*, and *Mycoplasma fermentans*) *(29–34)*. These ligands for TLR2 were confirmed by several studies. Lipoproteins or PGN were found to induce NF-κB activation in human embryonic kidney 293 cells or Chinese hamster macrophages transfected with TLR2. Furthermore, only anti-TLR2 antibody or a mutant TLR2 (TLR2-P681H, with inhibitory activity) were found to block TNF-α production induced by lipoprotein or whole *Mycobacterium tuberculosis* in human peripheral blood mononuclear cells or RAW264.7 macrophages. Akira's group also confirmed that macrophages form TLR2 knockout mice are hyposensitive to PGN derived from *S. aureus (30,35)*.

Interestingly, TLR2 can form functional pairs with TLR6 or TLR1 for recognition of diacylated or triacylated lipoprotein, respectively *(6,36–38)*. TLR1, TLR2, and TLR6 were therefore classified in the TLR2 subfamily. TLR1 and TLR6 exhibit 69.3% identity in overall amino acid sequence. They have similar genomic structures, and are located in tandem on the same chromosome, possibly the result of a gene duplication event *(6)*.

Other possible ligands of TLR2 are: lipoarabinomannan from Mycobacteria (e.g., *M. tuberculosis*) *(39,40)*; zymosan from fungi *(34)*; glycosylphosphatidylinositol (GPI) from parasite *Trypanosoma cruzi (41)*; glycolipid from *Treponema maltophilum* and *Treponema brennaborense (42)*; porins from Neisseria *(43)*; LPS from *Leptospira interrogans (44)*; phenol-soluble modulin from *Staphylococcus epidermidis* or zymosan (recognition by TLR2 and TLR6) *(45)*; and outer-surface lipoprotein from *Borrelia burgdorferi* (recognition by TLR2 and TLR1) *(46)*.

2.3. TLR5

Bacterial flagellin contributes to the virulence of pathogenic bacteria, because flagellin are required for bacterial movement, adhesion, invasion, and colonization of the intestinal mucosa of a host. In 2001, mammalian TLR5 was reported to be responsible for the recognition of flagellin from both Gram-positive and negative bacteria *(47,48)*. Using either reverse-phase chromatography or high-performance liquid chromatography to purify different components from *Listeria monocytogenes*, these researchers proved that flagellin-containing fractions induced NF-κB activation in TLR5-transfected Chinese hamster ovary (CHO) cells. Furthermore, deletion of the flagellin genes from *Salmonella typhimurium* abrogated their ability to stimulate TLR5. Another group utilized recombinant flagellin protein constructed from the *Salmonella dublin* flagellin gene, and demonstrated that this induced NF-κB activation and IL-8 production in Caco-2BBe human adenocarcinoma cells *(48)*. Recognition of bacteria flagellin by TLR5, therefore, plays an important role in host immune defense systems against flagellated bacteria.

2.4. TLR3

Double-stranded RNA (dsRNA) composes the genome of some viruses and is often also a product of viral replication inside host cells. NF-κB activation is induced when certain types of cells (e.g., RAW264.7 macrophages) are treated with synthetic dsRNA, polyinosine-polycytidylic acid [poly (I:C)] *(49)*. Cells derived from TLR3 knockout mice demonstrate impaired responses specifically to poly (I:C) *(49)*, but responses are not completely abolished. These results suggest that although dsRNA is one of the ligands triggering TLR3, other TLR3-independent pathways induced by dsRNA also exist.

2.5. TLR9

Roughly a decade ago, synthetic DNA fragments containing unmethylated CpG dinucleotides, CpG-oligodeoxynucleotide (CpG-ODN)—were reported to be useful immune modulators *(50)*. CpG-ODN is sufficient to induce immune cell responses, such as IL-6, TNF-α, IL-12, and interferon (IFN)-γ cytokine production and upregulation of costimulatory molecules. Administration of CpG-ODN is also sufficient to protect mice against lethal infection by *Francisella tularensis*, *Listeria monocytogenes*, or *Leishmania major*, possibly owing to the effect of CpG-ODN on the enhancement of Th1-responses *(51–53)*. The unmethylated CpG motifs are common in bacteria, but uncommon in the mammalian genome, which has a low frequency of CpG sequences that are mostly methylated. TLR9 recognizes not only bacterial DNA containing these unmethylated CpG motifs *(54,55)*, but also viral DNA (e.g., DNA from Herpes simplex virus-2) *(56)* and chromatin-IgG complexes found in patients with systemic autoimmune disease *(57)*.

Physiological functions of TLR9 were clearly demonstrated in mice with targeted disruption of the *tlr9* gene *(54)*. TLR9-deficient mice and cells are specifically impaired in responses (cytokine production and NF-κB activation) to CpG-ODN, but not to LPS or PGN. These data are consistent with the results that either *Escherichia coli* DNA or CpG-ODN is capable of inducing TNF-α, IL-8, IL-12 production, NF-κB activation, or B-cell proliferation in human B-cells, DC, or human TLR9-transfected 293 cells *(55)*.

TLR9 is expressed within endosomal membranes, and TLR9-mediated immune responses to CpG-ODN requires endosome acidification and/or maturation, because it is sensitive to inhibitors of endosomal acidification, e.g., chloroquine, ammonium chloride, concanamycin B, or balfilomycin A (the latter two drugs are inhibitors of the V-type ATPase which is responsible for the endosome and lysosome acidification) *(57–59)*.

2.6. TLR7 and TLR8

Synthetic ligands for TLR7 or TLR8 were initially discovered in 2002 *(60,61)*. These small imidazoquinoline compounds are guanosine nucleoside analogs, e.g., R-848 and imiquinod, and are potent antiviral or antitumor agents with the ability to induce inflammatory cytokines, especially interferons, and cellular immunity. Imidazoquinoline compounds activate NF-κB in human TLR7-, TLR8-, and murine TLR7-transfected 293 cells *(59–62)*. In addition, these agents fail to trigger TNF-α and IL-12 production or other immune responses in cells derived from TLR7-knockout mice *(60)*, suggesting that recognition of guanosine nucleoside analogs is dependent on TLR7.

Recently, two groups independently discovered that single-stranded RNA (ssRNA) is a potential natural ligand for TLR7 and TLR8. One group used influenza viral RNA, green fluorescent protein RNA, or polyU (mixed with RNA stabilizer polyethylenimine) *(63)*, and the other group utilized GU-rich HIV RNA oligonucleotides *(64)*. All of these ssRNA trigger inflammatory cytokine production or IFN-α production from wild type DC but not TLR7-deficient DC, again suggesting that recognition of ssRNA is TLR7-dependent. Paradoxically, HIV RNA oligonucleotides can trigger NF-κB activation in human TLR8-, but not human TLR7-transfected 293 cells.

TLR7 and TLR8 are homologous to TLR9 (in addition to sequence homology) for several reasons *(6,55,59,63)*. Firstly, they have similar genomic structures to TLR7, TLR8, and TLR9, which are encoded by two exons and are located close to each other on the X chromosome. Second, they all recognize DNA, RNA, or nucleoside analogs. Finally, they are all thought to be located in endosomal/lysosomal compartments because they are sensitive to chloroquine or bafilomycin A1. It is believed that mouse TLR8 is nonfunctional, which helps explain the prominent phenotypes of TLR7-knockout. The potential differences between human TLR7 and TLR8 concerning their ligands or tissue distributions are intriguing questions that require further investigation.

2.7. TLR11

TLR11 was recently identified by Ghosh's group using a sequence homology search with the conserved TIR domain *(11)*. TLR11 is expressed in macrophages and liver, kidney, and bladder epithelial cells. In contrast to nonpathogenic *E. coli* (e.g., BL21, DH5α), heat-killed uropathogenic *E. coli* (e.g., 8NU, NU14, HLK120, and AD110)

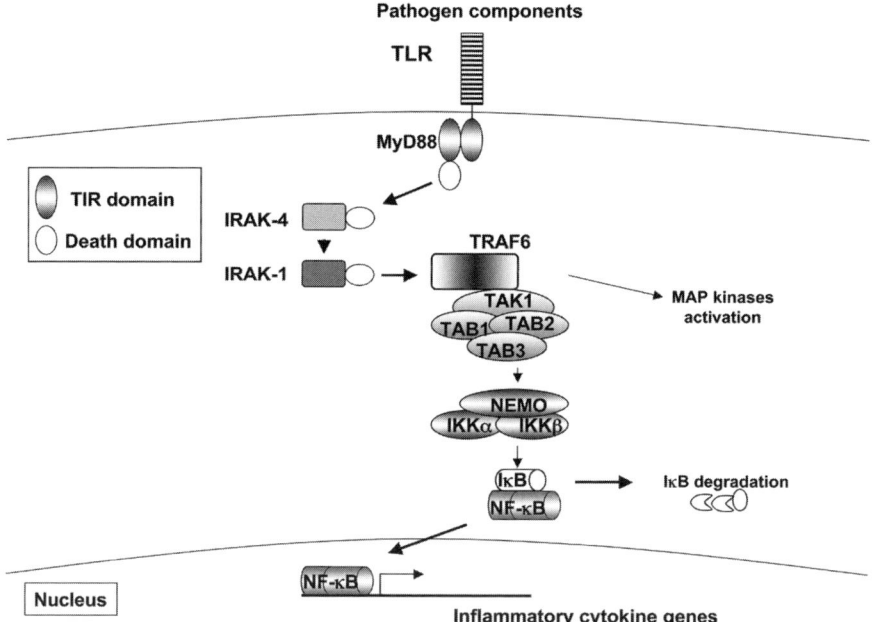

Fig. 5. MyD88-dependent TLR signaling pathway. After ligand binding, TLR recruits MyD88 through TIR domain–TIR domain interaction. MyD88 then interacts with IRAK4 via death domain–death domain interaction. IRAK4 autophosphorylates, phosphorylates IRAK1, and then recruits and activates TRAF6. This leads to TRAF6 activation of TAK1/TAB1/TAB2/TAB3 and IκB kinases IKKα, IKKβ, and their modulator NEMO (IKKγ). Finally, IκB kinases phosphorylate IκB leading to IκB degradation, freeing NF-κB to translocate into the nucleus to transcribe NF-κB-dependent antipathogen genes. TRAF6 can also trigger MAP kinase activation.

specifically activates NF-κB in TLR11-expressed 293 cells. In addition, TLR11-knockout mice are highly susceptible to kidney infection by uropathogenic *E. coli* 8NU. This study suggests that TLR11 provides an important sensor for uropathogenic *E. coli* on the surface of uroepithelial cells, and helps prevent infection of the host by these bacteria.

3. Signaling Cascades Induced by TLRs

Signal transduction mediated by a mammalian TIR domain-containing protein was first described in the IL-1R signaling pathway (Fig. 5). Myeloid differentiation primary response gene 88 (MyD88) is an adaptor protein that contains a TIR domain that associates with the IL-1R protein via direct TIR–TIR domain interaction. The death domain (DD) of MyD88, in turn, recruits through homologous DD interaction, a family of IL-1R-associated protein kinases (IRAKs) *(10,65–69)*. On ligand binding to the receptor, IRAK-1 is phosphorylated and dissociates from the receptor complex to associate with the signal transducer TNF receptor-associated protein 6 (TRAF6) *(70,71)*. TRAF6 then triggers downstream signaling pathways, presumably through TAK1/TAB protein complexes, resulting in the activation of NF-κB and various MAP kinases, including JNK and p38 MAPK. As mammalian TLRs share similar cytoplasmic domains with the IL-1 receptor family, they both utilize the same signaling pathway on ligand binding. The relatively simple cascade of MyD88→IRAK→TRAF6 therefore applies to most TLR signaling.

One outstanding question regarding the central signaling cascade is how IRAK-1 becomes phosphorylated and activated. One potential kinase that fits into this role and cascade is IRAK-4, which will be described below. Another intriguing phenomenon is that certain TLR signaling outcomes, e.g., antiviral responses and costimulatory functions induced by specific TLRs, is not accounted for by the MyD88-mediated signaling cascade. Indeed, several MyD88-like proteins that contain TIR domains have been discovered. Some of these proteins, as detailed in **Subheading 3.2.**, play important roles in the MyD88-independent signaling pathways.

3.1. MyD88-Dependent Signaling

With the exception of perhaps TLR3, all TLRs appear to contain a common MyD88-dependent signaling pathway (Fig. 5). We will first discuss the key signal transducers in this pathway.

3.1.1. MyD88

MyD88 was discovered to be an important adaptor linking the signal from IL-1 receptor to IRAK1 and TRAF6, both of which were known as pivotal mediators for IL-1-induced NF-κB activation. MyD88 is also the first adaptor identified with a carboxyl-terminal TIR domain for association with TLR receptors through direct TIR–TIR interactions. In addition, MyD88 contains an amino-terminal DD that was originally found in death receptors and their adaptors *(72)*, e.g., Fas and FADD, which are also involved in signal transduction. Overexpression of full-length MyD88 induces NF-κB activation. IL-1 or TLR4 overexpression-induced NF-κB activation can be blocked by a dominant-negative MyD88 mutant containing only the TIR domain. Many studies also suggest that MyD88 is the adaptor for the signaling pathways of TLR2/TLR6, TLR2/TLR1, TLR5, TLR9, TLR7, and TLR11 *(11,35,54,60)*.

Deficiency of MyD88 in mice results in severe defects in responses to IL-1 and many TLR ligands *(73–77)*, and causes resistance to LPS-induced septic shock. MyD88-deficient mice are more susceptible to various bacterial infections (e.g., *S. aureus*, mycobacteria), probably owing to the lack of innate immune responses. As mentioned in the introduction to this section, MyD88 does not control all TLR signaling pathways. For example, LPS/TLR4-induced NF-κB and MAPK activations are largely intact in MyD88-deficient cells. LPS-induced maturation of DCs, including upregulation of costimulatory molecules, does not require MyD88. MyD88 is also dispensable for interferon regulatory factor (IRF)3 activation and IFN-β production induced by TLR3 and TLR4 *(78)*. However, recent studies suggest that MyD88 does play a role in some antiviral responses and is in fact required for IRF7 activation and IFN-α production induced by TLR9 *(79,80)*.

3.1.2. IRAK-1 and IRAK-4

Of the four IRAK family members, prototypical IRAK-1 is the best characterized and plays a key role in IL-1R/TLR signaling pathway downstream of MyD88 *(66–68,70)*. However, IRAK-1-deficient mice and cells are only partially defective in immune responses induced by IL-1 and LPS, suggesting potential redundancy within IRAK family members or the presence of an alternative signaling pathway. In addition, IRAK-1 undergoes phosphorylation and activation on ligand binding, and autophosphorylation of IRAK-1 only accounts for part of the activation process.

One potential IRAK-1 kinase is IRAK-4. Both IRAK-4 and IRAK-1 are active kinases and homologous to *Drosophila* Pelle, the only IRAK homolog in fruit flies. IRAK-4 is capable of phosphorylating IRAK-1 in vitro *(81)*. We have generated IRAK-4 knockout mice and demonstrated that these mutant mice and cells are severely impaired in cytokine responses induced by LPS, PGN, poly(I:C), CpG-ODN, or IL-1 *(65)*. Like MyD88-deficient mice, IRAK-4 knockouts are resistant to LPS challenge and are highly susceptible to *S. aureus* infection. IRAK-4 is also dispensable for most MyD88-independent pathways, with the exception of TLR4-induced IFN-β and interferon-regulated genes, as well as certain TLR3 signals. The severity of IRAK-4 knockout phenotypes contrasts with the much milder phenotype observed in IRAK-1-deficient mice *(82)*, suggesting that other IRAK-4 substrates aside from IRAK-1 may also be involved in TLR signaling. We have also demonstrated that the kinase activity of IRAK-4 is at least partially required for IL-1-mediated signaling. Whether such kinase requirement applies to TLR signals is currently one focus of the follow-up studies.

3.1.3. TRAF6

TRAF6 is a key signal transducer in TLR signaling pathways (*see* Fig. 5). Biochemical studies have suggested that TRAF6 interacts with IRAK family proteins and probably plays a role downstream of IRAKs, as the TRAF6 dominant-negative mutant is capable of suppressing IRAK-induced NF-κB activation *(71)*. Indeed, TRAF6-deficient mice and cells fail to respond to IL-1 and LPS *(83)*. Although TRAF6 plays a prominent role in MyD88/IRAK-mediated signals, TRAF6 has also been found to interact with Toll/IL-1 receptor domain containing adaptor-inducing IFN-β (TRIF), which is an adaptor involved in MyD88-independent pathways (*see* **Subheading 3.2.1.**) *(84)*. This, along with studies using TRAF6-deficient DCs, also suggests a role for TRAF6 in certain MyD88-independent pathways. In addition, it is worth noting that TRAF6-mediated signals are not limited to TLRs, but are also important for signals induced by various tumor necrosis factor receptor family members, including CD40 and receptor activator of NF-κB (RANK) *(72,85–87)*. TRAF6-deficient mice are not responsive to CD40 activation and exhibit defects in tooth eruption, osteopetrosis, and lymph node organogenesis that are consistent with the lack of RANK signaling.

3.1.4. Mal/TIRAP

Mal (also called TIRAP) is the second TIR domain-containing cytoplasmic adaptor discovered to play a role in the TLR4 signaling pathway *(88)*. Mal knockout mice exhibit very similar phenotypes to those observed in MyD88-deficient mice, except that the defects are specifically associated with TLR4 and TLR2 *(88–90)*. Mal is thought to interact with MyD88 and likely mediates MyD88-dependent signals specifically downstream of TLR4 and TLR2 (Fig. 6).

3.2. MyD88-Independent Signaling

Studies of the MyD88-independent pathways have been concentrated on TLR3/dsRNA- and TLR4/LPS-induced IRF3 activation and production of type I interferons, and downstream interferon-regulated genes such as *IP-10* and *GARG16 (77)* (Fig. 6). It appears that two newly discovered TIR-containing adaptors, TRIF and TRIF-related adaptor molecule (TRAM), play critical roles in initiating this signaling cascade *(91–95)*. TRAFZ-associated kinase (T2K) and IκB kinase (IKKε) are recruited by TRIF and

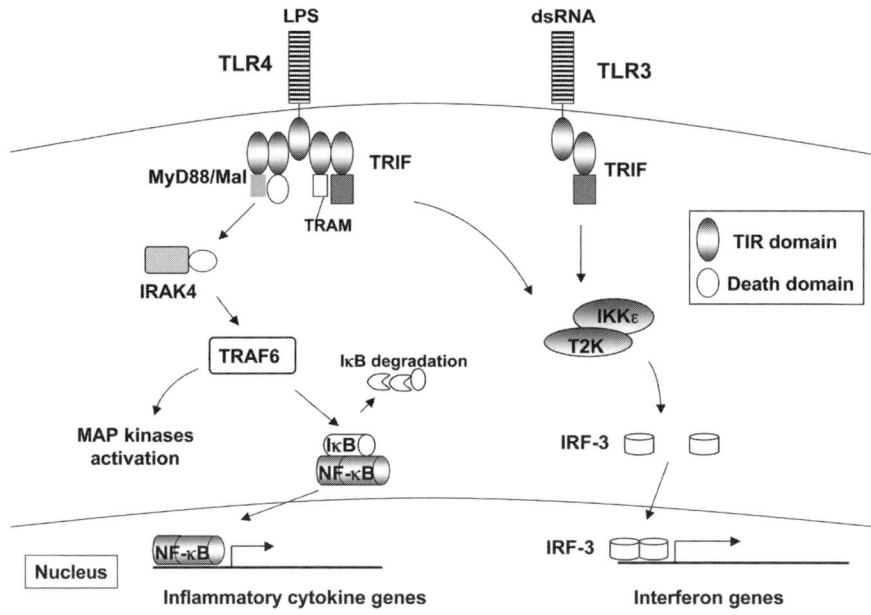

Fig. 6. TRIF-dependent TLR signaling pathway. After LPS binding, TLR4 can trigger both a MyD88-dependent signaling pathway (Fig. 5) and a TRIF-dependent pathway. The TRIF-dependent pathway leads to activation of IKKε and T2K (or called TBK1) and causes IRF3 dimerization. Dimerized IRF3 can translocate into the nucleus and result in transcription of IRF3-dependent genes (*IFN-*α or *IFN-*β). Upregulation of costimulatory molecules (Fig. 3) is also TRIF-dependent. In addition, there is only a TRIF-dependent pathway, but no MyD88-dependent pathway, for TLR3. Furthermore, TRAM is thought to be an adaptor only for TLR4.

then potentially interact with IRF3 directly to phosphorylate and activate IRF3 *(96,97)*. These molecules will be discussed in this Subheading. TLR3 and TLR4-induced upregulation of costimulatory molecules during DC maturation is also MyD88-independent *(98)*, and the search for signaling molecules involved in this response has also yielded some recent breakthroughs.

3.2.1. TRIF

TRIF (also called TICAM-1) was first discovered in 2002 as an important protein linking TLR to type I interferon production *(91–94)*. Recently, two studies using TRIF-deficient mouse models demonstrated that TRIF is indeed essential for TLR3-mediated signals and TLR4-induced IRF3 activation (Fig. 6). Using a reverse genetics approach, Beutler's group identified the mutation *lps2*, which displayed a defect in polyI:C- or LPS-induced interferon production, and confirmed the product of *lps2* gene as TRIF *(91)*. Intriguingly, these TRIF-defective mice also exhibited defects in LPS-induced inflammatory cytokines and resistance to LPS-induced endotoxic shock *(93)*. Although MyD88 knockout or TRIF-defective cells retain LPS-induced NF-κB activation, such signal is completely abolished in cells lacking both MyD88 and TRIF. These results suggest that TRIF not only is essential for the MyD88-independent pathway, but also cooperates with MyD88 in the branch of NF-κB and inflammatory signaling mediated by bacterial LPS. TRIF-knockout mice generated by Akira's group show the same fundamental phenotype as *lps2* mutant mice *(93,94)*. Furthermore, LPS-induced upregulation of costimulatory molecules such as CD40, CD80, and CD86 are also dependent on the TRIF-signaling pathway (Fig. 6) *(98)*.

3.2.2. TRAM

TRAM (also called TICAM2 or TIRP) is a closer homolog to TRIF than MyD88 or Mal in TIR domain homology *(92,95,99)*. Overexpression of TRAM in 293 cells induces NF-κB activation and also activates the genes encoding type I interferons. Interestingly, TRAM knockout mice show specifically impaired LPS responses, including cytokine production, B-cell activation, and costimulatory molecule upregulation. These data suggest a model where TRAM is involved in TRIF-dependent signaling, specifically in the LPS/TLR4 pathway. TRAM appears to directly interact with TLR4 and provides a signaling link between TLR4 and TRIF (Fig. 6).

3.2.3. T2K and IKKε

T2K (also called TBK1 or NAK) was originally isolated as a kinase associated with TRAF2, whereas IKKε (IKKi) appears to be inducible by inflammatory stimuli. T2K and IKKε are the most similar to each other and also share some homology with IKKα and IKKβ, particularly in the kinase domains. Initial characterizations of T2K and IKKε implicate them in NF-κB signaling *(100–102)*. Whereas IKKε-deficient mice are viable and grossly normal, T2K knockouts die during embryogenesis because of fetal liver apoptosis *(103)*. Similar to mice lacking RelA or IKKβ, which are both core components of NF-κB signaling, the embryonic lethality and liver apoptosis of T2K-deficient mice can be rescued by the elimination of tumor necrosis factor receptor-1 *(103)*. Intriguingly, NF-κB activation is largely preserved in T2K knockout cells, despite the fact that induction of a small subset of NF-κB-dependent target genes is impaired.

Recently, T2K and IKKε were identified as kinases responsible of phosphorylating and activating IRF3, the key transcription factor in the TRIF-dependent signaling pathway *(96,97)* (Fig. 6). Indeed, TRIF appears capable of interacting with T2K/IKKε, and TRIF-mediated NF-κB and IRF signals are impaired in T2K-deficient cells. In addition, T2K is required for a subset of virus-induced responses, consistent with the function of T2K along the line of IFN-β induction and IRF3 activation.

3.3. Negative Regulators of TLR Signals

Inflammatory responses are quite useful for the clearance of pathogens such as bacteria. However, these responses are intricately controlled to prevent tissue damage, septic shock, and autoimmune disorders resulting from excessive inflammation. In the following section we discuss examples of negative signal regulators relevant to TLR signals, as discovered by gene targeting studies.

3.3.1. IRAK-M

IRAK-M is an inducible member of the IRAK family and does not possess kinase activity. Expression of IRAK-M is somewhat restricted to monocytes/macrophages and is induced at least by LPS. Interestingly, IRAK-M knockout mice show enhanced inflammatory response on bacteria infection *(104)*. Cells lacking IRAK-M exhibit enhanced cytokine production and NF-κB and MAP kinase activation on stimulation using various TLR ligands or heat-inactivated bacteria (*E. coli* and *S. typhimurium*). A potential model for how IRAK-M may achieve negative signal regulation is through association with the MyD88/IRAK complex and inhibition of the association between active IRAK proteins and TRAF6.

3.3.2. SIGIRR and ST2

Two TIR-containing surface membrane proteins, SIGIRR and ST2, are implicated as negative signal regulators primarily because of knockout studies (105,106). SIGIRR- or ST2-deficient mice show enhanced sensitivity to LPS shock and lack of LPS tolerance, respectively. Cell surface expression of ST2 is induced after LPS treatment. ST2 knockout macrophages produce a heightened amount of inflammatory cytokines induced by LPS. Both SIGIRR and ST2 probably inhibit TLR signaling through interaction with TIR-containing cytoplasmic adaptors, e.g., MyD88 or Mal.

4. Conclusion and Clinical Perspectives

Innate immunity is an evolutionarily conserved response that antagonizes pathogen invasion, such as bacteria infection, in a swift and efficient way. To achieve this, recognition of pathogens is sorted by general molecular patterns and signals generated by such pattern recognition are then transmitted quickly to induce a legion of cytokines that amplify the inflammatory responses. This chapter focuses on the events of pathogen pattern recognition and signal transduction that lead to major downstream events initiating and regulating inflammation and pathogen control. As discussed in **Subheadings 2.** and **3.**, primarily based on information gleaned from various mouse studies, perturbations of TLRs and their signaling cascades can lead to severe, and sometimes specific, defects of innate immune responses.

Remarkably, TLRs or key signaling molecules mediating TLR signals, are rarely reported deleted or mutated in humans. TLR4 sequence polymorphism is potentially linked to susceptibility to bacterial infection or other diseases, but those observations remain to be further investigated (107–111). One possibility for the lack of identified human mutations may be that the mutations are generally silent and do not cause health problems. If this is the case, innate immune responses against regular doses of pathogens, at least in humans, may be well compensated by pattern recognition systems other than TLR signals. Along this line of discussion, it is interesting to note that there are in fact several cases of patients with recurrent pyogenic bacterial infections who harbor mutations in the *irak-4* gene that abolish expression of IRAK-4 (112,113). These patients suffered from repeated infections of *Streptococcus pneumoniae*, *Streptococcus intermedicus*, *S. aureus*, *Gemella morbillorum* (Gram-positive), and *Neisseria meningitidis* (Gram-negative) during their childhood, but gradually became resistant to pathogens as they grew older. Although adaptation of these patients supports the theory of non-TLR signal compensation, it is intriguing as to why IRAK-4 deficiency specifically draws infections from mostly Gram-positive bacteria. Further investigations of the patients, IRAK-4 signals, and other TLR signaling molecules will help address these issues.

Interestingly, recent investigation of Nod family proteins, which may function as intracellular pattern-recognition sensors, has yielded information relevant to clinical diseases. In particular, mutation of Nod2 is thought to be associated with Crohn's disease (114,115), which, along with ulcerative colitis, are the prominent inflammatory bowel disorders. Nod2 acts as a signal modulator alongside TLR signals, and Nod2 mutation found in Crohn's disease results in enhanced NF-κB and cytokine signals *(116–118)*. Further studies of Nod proteins will provide useful information that complements the investigation of TLR signal regulation outlined in this chapter.

References

1. Lemaitre, B., Nicolas, E., Michaut, L., Reichhart, J. M., and Hoffmann, J. A. (1996) The dorsoventral regulatory gene cassette spatzle/Toll/cactus controls the potent antifungal response in Drosophila adults. *Cell* **86,** 973–983.
2. Hoffmann, J. A., Kafatos, F. C., Janeway, C. A., and Ezekowitz, R. A. (1999) Phylogenetic perspectives in innate immunity. *Science* **284,** 1313–1318.
3. Michel, T., Reichhart, J. M., Hoffmann, J. A., and Royet, J. (2001) Drosophila Toll is activated by Gram-positive bacteria through a circulating peptidoglycan recognition protein. *Nature* **414,** 756–759.
4. Ligoxygakis, P., Pelte, N., Hoffmann, J. A., and Reichhart, J. M. (2002) Activation of Drosophila Toll during fungal infection by a blood serine protease. *Science* **297,** 114–116.
5. Tauszig-Delamasure, S., Bilak, H., Capovilla, M., Hoffmann, J. A., and Imler, J. L. (2002) Drosophila MyD88 is required for the response to fungal and Gram-positive bacterial infections. *Nat. Immunol.* **3,** 91–97.
6. Takeda, K., Kaisho, T., and Akira, S. (2003) Toll-like receptors. *Annu. Rev. Immunol.* **21,** 335–376.
7. Gomez-Gomez, L. and Boller, T. (2000) FLS2: an LRR receptor-like kinase involved in the perception of the bacterial elicitor flagellin in Arabidopsis. *Mol. Cell* **5,** 1003–1011.
8. Asai, T., Tena, G., Plotnikova, J., et al. (2002) MAP kinase signalling cascade in Arabidopsis innate immunity. *Nature* **415,** 977–983.
9. Janeway, C. A. Jr., Travers, P., Walport, M., and Shlomchik, M. (2001) Innate immunity, in *Immunobiology: The Immune System in Health and Disease*, Garland, NY, pp. 35–91.
10. Suzuki, N., Suzuki, S., and Yeh, W. C. (2002) IRAK-4 as the central TIR signaling mediator in innate immunity. *Trends Immunol.* **23,** 503–506.
11. Zhang, D., Zhang, G., Hayden, M. S., et al. (2004) A toll-like receptor that prevents infection by uropathogenic bacteria. *Science* **303,** 1522–1526.
12. Medzhitov, R., Preston-Hurlburt, P., and Janeway, C. A. Jr. (1997) A human homologue of the Drosophila Toll protein signals activation of adaptive immunity. *Nature* **388,** 394–397.
13. Poltorak, A., He, X., Smirnova, I., et al. (1998) Defective LPS signaling in C3H/HeJ and C57BL/10ScCr mice: mutations in Tlr4 gene. *Science* **282,** 2085–2088.
14. Qureshi, S. T., Lariviere, L., Leveque, G., et al. (1999) Endotoxin-tolerant mice have mutations in Toll-like receptor 4 (Tlr4). *J. Exp. Med.* **189,** 615–625.
15. Hoshino, K., Takeuchi, O., Kawai, T., et al. (1999) Cutting edge: Toll-like receptor 4 (TLR4)-deficient mice are hyporesponsive to lipopolysaccharide: evidence for TLR4 as the Lps gene product. *J. Immunol.* **162,** 3749–3752.
16. Shimazu, R., Akashi, S., Ogata, H., et al. (1999) MD-2, a molecule that confers lipopolysaccharide responsiveness on Toll-like receptor 4. *J. Exp. Med.* **189,** 1777–1782.
17. Visintin, A., Mazzoni, A., Spitzer, J. A., and Segal, D. M. (2001) Secreted MD-2 is a large polymeric protein that efficiently confers lipopolysaccharide sensitivity to Toll-like receptor 4. *Proc. Natl. Acad. Sci. USA* **98,** 12,156–12,161.
18. Nagai, Y., Akashi, S., Nagafuku, M., et al. (2002) Essential role of MD-2 in LPS responsiveness and TLR4 distribution. *Nat. Immunol.* **3,** 667–672.
19. Jiang, Q., Akashi, S., Miyake, K., and Petty, H. R. (2000) Lipopolysaccharide induces physical proximity between CD14 and toll-like receptor 4 (TLR4) prior to nuclear translocation of NF-kappa B. *J. Immunol.* **165,** 3541–3544.

20. Kurt-Jones, E. A., Popova, L., Kwinn, L., et al. (2000) Pattern recognition receptors TLR4 and CD14 mediate response to respiratory syncytial virus. *Nat. Immunol.* **1,** 398–401.

21. Bulut, Y., Faure, E., Thomas, L., et al. (2002) Chlamydial heat shock protein 60 activates macrophages and endothelial cells through Toll-like receptor 4 and MD2 in a MyD88-dependent pathway. *J. Immunol.* **168,** 1435–1440.

22. Rassa, J. C., Meyers, J. L., Zhang, Y., Kudaravalli, R., and Ross, S. R. (2002) Murine retro-viruses activate B cells via interaction with toll-like receptor 4. *Proc. Natl. Acad. Sci. USA* **99,** 2281–2286.

23. Kawasaki, K., Akashi, S., Shimazu, R., Yoshida, T., Miyake, K., and Nishijima, M. (2000) Mouse toll-like receptor 4.MD-2 complex mediates lipopolysaccharide-mimetic signal transduction by Taxol. *J. Biol. Chem.* **275,** 2251–2254.

24. Vabulas, R. M., Ahmad-Nejad, P., Ghose, S., Kirschning, C. J., Issels, R. D., and Wagner, H. (2002) HSP70 as endogenous stimulus of the Toll/interleukin-1 receptor signal pathway. *J. Biol. Chem.* **277,** 15,107–15,112.

25. Asea, A., Rehli, M., Kabingu, E., et al. (2002) Novel signal transduction pathway utilized by extracellular HSP70: role of toll-like receptor (TLR) 2 and TLR4. *J. Biol. Chem.* **277,** 15,028–15,034.

26. Termeer, C., Benedix, F., Sleeman, J., et al. (2002) Oligosaccharides of Hyaluronan activate dendritic cells via toll-like receptor 4. *J. Exp. Med.* **195,** 99–111.

27. Johnson, G. B., Brunn, G. J., Kodaira, Y., and Platt, J. L. (2002) Receptor-mediated monitoring of tissue well-being via detection of soluble heparan sulfate by Toll-like receptor 4. *J. Immunol.* **168,** 5233–5239.

28. Smiley, S. T., King, J. A., and Hancock, W. W. (2001) Fibrinogen stimulates macrophage chemokine secretion through toll-like receptor 4. *J. Immunol.* **167,** 2887–2894.

29. Takeuchi, O., Hoshino, K., Kawai, T., et al. (1999) Differential roles of TLR2 and TLR4 in recognition of gram-negative and gram-positive bacterial cell wall components. *Immunity* **11,** 443–451.

30. Takeuchi, O., Kaufmann, A., Grote, K., et al. (2000) Cutting edge: preferentially the R-stereoisomer of the mycoplasmal lipopeptide macrophage-activating lipopeptide-2 activates immune cells through a toll-like receptor 2- and MyD88-dependent signaling pathway. *J. Immunol.* **164,** 554–557.

31. Aliprantis, A. O., Yang, R. B., Mark, M. R., et al. (1999) Cell activation and apoptosis by bacterial lipoproteins through toll-like receptor-2. *Science* **285,** 736–739.

32. Schwandner, R., Dziarski, R., Wesche, H., Rothe, M., and Kirschning, C. J. (1999) Peptidoglycan- and lipoteichoic acid-induced cell activation is mediated by toll-like receptor 2. *J. Biol. Chem.* **274,** 17,406–17,409.

33. Lien, E., Sellati, T. J., Yoshimura, A., et al. (1999) Toll-like receptor 2 functions as a pattern recognition receptor for diverse bacterial products. *J. Biol. Chem.* **274,** 33,419–33,425.

34. Underhill, D. M., Ozinsky, A., Hajjar, A. M., et al. (1999) The Toll-like receptor 2 is recruited to macrophage phagosomes and discriminates between pathogens. *Nature* **401,** 811–815.

35. Takeuchi, O., Hoshino, K., and Akira, S. (2000) Cutting edge: TLR2-deficient and MyD88-deficient mice are highly susceptible to Staphylococcus aureus infection. *J. Immunol.* **165,** 5392–5396.

36. Takeda, K. and Akira, S. (2004) Microbial recognition by Toll-like receptors. *J. Dermatol. Sci.* **34,** 73–82.

37. Takeuchi, O., Sato, S., Horiuchi, T., et al. (2002) Cutting edge: role of Toll-like receptor 1 in mediating immune response to microbial lipoproteins. *J. Immunol.* **169,** 10–14.

38. Ozinsky, A., Underhill, D. M., Fontenot, J. D., et al. (2000) The repertoire for pattern recognition of pathogens by the innate immune system is defined by cooperation between toll-like receptors. *Proc. Natl. Acad. Sci. USA* **97,** 13,766–13,771.

39. Means, T. K., Wang, S., Lien, E., Yoshimura, A., Golenbock, D. T., and Fenton, M. J. (1999) Human toll-like receptors mediate cellular activation by Mycobacterium tuberculosis. *J. Immunol.* **163,** 3920–3927.

40. Means, T. K., Lien, E., Yoshimura, A., Wang, S., Golenbock, D. T., and Fenton, M. J. (1999) The CD14 ligands lipoarabinomannan and lipopolysaccharide differ in their requirement for Toll-like receptors. *J. Immunol.* **163,** 6748–6755.

41. Campos, M. A., Almeida, I. C., Takeuchi, O., et al. (2001) Activation of Toll-like receptor-2 by glycosylphosphatidylinositol anchors from a protozoan parasite. *J. Immunol.* **167,** 416–423.

42. Opitz, B., Schroder, N. W., Spreitzer, I., et al. (2001) Toll-like receptor-2 mediates Treponema glycolipid and lipoteichoic acid-induced NF-kappaB translocation. *J. Biol. Chem.* **276,** 22,041–22,047.

43. Massari, P., Henneke, P., Ho, Y., Latz, E., Golenbock, D. T., and Wetzler, L. M. (2002) Cutting edge: Immune stimulation by neisserial porins is toll-like receptor 2 and MyD88 dependent. *J. Immunol.* **168,** 1533–1537.

44. Werts, C., Tapping, R. I., Mathison, J. C., et al. (2001) Leptospiral lipopolysaccharide activates cells through a TLR2-dependent mechanism. *Nat. Immunol.* **2,** 346–352.

45. Hajjar, A. M., O'Mahony, D. S., Ozinsky, A., et al. (2001) Cutting edge: functional interactions between toll-like receptor (TLR) 2 and TLR1 or TLR6 in response to phenol-soluble modulin. *J. Immunol.* **166,** 15–19.

46. Alexopoulou, L., Thomas, V., Schnare, M., et al. (2002) Hyporesponsiveness to vaccination with Borrelia burgdorferi OspA in humans and in TLR1- and TLR2-deficient mice. *Nat. Med.* **8,** 878–884.

47. Hayashi, F., Smith, K. D., Ozinsky, A., et al. (2001) The innate immune response to bacterial flagellin is mediated by Toll-like receptor 5. *Nature* **410,** 1099–1103.

48. Eaves-Pyles, T. D., Wong, H. R., Odoms, K., and Pyles, R. B. (2001) Salmonella flagellin-dependent proinflammatory responses are localized to the conserved amino and carboxyl regions of the protein. *J. Immunol.* **167,** 7009–7016.

49. Alexopoulou, L., Holt, A. C., Medzhitov, R., and Flavell, R. A. (2001) Recognition of double-stranded RNA and activation of NF-kappaB by Toll-like receptor 3. *Nature* **413,** 732–738.

50. Wagner, H. (2001) Toll meets bacterial CpG-DNA. *Immunity* **14,** 499–502.

51. Krieg, A. M., Love-Homan, L., Yi, A. K., and Harty, J. T. (1998) CpG DNA induces sustained IL-12 expression in vivo and resistance to Listeria monocytogenes challenge. *J. Immunol.* **161,** 2428–2434.

52. Elkins, K. L., Rhinehart-Jones, T. R., Stibitz, S., Conover, J. S., and Klinman, D. M. (1999) Bacterial DNA containing CpG motifs stimulates lymphocyte-dependent protection of mice against lethal infection with intracellular bacteria. *J. Immunol.* **162,** 2291–2298.

53. Zimmermann, S., Egeter, O., Hausmann, S., et al. (1998) CpG oligodeoxynucleotides trigger protective and curative Th1 responses in lethal murine leishmaniasis. *J. Immunol.* **160,** 3627–3630.

54. Hemmi, H., Takeuchi, O., Kawai, T., et al. (2000) A Toll-like receptor recognizes bacterial DNA. *Nature* **408,** 740–745.

55. Bauer, S., Kirschning, C. J., Hacker, H., et al. (2001) Human TLR9 confers responsiveness to bacterial DNA via species-specific CpG motif recognition. *Proc. Natl. Acad. Sci. USA* **98,** 9237–9242.

56. Lund, J., Sato, A., Akira, S., Medzhitov, R., and Iwasaki, A. (2003) Toll-like receptor 9-mediated recognition of Herpes simplex virus-2 by plasmacytoid dendritic cells. *J. Exp. Med.* **198,** 513–520.

57. Leadbetter, E. A., Rifkin, I. R., Hohlbaum, A. M., Beaudette, B. C., Shlomchik, M. J., and Marshak-Rothstein, A. (2002) Chromatin-IgG complexes activate B cells by dual engagement of IgM and Toll-like receptors. *Nature* **416,** 603–607.

58. Ahmad-Nejad, P., Hacker, H., Rutz, M., Bauer, S., Vabulas, R. M., and Wagner, H. (2002) Bacterial CpG-DNA and lipopolysaccharides activate Toll-like receptors at distinct cellular compartments. *Eur. J. Immunol.* **32,** 1958–1968.

59. Heil, F., Ahmad-Nejad, P., Hemmi, H., et al. (2003) The Toll-like receptor 7 (TLR7)-specific stimulus loxoribine uncovers a strong relationship within the TLR7, 8 and 9 subfamily. *Eur. J. Immunol.* **33,** 2987–2997.

60. Hemmi, H., Kaisho, T., Takeuchi, O., et al. (2002) Small anti-viral compounds activate immune cells via the TLR7 MyD88-dependent signaling pathway. *Nat. Immunol.* **3,** 196–200.

61. Jurk, M., Heil, F., Vollmer, J., et al. (2002) Human TLR7 or TLR8 independently confer responsiveness to the antiviral compound R-848. *Nat. Immunol.* **3,** 499.

62. Lee, J., Chuang, T. H., Redecke, V., et al. (2003) Molecular basis for the immunostimulatory activity of guanine nucleoside analogs: activation of Toll-like receptor 7. *Proc. Natl. Acad. Sci. USA* **100,** 6646–6651.

63. Diebold, S. S., Kaisho, T., Hemmi, H., Akira, S., and Reis e Sousa, C. (2004) Innate antiviral responses by means of TLR7-mediated recognition of single-stranded RNA. *Science* **303,** 1529–1531.

64. Heil, F., Hemmi, H., Hochrein, H., et al. (2004) Species-specific recognition of single-stranded RNA via toll-like receptor 7 and 8. *Science* **303,** 1526–1529.

65. Suzuki, N., Suzuki, S., Duncan, G. S., et al. (2002) Severe impairment of interleukin-1 and Toll-like receptor signalling in mice lacking IRAK-4. *Nature* **416,** 750–756.

66. Adachi, O., Kawai, T., Takeda, K., et al. (1998) Targeted disruption of the MyD88 gene results in loss of IL-1- and IL-18-mediated function. *Immunity* **9,** 143–150.

67. Wesche, H., Henzel, W. J., Shillinglaw, W., Li, S., and Cao, Z. (1997) MyD88: an adapter that recruits IRAK to the IL-1 receptor complex. *Immunity* **7,** 837–847.

68. Muzio, M., Ni, J., Feng, P., and Dixit, V. M. (1997) IRAK (Pelle) family member IRAK-2 and MyD88 as proximal mediators of IL-1 signaling. *Science* **278,** 1612–1615.

69. Medzhitov, R., Preston-Hurlburt, P., Kopp, E., et al. (1998) MyD88 is an adaptor protein in the hToll/IL-1 receptor family signaling pathways. *Mol. Cell* **2,** 253–258.

70. Cao, Z., Henzel, W. J., and Gao, X. (1996) IRAK: a kinase associated with the interleukin-1 receptor. *Science* **271,** 1128–1131.

71. Cao, Z., Xiong, J., Takeuchi, M., Kurama, T., and Goeddel, D. V. (1996) TRAF6 is a signal transducer for interleukin-1. *Nature* **383,** 443–446.

72. Yeh, W. C., Hakem, R., Woo, M., and Mak, T. W. (1999) Gene targeting in the analysis of mammalian apoptosis and TNF receptor superfamily signaling. *Immunol. Rev.* **169,** 283–302.

73. Kaisho, T. and Akira, S. (2001) Dendritic-cell function in Toll-like receptor- and MyD88-knockout mice. *Trends Immunol.* **22,** 78–83.

74. Akira, S., Yamamoto, M., and Takeda, K. (2004) Toll-like receptor signalling. *Nat. Rev. Immunol.* **4,** 499–511.

75. Takeda, K. and Akira, S. (2004) TLR signaling pathways. *Semin. Immunol.* **16,** 3–9.

76. Kaisho, T., Takeuchi, O., Kawai, T., Hoshino, K., and Akira, S. (2001) Endotoxin-induced maturation of MyD88-deficient dendritic cells. *J. Immunol.* **166,** 5688–5694.

77. Kawai, T., Takeuchi, O., Fujita, T., et al. (2001) Lipopolysaccharide stimulates the MyD88-independent pathway and results in activation of IFN-regulatory factor 3 and the expression of a subset of lipopolysaccharide-inducible genes. *J. Immunol.* **167,** 5887–5894.

78. Smith, E. J., Marie, I., Prakash, A., Garcia-Sastre, A., and Levy, D. E. (2001) IRF3 and IRF7 phosphorylation in virus-infected cells does not require double-stranded RNA-dependent protein kinase R or Ikappa B kinase but is blocked by Vaccinia virus E3L protein. *J. Biol. Chem.* **276,** 8951–8957.

79. Honda, K., Yanai, H., Mizutani, T., et al. (2004) Role of a transductional-transcriptional processor complex involving MyD88 and IRF-7 in Toll-like receptor signaling. *Proc. Natl. Acad. Sci. USA* **101,** 15416–15421.

80. Kawai, T., Sato, S., Ishii, K. J., et al. (2004) Interferon-alpha induction through Toll-like receptors involves a direct interaction of IRF7 with MyD88 and TRAF6. *Nat. Immunol.* **5,** 1061–1068.

81. Li, S., Strelow, A., Fontana, E. J., and Wesche, H. (2002) IRAK-4: a novel member of the IRAK family with the properties of an IRAK-kinase. *Proc. Natl. Acad. Sci. USA* **99,** 5567–5572.

82. Kanakaraj, P., Schafer, P. H., Cavender, D. E., et al. (1998) Interleukin (IL)-1 receptor-associated kinase (IRAK) requirement for optimal induction of multiple IL-1 signaling pathways and IL-6 production. *J. Exp. Med.* **187,** 2073–2079.

83. Lomaga, M. A., Yeh, W. C., Sarosi, I., et al. (1999) TRAF6 deficiency results in osteopetrosis and defective interleukin-1, CD40, and LPS signaling. *Genes Dev.* **13,** 1015–1024.

84. Sato, S., Sugiyama, M., Yamamoto, M., et al. (2003) Toll/IL-1 receptor domain-containing adaptor inducing IFN-beta (TRIF) associates with TNF receptor-associated factor 6 and TANK-binding kinase 1, and activates two distinct transcription factors, NF-kappa B and IFN-regulatory factor-3, in the Toll-like receptor signaling. *J. Immunol.* **171,** 4304–4310.

85. Lomaga, M. A., Henderson, J. T., Elia, A. J., et al. (2000) Tumor necrosis factor receptor-associated factor 6 (TRAF6) deficiency results in exencephaly and is required for apoptosis within the developing CNS. *J. Neurosci.* **20,** 7384–7393.

86. Wu, H. and Arron, J. R. (2003) TRAF6, a molecular bridge spanning adaptive immunity, innate immunity and osteoimmunology. *Bioessays* **25,** 1096–1105.

87. Naito, A., Azuma, S., Tanaka, S., et al. (1999) Severe osteopetrosis, defective interleukin-1 signalling and lymph node organogenesis in TRAF6-deficient mice. *Genes Cells* **4,** 353–362.

88. Fitzgerald, K. A., Palsson-McDermott, E. M., Bowie, A. G., et al. (2001) Mal (MyD88-adapter-like) is required for Toll-like receptor-4 signal transduction. *Nature* **413,** 78–83.

89. Horng, T., Barton, G. M., Flavell, R. A., and Medzhitov, R. (2002) The adaptor molecule TIRAP provides signalling specificity for Toll-like receptors. *Nature* **420,** 329–333.

90. Yamamoto, M., Sato, S., Hemmi, H., et al. (2002) Essential role for TIRAP in activation of the signalling cascade shared by TLR2 and TLR4. *Nature* **420,** 324–329.

91. Hoebe, K., Du, X., Georgel, P., et al. (2003) Identification of Lps2 as a key transducer of MyD88-independent TIR signalling. *Nature* **424,** 743–748.

92. Yamamoto, M., Sato, S., Hemmi, H., et al. (2003) TRAM is specifically involved in the Toll-like receptor 4-mediated MyD88-independent signaling pathway. *Nat. Immunol.* **4,** 1144–1150.

93. Yamamoto, M., Sato, S., Hemmi, H., et al. (2003) Role of adaptor TRIF in the MyD88-independent toll-like receptor signaling pathway. *Science* **301,** 640–643.

94. Yamamoto, M., Sato, S., Mori, K., et al. (2002) Cutting edge: a novel Toll/IL-1 receptor domain-containing adapter that preferentially activates the IFN-beta promoter in the Toll-like receptor signaling. *J. Immunol.* **169,** 6668–6672.

95. Fitzgerald, K. A., Rowe, D. C., Barnes, B. J., et al. (2003) LPS-TLR4 signaling to IRF-3/7 and NF-kappaB involves the toll adapters TRAM and TRIF. *J. Exp. Med.* **198,** 1043–1055.

96. Fitzgerald, K. A., McWhirter, S. M., Faia, K. L., et al. (2003) IKKepsilon and TBK1 are essential components of the IRF3 signaling pathway. *Nat. Immunol.* **4,** 491–496.

97. Doyle, S., Vaidya, S., O'Connell, R., et al. (2002) IRF3 mediates a TLR3/TLR4-specific antiviral gene program. *Immunity* **17,** 251–263.

98. Hoebe, K., Janssen, E. M., Kim, S. O., et al. (2003) Upregulation of costimulatory molecules induced by lipopolysaccharide and double-stranded RNA occurs by Trif-dependent and Trif-independent pathways. *Nat. Immunol.* **4,** 1223–1229.

99. Bin, L. H., Xu, L. G., and Shu, H. B. (2003) TIRP, a novel Toll/interleukin-1 receptor (TIR) domain-containing adapter protein involved in TIR signaling. *J. Biol. Chem.* **278,** 24,526–24,532.

100. Hemmi, H., Takeuchi, O., Sato, S., et al. (2004) The roles of two IkappaB kinase-related kinases in lipopolysaccharide and double stranded RNA signaling and viral infection. *J. Exp. Med.* **199,** 1641–1650.

101. McWhirter, S. M., Fitzgerald, K. A., Rosains, J., Rowe, D. C., Golenbock, D. T., and Maniatis, T. (2004) IFN-regulatory factor 3-dependent gene expression is defective in Tbk1-deficient mouse embryonic fibroblasts. *Proc. Natl. Acad. Sci. USA* **101,** 233–238.

102. Perry, A. K., Chow, E. K., Goodnough, J. B., Yeh, W. C., and Cheng, G. (2004) Differential requirement for TANK-binding kinase-1 in type I interferon responses to toll-like receptor activation and viral infection. *J. Exp. Med.* **199,** 1651–1658.

103. Bonnard, M., Mirtsos, C., Suzuki, S., et al. (2000) Deficiency of T2K leads to apoptotic liver degeneration and impaired NF-kappaB-dependent gene transcription. *EMBO J.* **19,** 4976–4985.

104. Kobayashi, K., Hernandez, L. D., Galan, J. E., Janeway, C. A. Jr., Medzhitov, R., and Flavell, R. A. (2002) IRAK-M is a negative regulator of Toll-like receptor signaling. *Cell* **110,** 191–202.

105. Wald, D., Qin, J., Zhao, Z., et al. (2003) SIGIRR, a negative regulator of Toll-like receptor-interleukin 1 receptor signaling. *Nat. Immunol.* **4,** 920–927.

106. Brint, E. K., Xu, D., Liu, H., et al. (2004) ST2 is an inhibitor of interleukin 1 receptor and Toll-like receptor 4 signaling and maintains endotoxin tolerance. *Nat. Immunol.* **5,** 373–379.

107. Arbour, N. C., Lorenz, E., Schutte, B. C., et al. (2000) TLR4 mutations are associated with endotoxin hyporesponsiveness in humans. *Nat. Genet.* **25,** 187–191.

108. Boekholdt, S. M., Agema, W. R., Peters, R. J., et al. (2003) Variants of toll-like receptor 4 modify the efficacy of statin therapy and the risk of cardiovascular events. *Circulation* **107,** 2416–2421.

109. Smirnova, I., Mann, N., Dols, A., et al. (2003) Assay of locus-specific genetic load implicates rare Toll-like receptor 4 mutations in meningococcal susceptibility. *Proc. Natl. Acad. Sci. USA* **100,** 6075–6080.

110. Cunningham, P. N., Wang, Y., Guo, R., He, G., and Quigg, R. J. (2004) Role of Toll-like receptor 4 in endotoxin-induced acute renal failure. *J. Immunol.* **172,** 2629–2635.

111. Braun-Fahrlander, C., Riedler, J., Herz, U., et al. (2002) Environmental exposure to endotoxin and its relation to asthma in school-age children. *N. Engl. J. Med.* **347,** 869–877.

112. Picard, C., Puel, A., Bonnet, M., et al. (2003) Pyogenic bacterial infections in humans with IRAK-4 deficiency. *Science* **299,** 2076–2079.

113. Medvedev, A. E., Lentschat, A., Kuhns, D. B., et al. (2003) Distinct mutations in IRAK-4 confer hyporesponsiveness to lipopolysaccharide and interleukin-1 in a patient with recurrent bacterial infections. *J. Exp. Med.* **198,** 521–531.

114. Hugot, J. P., Chamaillard, M., Zouali, H., et al. (2001) Association of NOD2 leucine-rich repeat variants with susceptibility to Crohn's disease. *Nature* **411,** 599–603.

115. Ogura, Y., Bonen, D. K., Inohara, N., et al. (2001) A frameshift mutation in NOD2 associated with susceptibility to Crohn's disease. *Nature* **411,** 603–606.
116. Watanabe, T., Kitani, A., Murray, P. J., and Strober, W. (2004) NOD2 is a negative regulator of Toll-like receptor 2-mediated T helper type 1 responses. *Nat. Immunol.* **5,** 800–808.
117. Maeda, S., Hsu, L. C., Liu, H., et al. (2005) Nod2 mutation in Crohn's disease potentiates NF-kappaB activity and IL-1beta processing. *Science* **307,** 734–738.
118. Kobayashi, K. S., Chamaillard, M., Ogura, Y., et al. (2005) Nod2-dependent regulation of innate and adaptive immunity in the intestinal tract. *Science* **307,** 731–734.

5

Campylobacter

From Glycome to Pathogenesis

**John Kelly, Jean-Robert Brisson, N. Martin Young,
Harold C. Jarrell, and Christine M. Szymanski**

Summary

Genome sequencing of *Campylobacter jejuni* NCTC11168 identified an abundance of carbohydrate biosynthetic clusters comprising a large proportion of the genome. Many of these pathways were already under investigation including the lipooligosaccharide, flagellar *O*-linked protein glycosylation, and general *N*-linked protein glycosylation systems. Genome sequencing also identified a novel cluster of genes, which was subsequently shown to be involved in capsular polysaccharide biosynthesis. In order to fully understand the *C. jejuni* glycome, sophisticated analytical techniques were employed for functional characterization. We will describe these four important carbohydrate pathways highlighting the methods used to characterize these systems, the biological relevance the sugars play in campylobacter survival and pathogenesis, and the potential exploitation of the glycome for novel therapeutics against this common food-borne pathogen.

Key Words: Campylobacter; glycome; pathogenesis; mass spectrometry; NMR spectroscopy.

1. Introduction

1.1. Campylobacter

Since King first described human infection with a "related vibrio" in 1957, *Campylobacter jejuni* has become recognized as the leading cause of bacterial gastroenteritis worldwide and a significant cause of child morbidity in underdeveloped countries *(1)*. In addition, infection with *C. jejuni* is the most frequent antecedent to Guillain-Barré syndrome (GBS), the primary cause of neuroparalysis since the eradication of polio *(2)*. *C. jejuni*, from the epsilon group of Proteobacteria *(3)* and the closely related *Campylobacter coli*, account for as much as 95% of the reported human infections caused by campylobacters *(4)*. However, it is believed that other *Campylobacter* species are under-reported because of improper culturing techniques and/or lack of testing. For example, *Campylobacter upsaliensis* is another species that is receiving more attention owing to the reported high incidence among HIV-patients *(5)*, and the recent observation that it was the second most common species isolated among quinolone-resistant campylobacters *(6)*. Thus, the importance of campylobacter infection to human health is now being recognized more than ever because of the increase in public awareness, improvements in culturing and reporting techniques, investigation of other species,

From: *Bacterial Genomes and Infectious Diseases*
Edited by: V. L. Chan, P. M. Sherman, and B. Bourke © Humana Press Inc., Totowa, NJ

growing number of reports of campylobacter-induced autoimmunity and antimicrobial resistance, and the increasing public concern for food and water safety.

The inconsistency in these observations is that *C. jejuni* is a fastidious organism in the laboratory requiring restricted temperature and reduced oxygen for growth, yet it is able to persist through extreme environmental changes. Many domestic and wild animals are colonized with campylobacters in the mucus lining their gastrointestinal tracts where the organism remains highly motile *(7)*. Birds, particularly poultry, are considered the main reservoir, but *C. jejuni* has also been identified in several water-borne outbreaks *(8,9)*. Although *C. jejuni* contains a large number of characterized virulence factors, surprisingly the mechanisms by which the organism is able to persist and induce diarrheal disease are still unknown.

One of the best-characterized virulence factors are the bipolar flagella. Several recent reports are addressing the regulatory network and hierarchy of this complex system *(10–13)*. These organelles are not only essential for motility, colonization, and immune avoidance, but have recently been shown to play another role in the infection process. The flagellar channel serves as a primitive type III secretion system that is necessary for the export of the adhesive protein, FlaC, that influences human cell invasion *(14)* and export of the campylobacter invasion antigens, Cia proteins, that are upregulated when the organism comes into contact with host cells *(15)*. Similar to other mucosal pathogens, campylobacters also produce cytolethal-distending toxin, consisting of CdtABC, which is responsible for DNA damage leading to cell cycle arrest *(16)*. Recently, it was demonstrated that *C. jejuni* cytolethal distending toxin may have proinflammatory activity in vivo *(17)* suggesting that this toxin, in part, may contribute to human enteritis. Another recent report identified a type IV secretion system (TFSS) encoded by the pVir plasmid in certain isolates of *C. jejuni (18)*. Mutation of proteins forming the TFSS results in reduced adherence, invasion, competence, and ferret diarrheal disease *(18,19)*. Iron is also an essential component for survival of all human pathogens, and much work has been done in understanding iron acquisition and regulation in *C. jejuni (20,21)*.

The first genome sequence of *C. jejuni* NCTC11168 was published in 2000 and revealed that the organism contains multiple hypervariable sequences *(22)*. Genome sequencing confirmed that 23 genes contained variable homopolymeric tracts (and identified 9 possible others) that could potentially lead to 23! (2.6×10^{22}) different phenotypes being expressed in the population because of frameshift mutations. Dynamic populations with multiple variant cell types may contribute to the persistence of this fastidious organism and the ability to adapt to a wide range of environmental conditions. Although much work has been done to understand *C. jejuni* in the postgenomic era, it is still one of the least understood enteric pathogens. However, genome sequencing has provided "new insights into the biology of *C. jejuni* including . . . an unexpected capacity for polysaccharide production" *(22)*. This chapter will provide a summary of the ongoing efforts in understanding the role and relevance of polysaccharide production in this important mucosal pathogen.

1.2. Glycomics

The term glycome, first coined by V. N. Reinhold, was devised by analogy to proteome and describes the complete set of carbohydrate molecules synthesized by an organism,

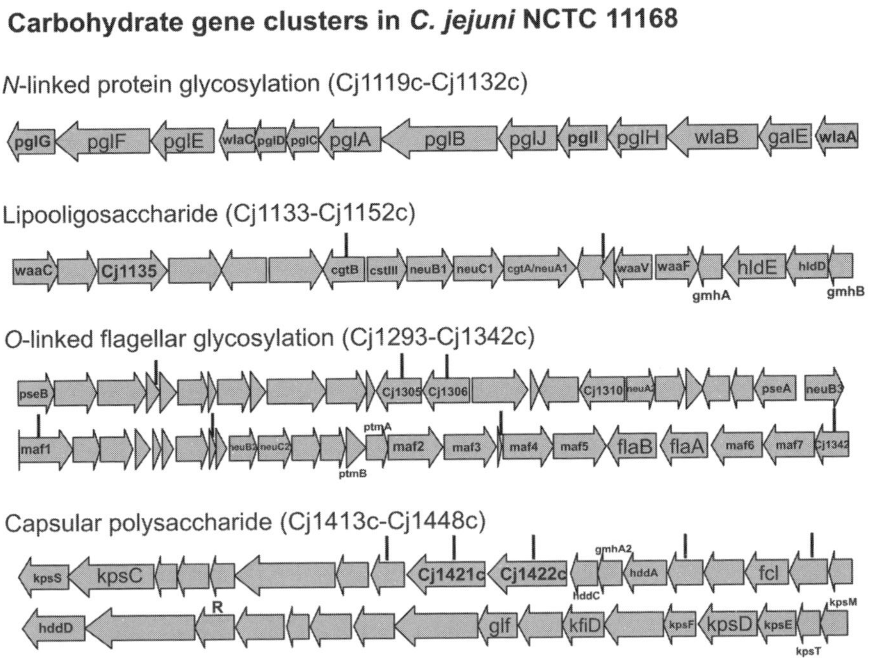

Carbohydrate gene clusters in *C. jejuni* NCTC 11168

N-linked protein glycosylation (Cj1119c-Cj1132c)

Lipooligosaccharide (Cj1133-Cj1152c)

O-linked flagellar glycosylation (Cj1293-Cj1342c)

Capsular polysaccharide (Cj1413c-Cj1448c)

Fig. 1. Gene schematic demonstrating the organization of the *Campylobacter jejuni* NCTC11168 carbohydrate loci. Genes that have not been described in the literature are represented as blank arrows. The loci are explained in more detail at: *N*-linked glycan *(67–69)*, lipooligosaccharide *(50,51,87)*, *O*-linked glycan *(67–69)* and capsular polysaccharide *(40,57,63)*. Vertical bars above the arrows indicate genes with variable homopolymeric nucleotide tracts *(22)*. R represents nine repeats of the amino acid sequence KIDLNNT. The figure was created using the CampyDB database (http://campy.bham.ac.uk/).

tissue, or cell. The term proteomics describes the methods and experiments employed to establish and investigate a proteome. However, carbohydrates are secondary gene products, i.e., they are derived from protein products encoded by genes. Hence, it is useful to extend glycomics to include the study of the relationship of a glycan's structure to the genes that govern it. From this standpoint, prior to the publication of the genome sequence of *C. jejuni (22)*, there had been little or no glycomic studies performed on this organism. Although several glycan structures had been determined, principally by Dr. G. O. Aspinall and colleagues (reviewed in ref. *23*), the genes responsible for these structures were largely unknown. Conversely, the carbohydrate structures made by the strains used in studies of pathogenesis, such as 81-176, were not known, and this was also true for the strain selected for genome sequencing, NCTC11168.

The *C. jejuni* 1.64 Mb genome sequence immediately clarified several general features of the glycome *(22)*. It was apparent that there were four distinct carbohydrate gene clusters, one of which contained two sets of genes transcribed in opposite directions (Fig. 1). Apart from the abundance of carbohydrate synthesis genes, the four loci differed from the remainder of the genome in three respects. First, they contained a large proportion of the identified genes prone to phase variation, a mechanism by which the structures of the cell-surface glycans can differ in daughter cells owing to frameshift mutations in their biosynthesis genes. Second, the majority of the genes in these

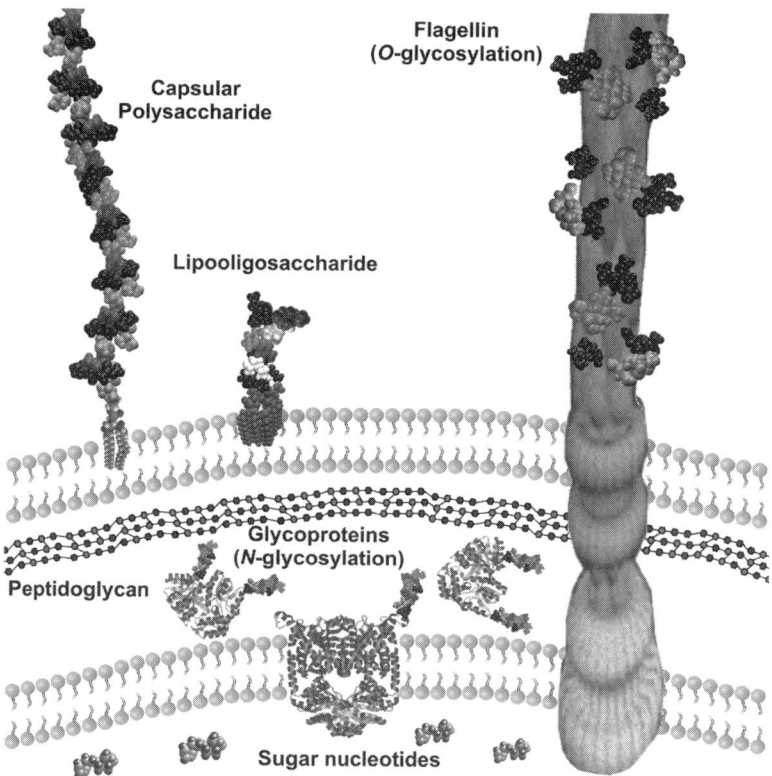

Fig. 2. Representation of the Gram-negative *Campylobacter jejuni* NCTC11168 cell glycome. The biosynthetic precursors (sugar nucleotides) are shown in the cytoplasm. *N*-linked glycoproteins are shown spanning the inner membrane and within the periplasm. The peptidoglycan, composed of a polymer containing *N*-acetyl-glucosamine and *N*-acetyl-muramic acid with several amino acids, is located within the periplasm. The flagella span both the inner and outer membranes, and are located at both cell poles. The lipooligosaccharides and capsular polysaccharides are attached to the outer leaflet of the outer membrane and surround the entire cell surface. The models are not drawn to scale.

loci were absent from the two sequenced genomes of the closely related organism, *Helicobacter pylori*. Notably, this organism does not show any clustering of its carbohydrate genes. Third, the guanine and cytosine content in the loci were slightly lower than the remainder of the genome, raising the possibility of gene transfer, particularly of the capsular and lipooligosaccharide (LOS) clusters. The whole genome contains approx 1630 genes, and the genes for carbohydrate biosynthesis total approx 100. Taking into account the genes for other glycans and for sugar nucleotide biosynthesis that are distributed throughout the genome, as much as one-tenth of the genome may be devoted to carbohydrate synthesis, making this organism an excellent model for glycomics studies.

The glycome of *C. jejuni* is summarized diagrammatically in Fig. 2. It is a set of five glycans: the peptidoglycan, the capsular polysaccharide (CPS), the LOS, and the two protein glycosylation systems, *N*- and *O*-linked. The peptidoglycan will not be described further here. The syntheses of the remaining glycans are directed by the glycosyltransferases of the four loci, which also contain the genes for the biosynthesis of the unusual sugars in each structure (Fig. 3). In their order within the genome, the loci encode the *N*-linked glycan and the LOS, the flagellin *O*-linked glycan, and the CPS (Fig. 1). Examining the breakdown of the genes in the loci according to the genome annotation demonstrates that they differ considerably in the proportion of transferases to sugar biosynthesis genes, and this mainly reflects the fact that the LOS is comprised largely of the common sugars such as glucose (Glc), galactose (Gal), and *N*-acetyl-galactosamine (GlcNAc) (Fig. 3), and thus the locus contains less biosynthetic genes. In contrast, the CPS and *N*- and *O*-linked glycans contain complex sugars (Fig. 3) requiring large biosynthetic gene clusters.

The genome information has been used to construct DNA microarrays with up to 95% coverage of the 1654 *C. jejuni* NCTC11168 open reading frames, and many arrays include additional genes from other strains. Studies have demonstrated extensive interstrain genomic variability, which was particularly evident in the genes encoding surface structures *(24–27)*. We have examined the variation among 75 strains (Fig. 4, upper panel) and showed several regions of hypervariability, with the LOS, flagellin, and CPS loci being prominent among them. However, enlargement of the *N*-glycan/LOS locus (Fig. 4, lower panel) shows a striking asymmetry with the *N*-glycan genes being almost entirely invariant (note that a function for *wlaJ* in this pathway has not been demonstrated), whereas the LOS genes involved in outer core synthesis are highly variable. Consistent with this variability plot, there are a wide variety of LOS, *O*-glycan, and CPS structures, although the *N*-glycan is conserved. Further analyses of these four pathways are described in detail in **Subheadings 2.–4.**

1.3. Analytical Tools to Examine Bacterial Glycomes

As science advances, there is a trend to merge multiple disciplines. For example, we need biology, chemistry, and physics to analyze complex biological problems. Advances in sequencing and high-throughput genomic technologies, such as microarrays and mass mutagenesis, are providing a wealth of data. This information is creating new challenges for bioinformatics, statistics, and mathematics. Mass spectrometry (MS) and nuclear magnetic resonance (NMR) spectroscopy are two analytical tools that are being relied on more heavily than before to generate functional information from the genetic data being generated.

1.3.1. Mass Spectrometry

Modern MS techniques, based on electrospray ionization (ESI) and/or MALDI, are exquisitely sensitive and can generate valuable information on tiny quantities of material (low femtomole). Therefore, it is no surprise that MS has become an essential element of any protocol to characterize novel glycoconjugates. A number of different MS techniques are used in a typical analysis strategy. This is illustrated very well in the examples described here which relate to the characterization of LOS, as well as the novel *N*- and *O*-linked glycan modifications on proteins from campylobacter.

All of the research work described here was performed on triple quadrupole and/or hybrid quadrupole time-of-flight (Q-TOF) mass spectrometers. ESI was used to ionize the samples and introduce them into the mass spectrometer. Typically, complex sample mixtures were separated by on-line high-performance liquid chromatography (HPLC) (i.e., protein digests) or capillary electrophoresis (CE) (i.e., LOS) prior to ionization.

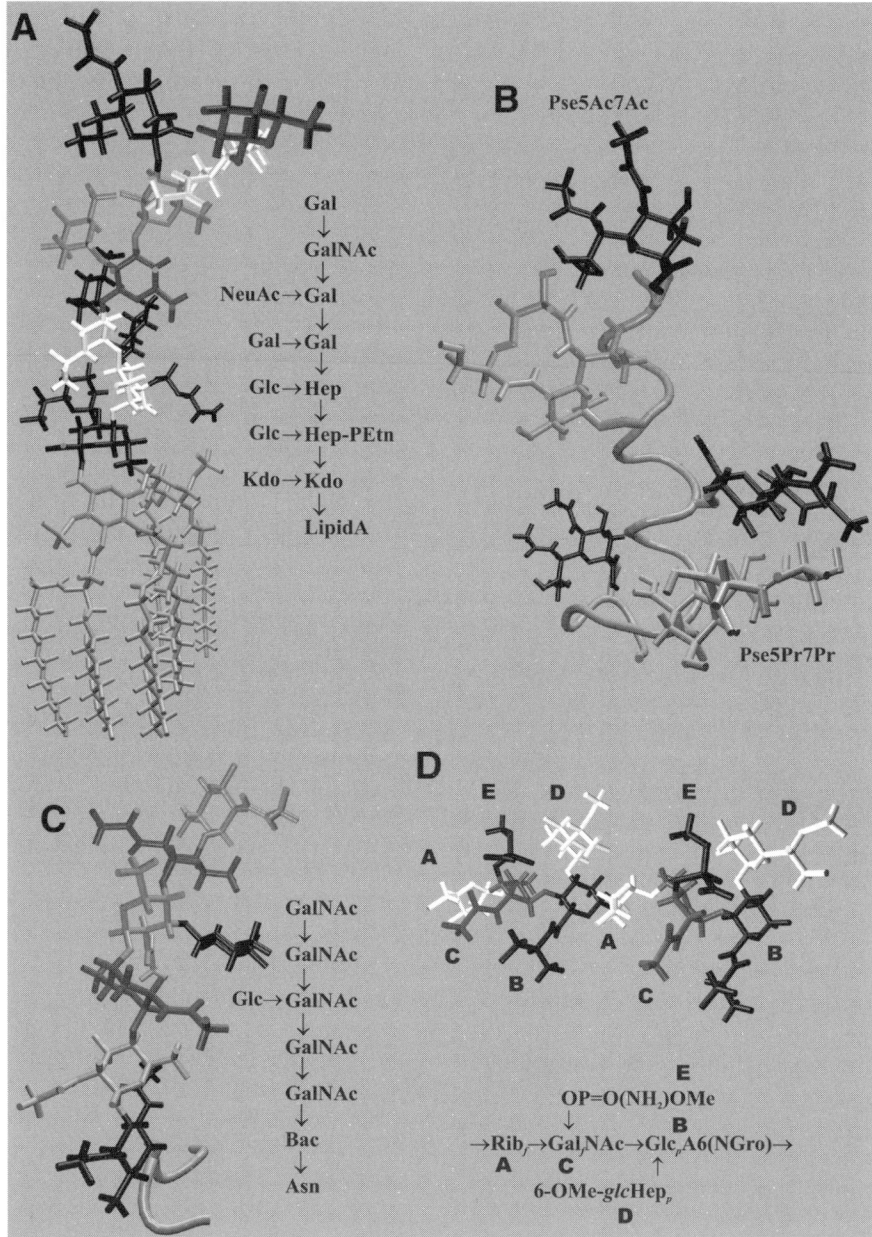

Fig. 3. Molecular models of the four *Campylobacter jejuni* glycans described. (**A**) Lipo-oligosaccharide structure *(57)*. (**B**) *O*-linked glycans showing pseudaminic acid (Pse5Ac7Ac) and its dihydroxypropionyl derivative (Pse5Pr7Pr) on the flagellin glycopeptide $_{392}$FTQNVSSI SAFMSAQGSGF$_{410}$ *(34)*. (**C**) *N*-linked glycan *(35)*. (**D**) Two repeating units of the capsular polysaccharide with phosphoramidate pendant groups *(57)*. All models where drawn with VMD *(88)*. Gal, galactose; GalNAc, *N*-acetyl-galactosamine; NeuAc, *N*-acetyl-neuraminic (sialic) acid; Glc, glucose; Hep, L-*glycero*-D-*manno*heptose; PEtn, phosphoethanolamine; KDO, 3-deoxy-D-*manno*-2-octulosonic acid; Bac, 2,4-diacetamido-2,4,6-trideoxy-D-*glucose* (bacillosamine); OP = O(NH$_2$)OMe, methyl phosphoramidate; Rib, ribose; Gal*f*Nac, 2-acetamido-2-deoxy-*galacto*-furanose; GlcA6(NGro), glucuronic acid with 2-amino-2-deoxyglycerol; and 6-*O*-Me-Hep, 6-*O*-methyl-D-*glycero*-α-L-*gluco*heptose.

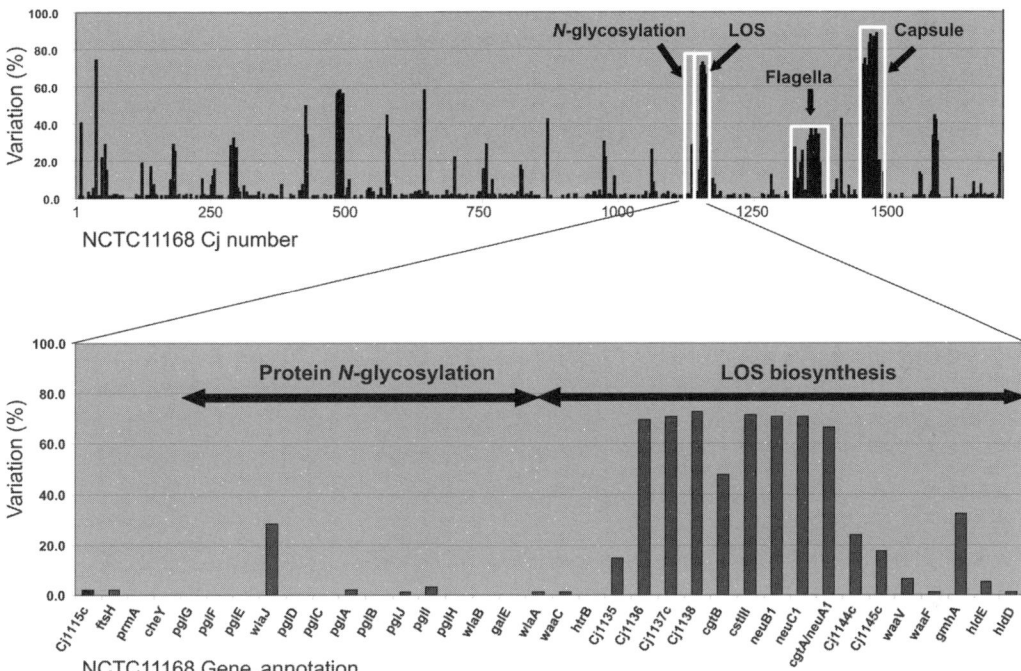

Fig. 4. Microarray comparison demonstrating gene variability and conservation in *Campylobacter jejuni* carbohydrate loci. Comparative genomic hybridizations were done for 75 *C. jejuni* strains as described in *(27)*. The variation was calculated for each individual gene in the genome strain NCTC11168, and is shown as the percentage of strains that show divergence at that gene (i.e., twofold or lower signal intensity than the genome strain). The genome region spanning all four carbohydrate loci is shown in the top panel. An expanded view of the variability in the *N*-linked protein glycosylation and lipooligosaccharide biosynthesis pathways is shown in the bottom panel.

The mass spectrometers were operated in different modes depending on the type of information that was required (Fig. 5). The masses of the intact ions (more correctly their mass/charge ratio, [m/z]) entering the mass spectrometer were determined in MS-only mode (Fig. 5A). In MS/MS mode (Fig. 5B) only ions of a specific m/z value were allowed to pass through the first quadrupole (Q_1), and were broken apart in the collision cell by collision with an inert gas, such as argon or nitrogen (a process called collision-induced dissociation). The resulting fragment ions were analyzed in the TOF (or third quadrupole in triple quadrupole instruments) and yielded important structural and sequence information. However, more detailed information was occasionally required on the fragment ions. This was achieved by fragmenting the precursor ions as they entered the mass spectrometer (a process called front-end collision-induced dissociation) and selecting one of the resulting fragment ions for MS/MS analysis (Fig. 5C). There are additional data acquisition modes that can yield useful information for a novel glycoconjugate (i.e., precursor ion scanning, neutral loss analysis, and others), but these will not be discussed. Furthermore, there are many different mass spectrometers (FTICR-MS, ion traps, hybrid instruments, and so on) available today each of which has its own particular set of strengths and weaknesses. However, only the techniques and instruments used in this work are described here.

A MS mode

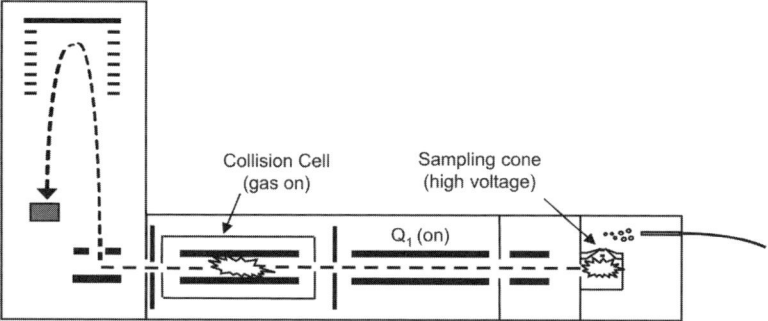

B MS/MS mode

C MS/MS with front end collision induced dissociation

Fig. 5. The different modes of mass spectrometry (MS) used in this research. The instrument represented here is a hybrid quadrupole time-of-flight mass spectrometer (Q-TOF). However, triple quadrupoles have a similar geometry except that the TOF tube is replaced with a second quadrupole similar to Q_1. In MS mode (**A**) all incoming ions pass intact through the quadrupole (Q_1) and the collision cell and are analyzed in the TOF. In MS/MS mode (**B**) only ions of a given m/z value are allowed to pass through Q_1 and enter the collision cell where they are fragmented by collision-induced dissociation. The resulting fragment ions are analyzed in the TOF. Second-generation MS/MS spectra can be acquired by increasing the orifice voltage such that incoming ions are fragmented in the source (**C**). The fragment ions of interest are then selected by Q_1, fragmented in the collision cell, and the second-generation fragment ions are analyzed in the TOF. It is worth noting that the design of some mass spectrometers (i.e., ion traps and FTICR-MS) makes it easier to acquire good quality second- and even third-generation MS/MS spectra.

1.3.2. NMR Spectroscopy and High-Resolution Magic Angle Spinning NMR

NMR has been used for decades to elucidate carbohydrate structures from all domains of life *(28–33)*. NMR is most often used in combination with MS and chemical methods for structural elucidation of glycan structures. For carbohydrates of biological origin, NMR experiments are done primarily for ^1H, ^{13}C, and ^{31}P nuclei. Proton homonuclear experiments are the most sensitive since ^1H is naturally abundant. NMR experiments for other nuclei are less sensitive and require more time or material. Protons can interact with each other through bonds or through space which provides information on the molecular distance and geometry used to elucidate the sugar type. Homonuclear experiments usually consist of correlated spectroscopy (COSY), total correlated spectroscopy (TOCSY), and nuclear overhauser effect spectroscopy (NOESY) experiments. In general, two-dimensional (2D) COSY and TOCSY are used to assign the proton resonances to each proton in the monosaccharide units, through H-C-H and H-C-C-H correlations. 2D NOESY correlations, which are dependent on interproton distances, are used to establish the sequence of sugars and locate pendant groups. The ^{13}C resonances are assigned with a heteronuclear multiple-quantum correlation (HMQC) or heteronuclear single-quantum correlation (HSQC) experiment that detects heteronuclear correlations. Long-range ^{13}C-^1H heteronuclear multiple bond correlations (HMBC) across the glycosidic bond (H-C-O-C-H) are used to establish the sequence of sugars. Chemical shifts and couplings through bonds are all dependent on the substitution patterns (H, C, O, or N) on the carbon atoms (electronic environment). Hence, the NMR parameters can be used to distinguish between all the various forms of monosaccharides possible, even if they have the same mass. Once, all resonances have been assigned, comparison of chemical shifts and coupling constants with model compounds (synthetic mono- or disaccharides) and known structures are used to confirm the structural analysis.

For high-resolution NMR, a soluble sample in sufficient quantity is required. For a glycan obtained from bacterial cells, this requires extensive purification methods that can be very time consuming. For example, to produce microgram to milligram amounts of the *O*-linked glycans from the flagellin, months of growing the appropriate amount of cells, isolation, and degradation of the flagellin, and isolation of pure tryptic glycopeptides were required *(34)*. For the *N*-linked glycopeptide, extensive purification followed by pronase digestion was used to isolate the glycan structures *(35)*. A nanoprobe, which allows NMR experiments to be done on nanomolar amounts of material, was then used to complete the structural elucidation of the glycan. For LOS, extensive purification is also required, along with cleavage of the lipid-A moiety to obtain a soluble sample *(36)*. For CPS, milligram quantities can be obtained, but viscosity owing to the high molecular weight (HMW) of the sample can lead to broad lines in the spectrum and can be problematic. In all cases, apart from impurities, heterogeneity in the glycan structure because of phase variation or loss of pendant groups or sugars during purification can further complicate the structural analysis.

In the analysis of the *O*-linked monosaccharide on *C. coli* flagellin (Fig. 6A), the glycan resonances could not be completely separated from the small peptide resonances in the partially purified pronase digest (Fig. 6B). In this case, selective 1D NMR methods were used to select resonances from the glycan structure that could be identified from a 2D COSY spectrum *(34)*. From the selective 1D TOCSY spectrum for the H-3e (equatorial) resonance (Fig. 6C), the H-3a (axial), H-4, and H-5 resonances could be observed.

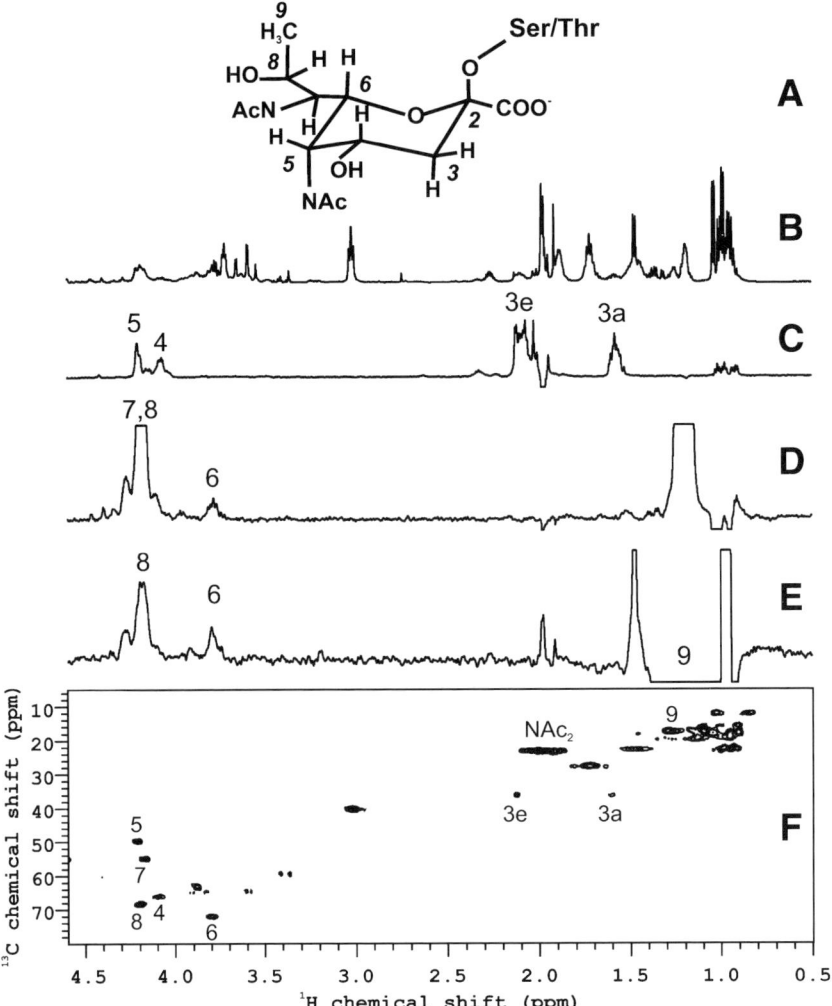

Fig. 6. Structure and nuclear magnetic resonance spectroscopy (NMR) spectra for the *O*-linked pseudaminic acid, α-Pse5Ac7Ac, found in the *Campylobacter coli* flagellin purified pronase digest. (**A**) Structure, (**B**) proton NMR spectrum, (**C**) 1D TOCSY for the H-3e resonance with a mixing time of 150 ms, (**D**) 1D TOCSY for the H-9 methyl resonance with a mixing time of 90 ms, (**E**) 1D NOESY for the H-9 methyl resonance with a mixing time of 800 ms, and (**F**) HMQC spectrum with assignment of the glycan cross-peaks. Experiments were performed at 600 MHz (^{1}H) using a nanoprobe (Varian).

From the selective 1D TOCSY spectrum for the H-9 methyl resonance (Fig. 6D), the H-8, H-7, and H-6 resonances were detected. Only spins that are coupled through consecutive H-C-H or H-C-C-H bonds were observed from a TOCSY experiment. From the selective 1D NOESY spectrum for the H-9 methyl resonance (Fig. 6E), the H-8 and H-6 resonances were detected because of the short interproton distances (<3 Å) between the C-9 methyl group and the H-8 and H-6 protons. Owing to overlap of resonances, the usual 2D COSY and TOCSY experiments were also performed to resolve any ambiguity.

It was possible to obtain a ^{13}C-^{1}H heteronuclear-correlated spectrum for this sample. This had not been previously possible for the tryptic digest *(34)*, because of less material being available and the higher molecular weights of the glycopeptides affecting the signal-to-noise ratio. From the HMQC spectrum in Fig. 6F, the ^{13}C chemical shifts of the glycan were obtained. The coupling constants were not obtained because the individual resonances exhibited heterogeneity in chemical shifts owing to multiple linkage sites to various peptides through serine (Ser) and threonine (Thr). A comparison of the ^{1}H and ^{13}C chemical shifts with model compounds and known structures confirmed that this sugar was Pse5Ac7Ac with the α anomeric configuration *(37)*. The absolute configuration (L or D) could not be determined from NMR and is usually determined by chemical methods which require a substantial amount of pure material. It was assumed to be the L configuration that is found in nature for this type of sugar *(37)*.

For glycomics, the genes encoding the components required for the synthesis of the glycan must be identified. Once the glycan structure has been determined from purified material, alterations in structure resulting from genomic modification need to be established without having to return to extensive glycan isolations. To this end, high-resolution magic angle spinning (HR-MAS) methods, which also use the nanoprobe, were utilized in order to directly observe glycans from intact bacterial cells and to provide a tool for geneticists to rapidly screen mutants for glycan alterations.

For HR-MAS NMR, a sample consists of 10^8–10^{10} bacterial cells suspended in aqueous buffer. To identify cellular components, such as the capsular polysaccharide, the widths of the individual resonances must be sufficiently narrow so that resonance overlap does not prevent critical resonance assignment. Ideally, measurement times for spectral acquisitions should be kept as short as possible to reduce the possibility that cellular degradation may influence experimental results. Because bacterial samples are relatively dilute, sensitivity becomes increasingly problematic with increasing linewidths leading to unacceptable measurement times. In liquids, rapid tumbling and isotropic (equal in all directions) molecular motion leads to sharp resonances. In bacteria, cellular components exhibit motions that are anisotropic with resulting broad lines. This residual line broadening can be reduced by spinning the sample rapidly around an axis that is inclined at an angle of 54.7°, the so-called "magic angle," with the direction of the magnetic field (Fig. 7). For some cellular components the resonance linewidths approach those achievable in liquid samples. Nearly 20 yr ago, Oldfield and coworkers demonstrated that model membrane samples composed of liposomal phospholipids could yield ^{1}H and ^{13}C NMR spectra with sufficient resolution under MAS, to allow individual resonance assignments *(38)*. In our hands, HR-MAS at spinning rates of 2000–3000 Hz can lead to linewidths of a few Hertz sufficient to acquire ^{1}H NMR spectra in a few minutes, to assign resonances to cellular components, and in fortuitous cases, to undertake structural studies *in situ* using both 2D and 1D NMR correlation experiments. It is also possible to perform heteronuclear correlation experiments such as ^{1}H-^{31}P and ^{1}H-^{13}C at natural abundance.

2. Campylobacter LOS

Lipopolysaccharides (LPS) coat the surfaces are all Gram-negative bacteria. They are composed of three domains: lipid A (also known as endotoxin) responsible for anchoring the sugars into the outer leaflet of the bacterial outer membrane, the core region that

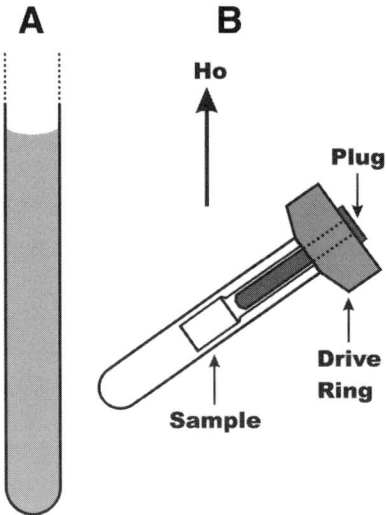

Fig. 7. Schematic diagram of nuclear magnetic resonance (NMR) spectroscopy tubes. (**A**) For high-resolution (HR)-NMR, a 5-mm NMR tube with 500 μL of soluble sample is typically used. (**B**) For HR-MAS NMR using a nanoprobe, the 40-μL sample is contained in a cylindrical rotor whose axis is inclined at an angle of 54.7° relative to the magnetic field (Ho). The sample is rotated about the rotor's axis at rates typically exceeding 2000 Hz. The tubes are drawn to scale relative to each other.

is further divided into inner core (heptoses and 3-deoxy-D-manno-oct-2-ulosonic acids [KDO], which display little variability between bacteria), and outer core (heterogeneous structures occasionally mimicking human antigens), and the HMW sugar repeats which extend beyond the bacterial cell surface. Several mucosal pathogens, such as *Neisseria* and *Haemophilus*, lack the sugar repeats and thus are said to express rough LPS or LOS. It was previously believed that campylobacters produced LPS and that these HMW structures were the serodeterminant of the heat-stable typing scheme developed by Penner *(39)*. However, even in the early studies reported by Aspinall, the attachment point for many of the sugar repeats was not defined. Subsequent genome sequencing demonstrated the unexpected presence of capsule genes and it was later shown that the HMW sugars produced by campylobacters were capsule (*see* **Subheading 3.**; *[22, 40,41]*). Further support that these HMW sugars are not attached to the core was provided by Fry et al. *(42)* and Oldfield et al. *(43)* when they analyzed deep LOS mutants and showed unaltered expression of the HMW polysaccharide. Thus, these organisms produce LOS, as shown in Fig. 2. Also similar to other mucosal pathogens, campylobacters express a phase-variable outer LOS core that is capable of mimicking human antigens. Many of these structures were solved by Aspinall et al. (reviewed in ref. *23*) and have been shown to be unique because they mimic human gangliosides in structure. It is this mimicry that was predicted to lead to the development of the antecedent neuropathies, GBS and Miller Fisher Syndrome, when antibodies that are generated against the LOS ganglioside mimics crossreact and attack host gangliosides *(44–47)*. Recently, Yuki and colleagues provided direct evidence that *C. jejuni* LOS can cause limb weakness in rabbits and, thus, carbohydrate mimicry can cause autoimmune disease *(48)*. It is speculated that *C. jejuni* expresses human ganglioside mimics to avoid host immunity, and that individuals susceptible to developing GBS would demonstrate a breakdown of immune tolerance *(49)*.

The LOS gene loci from multiple strains of *C. jejuni* have been sequenced and grouped into families based on the gene content and the resulting outer core structure *(50,51)*. The LOS cluster is the most thoroughly studied carbohydrate locus in *C. jejuni*. Through the work of Gilbert et al., many of the enzymes in the biosynthetic pathway have been characterized and the X-ray crystal structure of the first sialyltransferase, *C. jejuni* CstII, has recently been determined *(52)*. CstII has been demonstrated to be bifunctional and capable of adding both α2,3- and α2,8-linked sialic acids allowing for a greater variety of LOS structures *(50,51)*. *C. jejuni* LOS sialylation protects against complement-mediated killing, but does not play a role in adherence or invasion of host cells *(53,54)*. Recently, *galE*, which is located at the boundary between the LOS and *N*-linked glycosylation gene clusters (Fig. 1), was overexpressed and its predicted glucose-4-epimerase activity investigated *(55)*. Enzyme work demonstrated that the enzyme is actually bifunctional, exhibiting both uridine diphospho (UDP)-Glc to UDP-Gal, and UDP-GlcNAc to UDP-GalNAc conversion. Interestingly, derivatives of Gal and GalNAc are found in three *C. jejuni* NCTC11168 carbohydrate structures: LOS, CPS, and *N*-linked glycans (*see* Fig. 3) and remarkably mutation of *galE* affected all three structures *(55)*. Although enzymes involved in the biosynthesis of early sugar intermediates have been shown to be shared between pathways in other organisms, this was the first example of an enzyme being used for three separate carbohydrate pathways. Several of the genes in the LOS loci are prone to phase variation because of the presence of homopolymeric nucleotide tracts (*see* Fig. 1). Phase variation in *cgtB* (*wlaN*, galactosyltransferase *[56]*), and *cgtA* (*N*-acetylgalactosaminyltransferase *[54]*) have been linked to changes in LOS ganglioside mimicry. In addition, *cgtA* mutants exhibited increased bacterial attachment and invasion of human cells, as well as increased protection from complement mediated killing *(54)*.

Recently, we described the LOS structure for the genome sequenced strain NCTC11168, which exhibited both GM1a and GM2 ganglioside mimicry (Fig. 3; *[57]*). The variation in ganglioside mimicry was a result of the hypervariability of *cgtB* (*wlaN*) encoding the β-1,3-galactosyltransferase, which adds the terminal Gal residue to LOS. In order to more fully characterize the LOS structure in this strain, we analyzed the *O*-deacylated LOS by capillary electrophoresis–mass spectrometry (CE-MS) on a triple quadrupole MS, a method we developed to examine small quantities of these complex glycoconjugates *(58,59)*. Interestingly, we observed variability in sialylation, changes in the composition of the lipid A backbone, and variation in the relative proportions of phosphate and phosphoethanolamine modifications *(60)*. The *C. jejuni* lipid A region is unusual compared with other lipid A structures in that the glucosamine (GlcN) disaccharide backbone can be substituted with 3,4-diamino-Glc (GlcN3N) residues, and a mixture of all three forms: GlcN-GlcN, GlcN-GlcN3N, and GlcN3N-GlcN3N have been observed *(57,61)*. Changes in lipid A in other organisms have been demonstrated to affect immune response and antimicrobial activity, therefore, further examination of the effect of these changes and genes involved in *C. jejuni* is needed. Interestingly, unlike the other carbohydrate pathways described in this chapter, the genes required for lipid A and KDO biosynthesis are scattered throughout the genome *(22)*.

3. Campylobacter Capsular Polysaccharides

As mentioned in the chapter summary, capsular polysaccharides remained unnoticed until the genome sequencing project was initiated *(22)*. Subsequent mutation of capsule transport gene homologs demonstrated loss of the HMW polysaccharide and loss of the ability to be Penner typed *(40,41)*. It is now believed that all the HMW polysaccharide structures determined by Aspinall and colleagues are CPS and are the main serodeterminant of the Penner typing scheme. Further data supporting this conclusion was recently provided by Gilbert et al. when they identified a GBS isolate that serotyped as HS:2, but which had acquired an LOS gene cluster (and LOS structure) identical to the HS:19 serostrain *(62)*. Although many molecular typing techniques have been developed, the heat-stable typing scheme is still widely used and has demonstrated the clonality of certain *C. jejuni* isolates associated with GBS (i.e., HS:19 types in Japan and HS:41 types in South Africa). Recently, the *cps* loci from strains of multiple Penner serotypes were sequenced *(63)*. Karlyshev et al. demonstrated horizontal transfer of complex heptose gene clusters, as well as *cps* gene duplication, fusion, deletion, and contingency gene variation *(63)*. The CPS heptose pathway, which proceeds through GDP-linked intermediates rather than the common ADP-heptose pathway used primarily for LOS core biosynthesis, has also been identified in other pathogens, such as *Burkholderia*, *Yersinia*, and *Clostridium* species. Several of the genes involved in *C. jejuni* heptose biosynthesis of CPS (*hddA*, *hddC*, *hddD*) or shared with LOS (*gmhA*, *gmhA2*, *gmhB*) were verified through mutagenesis, followed by CE-MS analyses of the LOS and HR-MAS NMR analyses of the CPS *(63)*.

Sequencing of NCTC11168 also demonstrated that a large proportion of phase-variable genes are clustered within this locus (Fig. 1; *[22]*). The structure of the 11168 CPS was determined by us (Fig. 3) and shown to consist of β-D-Rib, β-D-Gal*f*NAc, α-D-GlcpA6(NGro)–a uronic acid with 2-amino-2-deoxyglycerol at C-6, and 6-*O*-methyl-D-*glycero*-α-L-*gluco*Hep as a side-branch *(57)*. Interestingly, in addition to aminoglycerol, ethanolamine substitutions were also detected in minor amounts during the analysis *(57)*. We optimized the HR-MAS NMR method described in **Subheading 1.3.2.** to examine intact *C. jejuni* cells. With this procedure, we were able to compare strains with known Penner type and unknown capsule structure to further demonstrate that strains with the same Penner type also showed similar capsule spectra *(60)*. HR-MAS NMR also provided us with an excellent method to analyze mutants (such as the heptose mutants) and to further examine the variation in 11168 CPS structure. We examined growth from individual colonies of wild-type 11168 by HR-MAS, silver staining, and antibody reactivity. This led to the identification of a variable 6-*O*-methyl group on the heptose branch, a population expressing high levels of the ethanolamine modification on glucuronic acid, and a unique phosphoramidate moiety on Gal*f*NAc that resembled man-made pesticides in structure *(60)*. This phosphoramidate has subsequently been detected on the HS:1 and HS:19 serostrains expressing different capsule structures and would suggest that the modification is added to different sugars in different linkages *(63)*. Interestingly, variants expressing the phosphoramidate did not stain with silver or react with Penner HS:2 typing sera *(60)*. In contrast, variants unable to express the 6-*O*-methyl group showed increased staining with silver and similar reactivity with Penner sera, whereas ethanolamine expressing variants showed similar silver-staining patterns but increased reactivity with antisera. Previous studies have demonstrated that intact CPS is required for ferret disease, chicken colonization, adherence and invasion of human epithelial cells, serum resistance, and surface charge *(41,64)*. However, the individual roles of these phase-variable modifications are currently unknown. Interestingly, early reports by Penner also described changes in HMW glycans after prolonged incubation in vitro and in vivo *(65,66)*. With the optimization of HR-MAS NMR used in our studies, we can now apply this method to examine bacteria recovered directly from representative models to further address these observations.

4. Campylobacter Protein Glycosylation Systems

Campylobacter is an excellent model system for examining bacterial glycomics because it contains two well characterized protein glycosylation systems. Similar to eukaryotes, *C. jejuni* is able to attach glycans to proteins through hydroxyl groups on Ser or Thr (*O*-linked glycosylation) or through amino groups on asparagine residues (*N*-linked glycosylation). These two pathways in *C. jejuni* have recently been reviewed *(67–69)*, so we will only briefly describe these two systems and instead describe the methodologies used in their characterization.

4.1. O-Linked Glycosylation of Flagellin

The structures of the *O*-linked sugars on *C. jejuni* 81-176 and *C. coli* VC167 flagella have recently been described (Fig. 3; *[34,70]*). The predominant glycoform is a sugar similar to sialic acid known as pseudaminic acid (Pse), a nine carbon sugar and member of the 5,7-diamino-3,5,7,9-tetradeoxynon-2-ulosonic acids. Pse has also been identified on the pilin of *Pseudomonas aeruginosa* and on the flagellin of *Helicobacter pylori* *(71,72)*. Interestingly, in both *Pseudomonas* and *Helicobacter*, the *O*-linked glycans are invariant. However, the sugars on *Campylobacter* flagella are very heterogeneous, and this is reflected in the polymorphic flagellar gene clusters (*see* Fig. 4; *[67]*). In addition to Pse, *C. jejuni* 81-176 also attaches derivatives of Pse modified with acetamidino (PseAm), *O*-acetylacetamidino, and dihydroxypropionyl to its flagellin *(34)*. In contrast, *C. coli* VC167 flagellin is modified with Pse, PseAm, and with a disaccharide structure containing these Pse derivatives possibly attached to deoxypentose *(70)*. Furthermore, the acetamidino group of *C. coli* PseAm is presumably attached at the alternate carbon, because the sugar fragments differently by MS and reacts differently with glycan specific antisera in comparison with the *C. jejuni* PseAm *(70)*. These results are consistent with differences in the composition of carbohydrate biosynthetic genes found in the corresponding flagellar loci *(68,70)*. The reason for the diversity in *O*-linked structures is currently unknown, but because this protein is the immunodominant antigen during infection, it may be advantageous for the organism to vary the carbohydrate structures. It has also been demonstrated that disruption of the Pse biosynthetic pathway results in loss of flagellar filament assembly and the subsequent loss of motility *(73)*. The *O*-linked glycans on *Helicobacter* flagellin are also needed for filament assembly and motility *(72)*, but are not required for pilin assembly in *Pseudomonas (74)*.

4.1.1. MS Analysis of Intact Flagellin Protein From C. jejuni 81-176

Analyzing the intact protein by MS is usually the first step in characterizing its posttranslational modifications (PTMs), especially those that are novel. The analysis provides a molecular weight which must all be accounted for (i.e., the combined mass of the amino acids plus the PTMs must equal the mass of the intact protein). Furthermore, a protein may exist in a variety of modified forms, and this can often be determined easily from the mass spectrum of the intact protein. In the example given here, the flagellin from *C. jejuni* 81-176 was purified, dialyzed to remove salt, and analyzed by ESI-MS on a Q-TOF mass spectrometer *(34)*. The resulting mass spectrum (Fig. 8A) is composed of a series of multiply charged ions each of which represents a different charge state for the intact flagellin. Reconstructing the molecular mass profile from this spectrum yielded a broad peak stretching over 650 Da that was dominated by two principal components at 65,766 and 65,841 Da, respectively (inset in Fig. 8A). The observed mass profile is approx 6.5 kDa larger than that predicted from the amino acid sequence alone (59,240 Da), suggesting extensive modification of the protein. Furthermore, the broadness of the mass profile indicates some variation in the extent of modification.

4.1.2. Characterization of the Novel Glycan Modifications on the Flagellin of C. jejuni 81-176

The next step in the analysis procedure was to digest the flagellin protein with trypsin and to analyze the resulting tryptic peptides by capillary HPLC-MS/MS. Approximately 250 ng of the digest was separated on an HPLC column that was linked directly to the ESI source of the Q-TOF mass spectrometer. The mass spectrometer was set to operate in automatic MS/MS acquisition mode, and MS/MS spectra were acquired on multiply charged ions only (i.e., peptide ions are usually multiply charged, whereas the background ions arising from the HPLC solvent are mostly singly charged). The MS/MS spectrum of one of the modified flagellin tryptic peptides ($T^{200-222}$) is given in Fig. 8B. This spectrum was dominated by ions in the low mass region of the spectra that could not be assigned to fragmentation of the peptide itself (m/z 317.1 and 299.1) and were assumed, correctly as it turned out, to be the oxonium ions of unusual glycan modifications (i.e., the ion formed by cleavage of a glycosidic bond). In total, three novel glycan ions were observed in the MS/MS spectra of all of the modified peptides in the tryptic digest of this flagellin (m/z 316.1, 317.1, and 409.1, respectively). Accurate mass analysis was carried out on the glycan ion at m/z 317.1 using the predicted masses of the surrounding peptide fragment ions as internal calibrants (Fig. 8B). Thus, the accurate mass of the neutral glycan residue was determined to be 316.122 ± 0.004. The most plausible empirical formula that satisfied this mass constraint was determined to be $C_{13}H_{20}O_7N_2$.

Further information regarding the structure of this unusual glycan modification was obtained by second-generation MS/MS analysis (as described in Fig. 5C). The glycopeptide precursor ion was fragmented as it entered the mass spectrometer by increasing the orifice voltage to 100 V (normally the orifice voltage is set to 30 V), and conventional MS/MS analysis was carried out on the glycan oxonium ion at m/z 317.1 (20 V collision activation). The MS/MS spectrum was dominated by the loss of small neutral molecules (water, formic acid, ketene) arising principally from the ligand functionalities appended to the glycan. This information, taken together with the empirical formula derived from the accurate mass analysis, indicated that the glycan was a diamino sugar containing two *N*-acetyl moieties and a carboxylic group, and could therefore be a pseudaminic acid. Subsequent analysis by NMR (*see* **Subheading 1.3.2.**) clearly identified the glycan as 5,7-diacetamido-3,5,7,9-tetradeoxy-ʟ-glycero-ʟ-manno-nonulosonic acid (pseudaminic acid, Pse5Ac7Ac). Second generation MS/MS analysis of the glycan ion at m/z 316.1

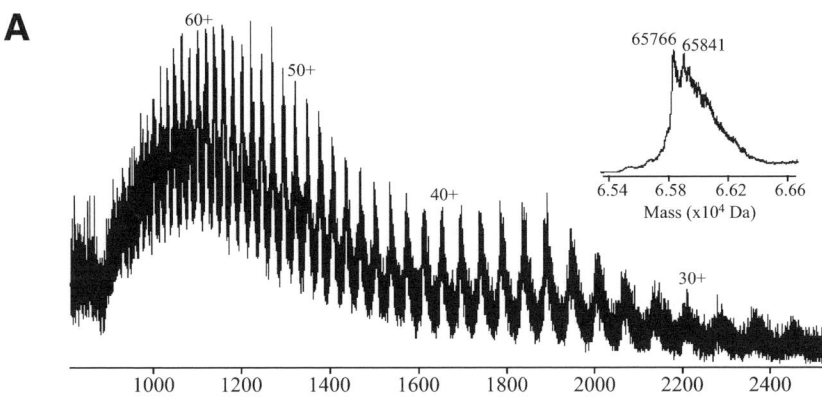

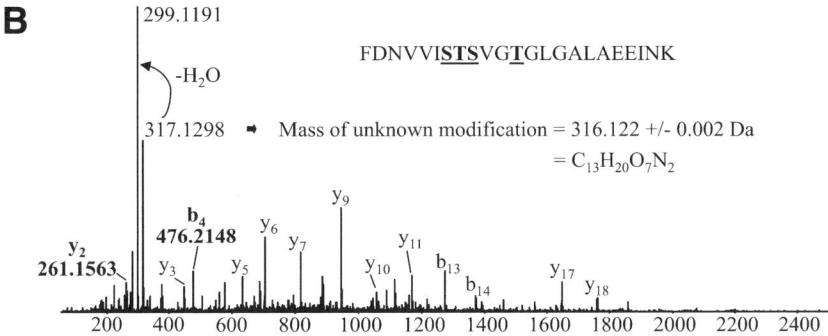

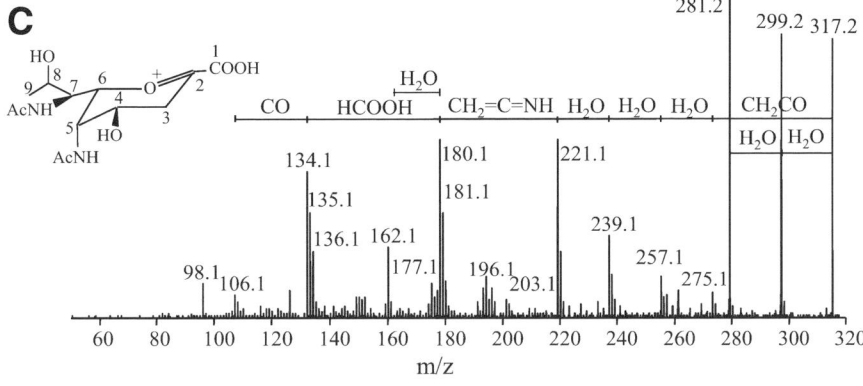

Fig. 8. Mass spectrometry (MS) characterization of the *O*-linked glycan on the flagellin of *Campylobacter jejuni* 81-176. (**A**) The ESI-MS mass spectrum of the intact flagellin. The reconstructed mass profile presented in the inset was derived from this spectrum. (**B**) MS/MS spectrum of the flagellin tryptic peptide, T[200-222]. The fragment ions derived from this peptide (sequence presented in the inset) are annotated in the spectrum. The exact masses of the glycan oxonium ion at m/z 317.1298, and the related dehydration product at 299.1191, were determined using the theoretical masses of the y_2 and b_4 peptide fragment ions as internal standards. (**C**) Second generation MS/MS spectrum of the pseudaminic (Pse5Ac7Ac) oxonium ion at m/z 317.1. The orifice voltage on the mass spectrometer was raised to 100 V in order to fragment the glycopeptide ion in the source. Standard MS/MS analysis was then carried out on the m/z 317.1 oxonium ion (20V collisional offset). The observed losses are annotated and the structure of the pseudaminic acid oxonium ion is presented in the inset.

(data not shown) suggested that one of the two acetamido groups was substituted with an acetamidino group (Pse5Am7Ac). A similar analysis of the less abundant glycan ion at m/z 409.1 suggested that the two acetamido groups were substituted with two N-2,3-dihydroxypropionyl groups (Pse5Pr7Pr).

All the evidence indicated that these glycans are O-linked to the hydroxyl group of Ser/Thr residues. However, because of the labile nature of this bond, the sites of modification could not be identified from the MS/MS spectra. Therefore, purified fractions of the tryptic glycopeptides were subjected to base-catalyzed β-elimination (incubation overnight at room temperature in 25% ammonium hydroxide), which removes O-linked glycans and converts the side chain of the linking amino acid to an amino group. The linking amino acid could then be easily identified by MS/MS analysis, as its mass was one Da less than the unmodified amino acid from which it was derived. In all, a total of 19 sites of modification were identified on the flagellin protein of *C. jejuni* 81-176. All sites were fully occupied, although the type of glycan attached at most of the sites varied somewhat. Similar results were obtained for *C. coli* VC-167 in which a total of 16 sites were identified *(70)*. The two principal glycan modifications on the *C. coli* flagellin had masses of 315 and 316 Da, respectively. MS/MS analysis indicated that the 316 Da glycan was Pse5Ac7Ac. However, MS/MS analysis of the 315 Da glycan suggested that it was an acetamidino version of pseudaminic acid (PseAm), but one that was structurally distinct from that produced by *C. jejuni* 81-176. This indicates that PseAm is synthesized by alternate mechanisms in the two strains.

4.2. N-*Linked Glycosylation of Multiple Proteins*

Since the discovery of the first Bacterial N-linked glycosylation system in *C. jejuni*, the field has moved with remarkable progress. In contrast to the gene clusters involved in LOS, CPS, and O-linked flagellar glycosylation (Fig. 1), the genes required for the general N-linked protein glycosylation pathway are remarkably conserved (Fig. 3; *[69]*) suggesting that the resulting glycans would also have the same structure. We described the complete structure of the N-linked glycan: GalNAc-α1,4-GalNAc-α1,4-[Glcβ1,3]GalNAc-α1,4-GalNAc-α1,4-GalNAc-α1,3-Bac-β1,N-Asn-Xaa, where Bac is bacillosamine, 2,4-diacetamido-2,4,6-trideoxyglucopyranose *(35)*. Using HR-MAS NMR, we observed small resonances that were conserved in all *C. jejuni* strains examined. Further analyses of the whole cells using NOESY and TOCSY experiments was possible and confirmed that we were detecting the common N-linked heptasaccharide *(60)*. This was the first demonstration of the ability to detect glycans attached to glycoproteins within intact bacterial cells. Thus, we used this method to screen other strains of *C. jejuni* and *C. coli*, and showed that the resonances exhibited similar chemical shifts indicative of the same carbohydrate structure. The genes involved in the biosynthesis of the N-linked glycan are named *pgl* for protein glycosylation (Fig. 1; *[75]*). Mutation of the *pgl* genes affects the glycosylation of numerous proteins and, therefore, has multiple effects including decreases in bacterial attachment and invasion of human epithelial cells, mouse and chick colonization, and protein reactivity to both human immune and Penner typing antisera *(68,75–78)*. Recently, one of the components of the TFSS in 81-176, VirB10, was shown to be modified by this pathway *(79)*. Amino acid substitution studies demonstrated for the first time, a direct role for the N-linked glycan in protein assembly/stability, which resulted in the reduction in natural competence *(79)*. Another recent study demonstrated the remarkable ability to functionally transfer the *pgl* locus into *E. coli* to glycosylate recombinant *C. jejuni* proteins, such as PEB3 (major antigenic protein) and AcrA (component of the multi-drug efflux pump) opening up enormous possibilities for glycoengineering in bacteria *(80)*.

4.2.1. MS Analysis of the Intact C. jejuni PEB3

Periplasmic and loosely associated cell surface proteins were isolated by gentle extraction of *C. jejuni* NCTC11168 cells (wild-type, as well as the oligosaccharyltransferase mutant, *pglB*) with glycine-HCl buffer, pH 2.2 *(35)*. The proteins were fractionated by cation-exchange chromatography (CEC) and those fractions thought to contain PEB3 (two bands by 1D-gel electrophoresis separated by ~1500 Da) were dialyzed and analyzed by ESI-MS in a manner similar to that described for the flagellin protein (**Subheading 4.1.1.**). Three protein peaks were observed in the reconstructed mass profile of the principal PEB3-containing fraction (inset in Fig. 9A). The peaks at 25,454 and 28,376 Da corresponded well with the expected masses of PEB3 (25,453 Da) and PEB4 (28,377 Da) when the signal peptides have been removed. The peak at 26,861 Da could not be identified immediately.

4.2.2. Capillary HPLC-MS/MS Analysis of PEB3

The PEB3-containing CEC fraction was digested with trypsin and analyzed by capillary HPLC-MS/MS on the Q-TOF mass spectrometer in a manner similar to that described in **Subheading 4.1.2.** All but one of the peptides in the digest of the sample could be assigned to the sequence of PEB3 or PEB4. The presence of a strong *N*-acetyl-hexosamine (HexNAc) oxonium ion at m/z 204.1 in the MS/MS spectrum of the unknown ion together with the consecutive losses of HexNAc and hexose (Hex) in the higher m/z regions clearly identified it as a glycopeptide (Fig. 9). In addition, there was sufficient peptide fragment information in this MS/MS spectrum to identify the tryptic peptide as [68]DFNVSK[73] from PEB3. The oligosaccharide is composed of 5 HexNAcs, 1 Hex and 1 unusual sugar with a residue mass of 228 Da. Later analysis by NMR identified this sugar as 2,4-diacetamido-2,4,6-trideoxyglucopyranose (Bac). The mass of the oligosaccharide (1406 Da) corresponds very well with the differences in molecular weights between PEB3 and the unknown protein peak at 26,861 Da (1407 Da). Taken all together this indicates that PEB3 is approx 50% substituted with the novel oligosaccharide. Furthermore, the oligosaccharide is linked to the peptide via the Bac residue.

This PEB3 tryptic peptide contains sites for both *N*- and *O*-linkage (i.e., Asn[70] and Ser[72], respectively). However, the actual site of glycan linkage could not be readily determined from the MS/MS spectrum. Therefore, second generation MS/MS analysis was carried out on the m/z 937.4 fragment ion (the ion comprising the tryptic peptide plus the linking Bac residue) in a manner similar to that described in **Subheading 4.1.2.** The resulting MS/MS spectrum (Fig. 9C) contained an ion at m/z 605.3, which was assigned to the b_3 peptide fragment containing the Bac glycan residue indicating that the oligosaccharide is *N*-linked to Asn[70]. Furthermore, the failure to remove the glycan by β-elimination (a procedure that removes *O*-linked modifications only, *see* **Subheading 4.1.2.**) substantiated this finding.

A

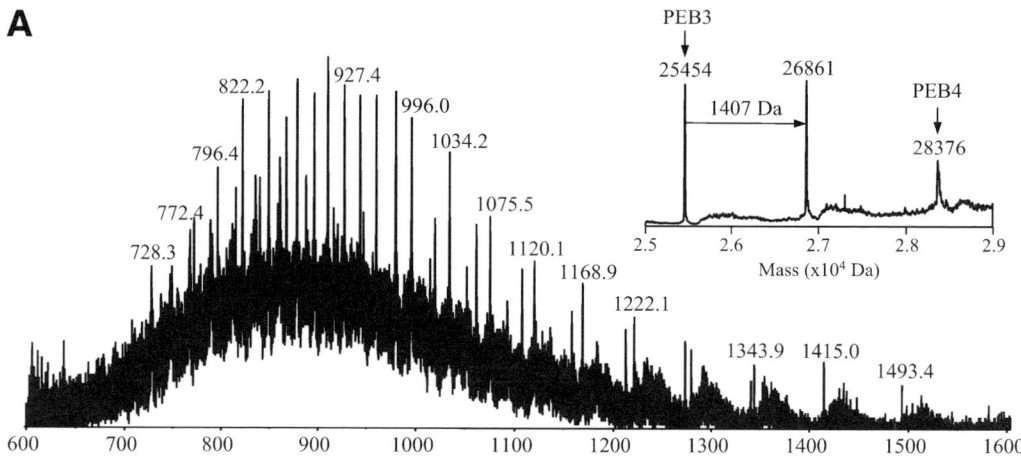

B

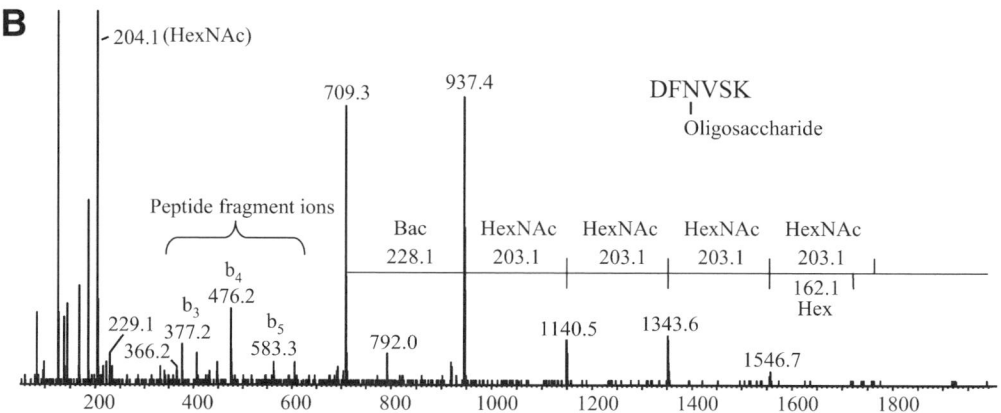

C

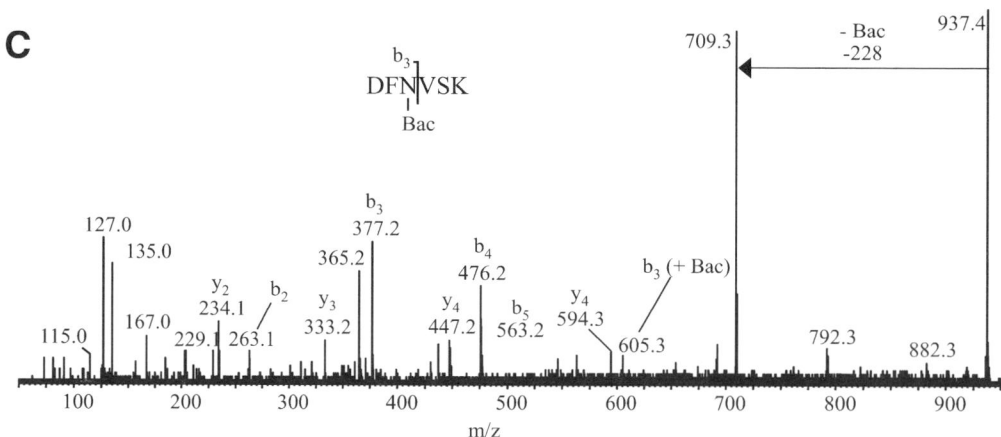

Fig. 9. Mass spectrometry (MS) characterization of the *N*-linked glycan on PEB3 from *Campylobacter jejuni* NCTC11168: (**A**) electrospray ionization (ESI)-MS mass spectrum of a cation-exchange chromatography (CEC) fraction of the glycine extract of *C. jejuni* proteins. The reconstructed mass profile derived from this spectrum is provided in the inset. The fraction was determined to contain PEB3 (25,454 Da) and PEB4 (28,376 Da), plus an unknown protein (26,861 Da) that was later determined to be glycosylated PEB3. (**B**) MS/MS analysis of the PEB3 tryptic glycopeptide,

4.2.3. Characterization of the N-Linked Glycoproteome From C. jejuni NCTC11168

A mixture of possible *N*-linked glycoproteins was isolated from the glycine-HCl extract of *C. jejuni* NCTC11168 by SBA (GalNAc specific lectin; *[81]*) affinity chromatography. The enriched glycoproteins were separated by 1D- and 2D-PAGE. Approximately 13 proteins were identified by in-gel tryptic digestion and nanoHPLC-MS/MS of the 1D-bands. However, the true complexity of the isolate was revealed by 2D-PAGE. Over 100 protein spots were resolved by 2D-PAGE representing 32 proteins. Many proteins exhibited vertical trains of spots with identical isoelectric points, indicating that these proteins have multiple sites of glycosylation, some of which are partially occupied. Indeed, almost all of the proteins identified have one or more *N*-linked sequon sites (Asn-Xaa-Ser/Thr). In total, 39 proteins are now known to be glycosylated, 24 of which are annotated as periplasmic proteins (Kelly and Young, unpublished observation; *[35]*). It is noteworthy that nearly all the glycoproteins are of unknown function. Glycopeptides were often detected by nanoLC-MS/MS in the tryptic digests of the protein bands/ spots, and in every case the glycan was identical to that observed on PEB3. This is in marked contrast to the varied *O*-linked glycans observed on the *C. jejuni* flagellin protein. Furthermore, the identified *O*-linked flagellin glycosylation sites are always fully occupied, but this is not the case for the *N*-linked glycosylation sites.

5. Biological Roles for Surface Carbohydrates and Potential Uses

Because polysaccharides represent the predominant structures on the bacterial cell surface, they are important in the interaction between the pathogen, host, and environment. To summarize the reports described in **Subheadings 2.–4.**, *C. jejuni* carbohydrates are involved in many cellular functions, including assembly of the flagellar filament *(73)*, which we now realize is a type III-like secretory apparatus necessary for the release of invasion/colonization proteins *(14,15)*, motility *(73)*, assembly of the TFSS that affects DNA uptake and natural transformation *(79)*, adherence and invasion in vitro *(41,54,76)*, colonization, and disease in vivo *(41,64,68,76–78)*, molecular mimicry of gangliosides *(82)*, autoimmunity leading to GBS *(48)*, maintenance of cell surface charge *(41)*, serum resistance *(41,53)*, antigenic *(53,60,83)* and phase variation *(41,54,57,83,84)*. Considering the importance of glycoconjugates in *C. jejuni* biology and based on the hypothesis that glycoconjugates have evolved in bacteria to enable them to become established in their preferred environmental niches, it is not surprising that these sugars play critical roles in *C. jejuni* survival and persistence.

Further characterization of the conserved campylobacter *N*-linked glycosylation pathway may lead to the development of novel therapeutic agents directed against the common bacterial heptasaccharide, or against enzymes involved in its biosynthesis. Because inactivation of the pathway also inhibits colonization, we may identify potential

---

Fig. 9. *(Continued)* T[68-73]. The fragment ions originating from the sequential loss of the oligosaccharide residues are indicated in the spectrum, as are the peptide fragment ions. The oligosaccharide is linked to the peptide via the unusual 288.1 Da glycan that was later identified as bacillosamine (Bac). (**C**) Second-generation MS/MS analysis of the glycopeptide fragment ion at m/z 937.4. The observation of the b[3]+288.1 fragment ion at m/z 605.3 clearly indicated that the oligosaccharide is *N*-linked to Asn[70].

targets for disease intervention, or crucial processes necessary for host–pathogen inter-actions. Studies may also identify unique glycosyltransferases (such as PglB) with poten-tial commercial applications, including the enormous potential for glycoengineering. Further studies using campylobacter as the model system will aid in understanding the importance of protein glycosylation, and the pathways involved for biosynthesis that will have applications in both bacterial and eukaryotic glycobiology and infectious diseases.

Although there is more variability in the flagellin *O*-linked glycans, this pathway is also an attractive therapeutic target for two obvious reasons: (1) Pse is common to two other important human pathogens (*H. pylori* and *P. aeruginosa*) and (2) inhibition of Pse biosynthesis results in inhibition of flagellar assembly and motililty in both campylo-bacter and helicobacter. In addition, further elucidation of the biosynthetic pathway will identify novel enzymes involved in Pse biosynthesis. Indeed, enzymes identified from all four pathways: *N*- and *O*-linked protein glycosylation, LOS, and CPS may be useful in industrial applications including glycoengineering. For example, there is an interest to produce synthetic gangliosides using *C. jejuni* enzymes for use in cancer ther-apy *(85)*. Further study of these pathways can also lead to novel strategies for the detec-tion and identification of these pathogens.

6. Conclusions and Future Directions

Genome sequencing demonstrated that *C. jejuni* contains multiple carbohydrate loci including both *N*- and *O*-linked protein glycosylation pathways. *C. jejuni* is an excel-lent model system for development of new techniques that can be applied to other organisms, and for understanding the biological relevance of carbohydrates and their importance in bacterial survival. Carbohydrates are required for multiple cellular func-tions and, thus, should provide multiple targets to be further exploited for the reduction of *C. jejuni* and related pathogens.

However, there are still a large number of questions to be answered. What is the func-tion of the common *N*-linked glycan that is added to so many proteins? Where are the glycosylation pathways located within the cell? What other pathways and/or proteins interact with the glycosylation systems and how are they regulated? How do the *O*-linked sugars assist with flagellar filament assembly and influence host immunity? Why do only a limited number of individuals infected with *C. jejuni* develop GBS, and what is the intended role for these LOS ganglioside mimics in host–pathogen interactions? Why do campylobacters express such a diverse assortment of CPS structures and phase-variable modifications?

Understanding these questions will allow us to gain a better awareness of the impor-tance of glycosylation in campylobacter survival and pathogenesis. This will require further model development and optimization of available analytical techniques. For example, fresh *C. jejuni* isolates (10^8–10^{10} cells) can now be examined directly by HR-MAS NMR to look for changes in *N*-linked glycan and CPS expression without labora-tory culturing or further purification. LOS structural changes (using 10^7–10^8 cells) can also be examined by using on-line preconcentration methods of capillary electrophore-sis electrospray (CE-ES) MS *(59)*. When structures need to be determined, purified car-bohydrates can be analyzed in cryoprobes to increase sensitivity and therefore decrease the amount of sample needed for NMR analysis. The field of metabolomics (analysis of small metabolites) is also rapidly expanding through methods, such as selective precursor ion scanning coupled to CE-ES MS, and Fourier transform ion cyclotron resonance MS. Indeed, precursor ion scanning MS has already identified several biosynthetic intermediates in the synthesis of Pse in *C. jejuni (86)*. High-throughput techniques such as mass mutagenesis, microarrays, and proteomics, as well as new methods such as isotope-coded affinity tagging MS and real-time method development will not only advance the understanding of glycobiology, but will move the field of bacterial pathogenesis forward in leaps. We are currently in an exciting era in microbiology!

Acknowledgments

The authors would like to acknowledge Warren Wakarchuk and Tom Devecseri for the bacterial cell image, Nam Huan Khieu for molecular modeling, Eduardo Taboada for the microarray image, and Roy Chaudhuri and Mark Pallen who created the CampyDB database. We would also like to thank Susan Logan, Michel Gilbert, Warren Wakarchuk, Jianjun Li, Brendan Wren, and Patricia Guerry for stimulating discussions and initiating many of the projects described. The funding for this work was through the NRC–Genomics and Health Initiative.

Dedication

We would like to dedicate this chapter to Gerald O. Aspinall and John L. Penner whose pioneering work on campylobacter carbohydrates has laid the foundation for several of the studies described.

References

1. King, E. O. (1957) Human infections with *Vibrio fetus* and a closely related vibrio. *J. Infect. Dis.* **101,** 119–128.
2. Jacobs, B. C., Rothbarth, P. H., van der Meche, F. G., et al. (1998) The spectrum of antecedent infections in Guillain-Barré syndrome: a case-control study. *Neurology* **51,** 1110–1115.
3. Walz, S. E., Baqar, S., Beecham, H. J., et al. (2001) Pre-exposure anti-*Campylobacter jejuni* immunoglobulin a levels associated with reduced risk of *Campylobacter* diarrhea in adults traveling to Thailand. *Am. J. Trop. Med. Hyg.* **65,** 652–656.
4. Park, S. F. (2002) The physiology of *Campylobacter* species and its relevance to their role as foodborne pathogens. *Int. J. Food Microbiol.* **74,** 177–188.
5. Jenkin, G. A. and Tee, W. (1998) *Campylobacter upsaliensis*-associated diarrhea in human immunodeficiency virus-infected patients. *Clin. Infect. Dis.* **27,** 816–821.
6. Labarca, J. A., Sturgeon, J., Borenstein, L., et al. (2002) *Campylobacter upsaliensis*: Another pathogen for consideration in the United States. *Clin. Infect. Dis.* **34,** E59–E60.
7. Szymanski, C. M., King, M., Haardt, M., and Armstrong, G. D. (1995) *Campylobacter jejuni* motility and invasion of Caco-2 cells. *Infect. Immun.* **63,** 4295–4300.
8. Sharma, S., Sachdeva, P., and Virdi, J. S. (2003) Emerging water-borne pathogens. *Appl. Microbiol. Biotechnol.* **61,** 424–428.
9. Hrudey, S. E., Payment, P., Huck, P. M., Gillham, R. W., and Hrudey, E. J. (2003) A fatal waterborne disease epidemic in Walkerton, Ontario: comparison with other waterborne outbreaks in the developed world. *Water Sci. Technol.* **47,** 7–14.
10. Hendrixson, D. R. and DiRita, V. J. (2003) Transcription of sigma54-dependent but not sigma28-dependent flagellar genes in *Campylobacter jejuni* is associated with formation of the flagellar secretory apparatus. *Mol. Microbiol.* **50,** 687–702.

11. Wosten, M. M., Wagenaar, J. A., and van Putten, J. P. (2004) The FlgS/FlgR two-component signal transduction system regulates the *fla* regulon in *Campylobacter jejuni. J. Biol. Chem.* **279,** 16,214–16,222.

12. Jagannathan, A., Constantinidou, C., and Penn, C. W. (2001) Roles of *rpoN, fliA,* and *flgR* in expression of flagella in *Campylobacter jejuni. J. Bacteriol.* **183,** 2937–2942.

13. Carrillo, C. D., Taboada, E., Nash, J. H., et al. (2004) Genome-wide expression analyses of *Campylobacter jejuni* NCTC11168 reveals coordinate regulation of motility and virulence by *flhA. J. Biol. Chem.* **279,** 20,327–20,338.

14. Song, Y. C., Jin, S., Louie, H., et al. (2004) FlaC, a protein of *Campylobacter jejuni* TGH9011 (ATCC43431) secreted through the flagellar apparatus, binds epithelial cells and influences cell invasion. *Mol. Microbiol.* **53,** 541–553.

15. Konkel, M. E., Klena, J. D., Rivera-Amill, V., et al. (2004) Secretion of virulence proteins from *Campylobacter jejuni* is dependent on a functional flagellar export apparatus. *J. Bacteriol.* **186,** 3296–3303.

16. Lee, R. B., Hassane, D. C., Cottle, D. L., and Pickett, C. L. (2003) Interactions of *Campylobacter jejuni* cytolethal distending toxin subunits CdtA and CdtC with HeLa cells. *Infect. Immun.* **71,** 4883–4890.

17. Fox, J. G., Rogers, A. B., Whary, M. T., et al. (2004) Gastroenteritis in NF-kappaB-deficient mice is produced with wild-type *Camplyobacter jejuni* but not with *C. jejuni* lacking cytolethal distending toxin despite persistent colonization with both strains. *Infect. Immun.* **72,** 1116–1125.

18. Bacon, D. J., Alm, R. A., Burr, D. H., et al. (2000) Involvement of a plasmid in virulence of *Campylobacter jejuni* 81-176. *Infect. Immun.* **68,** 4384–4390.

19. Bacon, D. J., Alm, R. A., Hu, L., et al. (2002) DNA sequence and mutational analyses of the pVir plasmid of *Campylobacter jejuni* 81-176. *Infect. Immun.* **70,** 6242–6250.

20. van Vliet, A. H., Ketley, J. M., Park, S. F., and Penn, C. W. (2002) The role of iron in *Campylobacter* gene regulation, metabolism and oxidative stress defense. *FEMS Microbiol. Rev.* **26,** 173–186.

21. Palyada, K., Threadgill, D., and Stintzi, A. (2004) Iron acquisition and regulation in *Campylobacter jejuni. J. Bacteriol.* **186,** 4714–4729.

22. Parkhill, J., Wren, B. W., Mungall, K., et al. (2000) The genome sequence of the food-borne pathogen *Campylobacter jejuni* reveals hypervariable sequences. *Nature* **403,** 665–668.

23. Moran, A. P., Penner, J. L., and Aspinall, G. O. (2000) *Campylobacter* lipopolysaccharides, in *Campylobacter* (Nachamkin, I. and Blaser, M. J., eds.). American Society for Microbiology, Washington, D.C., pp. 241–257.

24. Dorrell, N., Mangan, J. A., Laing, K. G., et al. (2001) Whole genome comparison of *Campylobacter jejuni* human isolates using a low-cost microarray reveals extensive genetic diversity. *Genome Res.* **11,** 1706–1715.

25. Leonard, E. E., Takata, T., Blaser, M. J., Falkow, S., Tompkins, L. S., and Gaynor, E. C. (2003) Use of an open-reading frame-specific *Campylobacter jejuni* DNA microarray as a new genotyping tool for studying epidemiologically related isolates. *J. Infect. Dis.* **187,** 691–694.

26. Pearson, B. M., Pin, C., Wright, J., I'Anson, K., Humphrey, T., and Wells, J. M. (2003) Comparative genome analysis of *Campylobacter jejuni* using whole genome DNA microarrays. *FEBS Lett.* **554,** 224–230.

27. Taboada, E. N., Acedillo, R. R., Carrillo, C. D., et al. (2004) Large-scale comparative genomics meta-analysis of *Campylobacter jejuni* isolates reveals low level of genome plasticity. *J. Clin. Microbiol.* **42,** 4566–4576.

28. Brisson, J. R., Sue, S. C., Wu, W. G., McManus, G., Nghia, P. T., and Uhrin, D. (2002) NMR of carbohydrates: 1D homonuclear selective methods, in *NMR Spectroscopy of Glyco-*

conjugates (Jimenez-Barbero, J. and Peters, T., eds.). Wiley-VCH, Weinhem, Germany, pp. 59–93.

29. Duus, J. O., Gotfredsen, C. H., and Bock, K. (2000) Carbohydrate structural determination by NMR spectroscopy: modern methods and limitations. *Chem. Rev.* **100,** 4589–4614.

30. Kogan, G. and Uhrin, D. (2000) Current NMR methods in the structural elucidation of polysaccharides, in *New Advances in Analytical Chemistry* (Atta-ur-Rahman, ed.). Harwood Academic, Amsterdam, pp. 73–134.

31. Uhrin, D. and Brisson, J. R. (2000) Structure determination of microbial polysaccharides by high resolution NMR spectroscopy, in *NMR in Microbiology: Theory and Applications* (Barbotin, J. N. and Portais, J. C., eds.). Horizon Scientific Press, Wymondham, UK, pp. 165–210.

32. van Halbeek, H. (1994) ^{1}H nuclear magnetic resonance spectroscopy of carbohydrate chains of glycoproteins. *Methods Enzymol.* **230,** 132–168.

33. Agrawal, P. K. (1992) NMR spectroscopy in the structural elucidation of oligosaccharides and glycosides. *Phytochemistry* **31,** 3307–3330.

34. Thibault, P., Logan, S. M., Kelly, J. F., et al. (2001) Identification of the carbohydrate moieties and glycosylation motifs in *Campylobacter jejuni* flagellin. *J. Biol. Chem.* **276,** 34,862–34,870.

35. Young, N. M., Brisson, J. R., Kelly, J., et al. (2002) Structure of the *N*-linked glycan present on multiple glycoproteins in the gram-negative bacterium, *Campylobacter jejuni. J. Biol. Chem.* **277,** 42,530–42,539.

36. Brisson, J. R., Crawford, E., Khieu, N. H., Perry, M. B., and Richards, J. C. (2002) The core oligosaccharide component from *Mannheimia (Pasteurella) haemolytica* serotype A1 lipopolysaccharide contains L-*glycero*-D-*manno*- and D-*glycero*-D-*manno*-heptoses. Analysis of the structure and conformation by high resolution NMR spectroscopy. *Can. J. Chem.* **80,** 949–963.

37. Knirel, Y. A., Shashkov, A. S., Tsvetkov, Y. E., Jansson, P. E., and Zahringer, U. (2003) 5,7-diamino-3,5,7,9-tetradeoxynon-2-ulosonic acids in bacterial glycopolymers: chemistry and biochemistry. *Adv. Carbohydr. Chem. Biochem.* **58,** 371–417.

38. Oldfield, E., Bowers, J. L., and Forbes, J. (1987) High-resolution proton and carbon-13 NMR of membranes: why sonicate? *Biochemistry* **26,** 6919–6923.

39. Penner, J. L. and Hennessy, J. N. (1980) Passive hemagglutination technique for serotyping *Campylobacter fetus* subsp. *jejuni* on the basis of soluble heat-stable antigens. *J. Clin. Microbiol.* **12,** 732–737.

40. Karlyshev, A. V., Linton, D., Gregson, N. A., Lastovica, A. J., and Wren, B. W. (2000) Genetic and biochemical evidence of a *Campylobacter jejuni* capsular polysaccharide that accounts for Penner serotype specificity. *Mol. Microbiol.* **35,** 529–541.

41. Bacon, D. J., Szymanski, C. M., Burr, D. H., Silver, R. P., Alm, R. A., and Guerry, P. (2001) A phase-variable capsule is involved in virulence of *Campylobacter jejuni* 81-176. *Mol. Microbiol.* **40,** 769–777.

42. Fry, B. N., Feng, S., Chen, Y. Y., Newell, D. G., Coloe, P. J., and Korolik, V. (2000) The *galE* gene of *Campylobacter jejuni* is involved in lipopolysaccharide synthesis and virulence. *Infect. Immun.* **68,** 2594–2601.

43. Oldfield, N. J., Moran, A. P., Millar, L. A., Prendergast, M. M., and Ketley, J. M. (2002) Characterization of the *Campylobacter jejuni* heptosyltransferase II gene, *waaF*, provides genetic evidence that extracellular polysaccharide is lipid A core independent. *J. Bacteriol.* **184,** 2100–2107.

44. Yuki, N., Taki, T., Inagaki, F., et al. (1993) A bacterium lipopolysaccharide that elicits Guillain-Barré syndrome has a GM1 ganglioside-like structure. *J. Exp. Med.* **178,** 1771–1775.

45. Salloway, S., Mermel, L. A., Seamans, M., et al. (1996) Miller-Fisher syndrome associated with *Campylobacter jejuni* bearing lipopolysaccharide molecules that mimic human ganglioside GD3. *Infect. Immun.* **64,** 2945–2949.

46. Jacobs, B. C., Endtz, H., van der Meche, F. G., Hazenberg, M. P., Achtereekte, H. A., and van Doorn, P. A. (1995) Serum anti-GQ1b IgG antibodies recognize surface epitopes on *Campylobacter jejuni* from patients with Miller Fisher syndrome. *Ann. Neurol.* **37,** 260–264.

47. Ang, C. W., Laman, J. D., Willison, H. J., et al. (2002) Structure of *Campylobacter jejuni* lipopolysaccharides determines antiganglioside specificity and clinical features of Guillain-Barré and Miller Fisher patients. *Infect. Immun.* **70,** 1202–1208.

48. Yuki, N., Susuki, K., Koga, M., et al. (2004) Carbohydrate mimicry between human ganglioside GM1 and *Campylobacter jejuni* lipooligosaccharide causes Guillain-Barré syndrome. *Proc. Natl. Acad. Sci. USA* **101,** 11,404–11,409.

49. Bowes, T., Wagner, E. R., Boffey, J., et al. (2002) Tolerance to self gangliosides is the major factor restricting the antibody response to lipopolysaccharide core oligosaccharides in *Campylobacter jejuni* strains associated with Guillain-Barré syndrome. *Infect. Immun.* **70,** 5008–5018.

50. Gilbert, M., Karwaski, M. F., Bernatchez, S., et al. (2002) The genetic bases for the variation in the lipo-oligosaccharide of the mucosal pathogen, *Campylobacter jejuni*. Biosynthesis of sialylated ganglioside mimics in the core oligosaccharide. *J. Biol. Chem.* **277,** 327–337.

51. Gilbert, M., Godschalk, P. C. R., Parker, C. T., Endtz, H., and Wakarchuk, W. W. (2004) Genetic basis for the variation in the lipooligosaccharide outer core of *Campylobacter jejuni* and possible association of glycosyltransferase genes with post-infectious neurophathies, in *Campylobacter jejuni: New Perspectives in Molecular and Cellular Biology* (Ketley, J. and Konkel, M. E., eds.). Horizon Scientific Press, Norfolk, UK, pp. 219–248.

52. Chiu, C. P., Watts, A. G., Lairson, L. L., et al. (2004) Structural analysis of the sialyltransferase CstII from *Campylobacter jejuni* in complex with a substrate analog. *Nat. Struct. Mol. Biol.* **11,** 163–170.

53. Guerry, P., Ewing, C. P., Hickey, T. E., Prendergast, M. M., and Moran, A. P. (2000) Sialylation of lipooligosaccharide cores affects immunogenicity and serum resistance of *Campylobacter jejuni*. *Infect. Immun.* **68,** 6656–6662.

54. Guerry, P., Szymanski, C. M., Prendergast, M. M., et al. (2002) Phase variation of *Campylobacter jejuni* 81-176 lipooligosaccharide affects ganglioside mimicry and invasiveness in vitro. *Infect. Immun.* **70,** 787–793.

55. Bernatchez, S., Szymanski, C. M., Ishiyama, N., et al. (2005) A single bifunctional UDP-GlcNAc/Glc 4-epimerase supports the synthesis of three cell surface glycoconjugates in *Campylobacter jejuni*. *J. Biol. Chem.* **280,** 4792–4802.

56. Linton, D., Gilbert, M., Hitchen, P. G., et al. (2000) Phase variation of a beta-1,3 galactosyltransferase involved in generation of the ganglioside GM1-like lipo-oligosaccharide of *Campylobacter jejuni*. *Mol. Microbiol.* **37,** 501–514.

57. St. Michael, F., Szymanski, C. M., Li, J., et al. (2002) The structures of the lipooligosaccharide and capsule polysaccharide of *Campylobacter jejuni* genome sequenced strain NCTC 11168. *Eur. J. Biochem.* **269,** 5119–5136.

58. Kelly, J., Masoud, H., Perry, M. B., Richards, J. C., and Thibault, P. (1996) Separation and characterization of O-deacylated lipooligosaccharides and glycans derived from *Moraxella catarrhalis* using capillary electrophoresis-electrospray mass spectrometry and tandem mass spectrometry. *Anal. Biochem.* **233,** 15–30.

59. Li, J., Thibault, P., Martin, A., Richards, J. C., Wakarchuk, W. W., and vander Wilp, W. (1998) Development of an on-line preconcentration method for the analysis of pathogenic lipopolysaccharides using capillary electrophoresis-electrospray mass spectrometry. *J. Chromatogr. A* **817,** 325–336.

60. Szymanski, C. M., Michael, F. S., Jarrell, H. C., et al. (2003) Detection of conserved *N*-linked glycans and phase-variable lipooligosaccharides and capsules from *Campylobacter* cells by mass spectrometry and high resolution magic angle spinning NMR spectroscopy. *J. Biol. Chem.* **278,** 24,509–24,520.

61. Moran, A. P., Zahringer, U., Seydel, U., Scholz, D., Stutz, P., and Rietschel, E. T. (1991) Structural analysis of the lipid A component of *Campylobacter jejuni* CCUG 10936 (serotype O:2) lipopolysaccharide. Description of a lipid A containing a hybrid backbone of 2-amino-2-deoxy-D-glucose and 2,3-diamino-2,3-dideoxy-D-glucose. *Eur. J. Biochem.* **198,** 459–469.

62. Gilbert, M., Godschalk, P. C., Karwaski, M. F., et al. (2004) Evidence for acquisition of the lipooligosaccharide biosynthesis locus in *Campylobacter jejuni* GB11, a strain isolated from a patient with Guillain-Barré syndrome, by horizontal exchange. *Infect. Immun.* **72,** 1162–1165.

63. Karlyshev, A. V., Champion, O. L., Churcher, C., et al. Analysis of *Campylobacter jejuni* capsular loci reveals multiple mechanisms for the generation of genetic diversity and the ability to form complex heptoses. *Mol. Microbiol.* **55,** 90–103.

64. Jones, M. A., Marston, K. L., Woodall, C. A., et al. (2004) Adaptation of *Campylobacter jejuni* NCTC11168 to high-level colonization of the avian gastrointestinal tract. *Infect. Immun.* **72,** 3769–3776.

65. Mills, S. D., Kurjanczyk, L. A., Shames, B., Hennessy, J. N., and Penner, J. L. (1991) Antigenic shifts in serotype determinants of *Campylobacter coli* are accompanied by changes in the chromosomal DNA restriction endonuclease digestion pattern. *J. Med. Microbiol.* **35,** 168–173.

66. Mills, S. D., Kuzniar, B., Shames, B., Kurjanczyk, L. A., and Penner, J. L. (1992) Variation of the O antigen of *Campylobacter jejuni in vivo. J. Med. Microbiol.* **36,** 215–219.

67. Szymanski, C. M., Logan, S. M., Linton, D., and Wren, B. W. (2003) *Campylobacter*: a tale of two protein glycosylation systems. *Trends Microbiol.* **11,** 233–238.

68. Szymanski, C. M., Goon, S., Allan, B., and Guerry, P. (2005) Campylobacter protein glycosylation, in C*ampylobacter: New Perspectives in Molecular and Cellular Biology* (Ketley, J. and Konkel, M. eds.). Horizon Scientific and Caister Academic Press, Norfolk, UK, pp. 259–273.

69. Szymanski, C. M. and Wren, B. W. (2005) Protein glycosylation in bacterial mucosal pathogens. *Nat. Rev. Microbiol.* **3,** 225–237.

70. Logan, S. M., Kelly, J. F., Thibault, P., Ewing, C. P., and Guerry, P. (2002) Structural heterogeneity of carbohydrate modifications affects serospecificity of *Campylobacter* flagellins. *Mol. Microbiol.* **46,** 587–597.

71. Castric, P., Cassels, F. J., and Carlson, R. W. (2001) Structural characterization of the *Pseudomonas aeruginosa* 1244 pilin glycan. *J. Biol. Chem.* **276,** 26,479–26,485.

72. Schirm, M., Soo, E. C., Aubry, A. J., Austin, J., Thibault, P., and Logan, S. M. (2003) Structural, genetic and functional characterization of the flagellin glycosylation process in *Helicobacter pylori. Mol. Microbiol.* **48,** 1579–1592.

73. Goon, S, Kelly, J., Logan, S. M., Ewing, C. P., and Guerry, P. (2003) Pseudaminic acid, the major modification on *Campylobacter* flagellin, is synthesized via the Cj1293 gene. *Mol. Microbiol.* **50,** 659–671.

74. DiGiandomenico, A., Matewish, M. J., Bisaillon, A., Stehle, J. R., Lam, J. S., and Castric, P. (2002) Glycosylation of Pseudomonas aeruginosa 1244 pilin: glycan substrate specificity. *Mol. Microbiol.* **46,** 519–530.

75. Szymanski, C. M., Yao, R., Ewing, C. P., Trust, T. J., and Guerry, P. (1999) Evidence for a system of general protein glycosylation in *Campylobacter jejuni. Mol. Microbiol.* **32,** 1022–1030.

76. Szymanski, C. M., Burr, D. H., and Guerry, P. (2002) *Campylobacter* protein glycosylation affects host cell interactions. *Infect. Immun.* **70,** 2242–2244.
77. Hendrixson, D. R. and DiRita, V. J. (2004) Identification of *Campylobacter jejuni* genes involved in commensal colonization of the chick gastrointestinal tract. *Mol. Microbiol.* **52,** 471–484.
78. Karlyshev, A. V., Everest, P., Linton, D., Cawthraw, S., Newell, D. G., and Wren, B. W. (2004) The *Campylobacter jejuni* general glycosylation system is important for attachment to human epithelial cells and in the colonization of chicks. *Microbiology* **150,** 1957–1964.
79. Larsen, J. C., Szymanski, C., and Guerry, P. (2004) *N*-linked protein glycosylation is required for full competence in *Campylobacter jejuni* 81-176. *J. Bacteriol.* **186,** 6508–6514.
80. Wacker, M., Linton, D., Hitchen, P. G., et al. (2002) *N*-linked glycosylation in *Campylobacter jejuni* and its functional transfer into *E. coli. Science* **298,** 1790–1793.
81. Linton, D., Allan, E., Karlyshev, A. V., Cronshaw, A. D., and Wren, B. W. (2002) Identification of N-acetylgalactosamine-containing glycoproteins PEB3 and CgpA in *Campylobacter jejuni. Mol. Microbiol.* **43,** 497–508.
82. Aspinall, G. O., Fujimoto, S., McDonald, A. G., Pang, H., Kurjanczyk, L. A., and Penner, J. L. (1994) Lipopolysaccharides from *Campylobacter jejuni* associated with Guillain-Barré, syndrome patients mimic human gangliosides in structure. *Infect. Immun.* **62,** 2122–2125.
83. Guerry, P., Alm, R., Szymanski, C. M., and Trust, T. J. (2000) Structure, function, and antigenicity of *Campylobacter* flagella, in *Campylobacter* (Nachamkin, I. and Blaser, M. J., eds.). American Society for Microbiology, Washington, DC, pp. 405–421.
84. Brooks, B. W., Robertson, R. H., Lutze-Wallace, C. L., and Pfahler, W. (2001) Identification, characterization, and variation in expression of two serologically distinct O-antigen epitopes in lipopolysaccharides of *Campylobacter fetus* serotype A strains. *Infect. Immun.* **69,** 7596–7602.
85. Antoine, T., Priem, B., Heyraud, A., et al. (2003) Large-scale in vivo synthesis of the carbohydrate moieties of gangliosides GM1 and GM2 by metabolically engineered *Escherichia coli. Chembiochem.* **4,** 406–412.
86. Soo, E. C., Aubry, A. J., Logan, S. M., et al. (2004) Selective detection and identification of sugar nucleotides by CE-electrospray-MS and its application to bacterial metabolomics. *Anal. Chem.* **76,** 619–626.
87. Valvano, M. A., Messner, P., and Kosma, P. (2002) Novel pathways for biosynthesis of nucleotide-activated *glycero-manno*-heptose precursors of bacterial glycoproteins and cell surface polysaccharides. *Microbiology* **148,** 1979–1989.
88. Humphrey, W., Dalke, A., and Schulten, K. (1996) VMD: visual molecular dynamics. *J. Mol. Graph.* **14,** 33–38.

6

Genomics of *Helicobacter* Species

Zhongming Ge and David B. Schauer

Summary

Helicobacter pylori was the first bacterial species to have the genome of two independent strains completely sequenced. Infection with this pathogen, which may be the most frequent bacterial infection of humanity, causes peptic ulcer disease and gastric cancer. Other *Helicobacter* species are emerging as causes of infection, inflammation, and cancer in the intestine, liver, and biliary tract, although the true prevalence of these enterohepatic *Helicobacter* species in humans is not yet known. The murine pathogen *Helicobacter hepaticus* was the first enterohepatic *Helicobacter* species to have its genome completely sequenced. Here, we consider functional genomics of the genus *Helicobacter*, the comparative genomics of the genus *Helicobacter*, and the related genera *Campylobacter* and *Wolinella*.

Key Words: Cytotoxin-associated gene; ε-Proteobacteria; gastric cancer; genomic evolution; genomic island; hepatobiliary; peptic ulcer disease; type IV secretion system.

1. Introduction

The genus *Helicobacter* belongs to the family Helicobacteriaceae, order Campylobacterales, and class ε-Proteobacteria, which is also known as the ε subdivision of the phylum Proteobacteria. The ε-Proteobacteria comprise of a relatively small and recently recognized line of descent within this extremely large and phenotypically diverse phylum. Other genera that colonize and/or infect humans and animals include *Campylobacter*, *Arcobacter*, and *Wolinella*. These organisms are all microaerophilic, chemoorganotrophic, nonsaccharolytic, spiral shaped or curved, and motile with a corkscrew-like motion by means of polar flagella. Increasingly, free living ε-Proteobacteria are being recognized in a wide range of environmental niches, including seawater, marine sediments, deep-sea hydrothermal vents, and even as symbionts of shrimp and tubeworms in these environments. To date, genome sequencing has been done on only a handful of ε-Proteobacteria (Table 1). At the time of writing, six ε-Proteobacteria genomes have been completed and six more are in progress (Table 2).

2. Diversity in the Genus *Helicobacter*

2.1. Helicobacter pylori

The genus *Helicobacter* comprises a remarkable group of bacterial species that thrive in the surface mucus overlaying epithelia in the gastrointestinal tract of humans and animals. Bacteria resembling the *Helicobacter* species were first observed in gastric

From: *Bacterial Genomes and Infectious Diseases*
Edited by: V. L. Chan, P. M. Sherman, and B. Bourke © Humana Press Inc., Totowa, NJ

Table 1

Characteristics of ε-Proteobacteria With Completed or in Progress Genome Sequences From the NCBI Entrez Genome Project[a]

Organism	Shape	Motility	Oxygen req.	Habitat	Disease
Campylobacter coli RM2228	Spiral	Yes	Microaerophilic	Host-associated; intestine	Enteritis and septicemia (humans and animals)
Campylobacter fetus	Spiral	Yes	Microaerophilic	Host-associated; intestine, reproductive tract, bloodstream	Infertility, abortions, septicemia, meningitis (humans and animals)
Campylobacter jejuni RM1221	Spiral	Yes	Microaerophilic	Host-associated; intestine	Enteritis (humans and animals)
C. jejuni subsp. *jejuni* NCTC 11168	Spiral	Yes	Microaerophilic	Host-associated; intestine	Enteritis (humans and animals)
Campylobacter lari RM2100	Spiral	Yes	Microaerophilic	Host-associated; intestine, bloodstream	Gastroenteritis and diarrhea (humans and animals)
Campylobacter upsaliensis RM3195	Spiral	Yes	Microaerophilic	Host-associated; intestine, bloodstream	Gastroenteritis and diarrhea (humans and animals)
Helicobacter hepaticus ATCC 51449	Spiral	Yes	Microaerophilic	Host-associated; intestine, liver, biliary tract	Hepatitis, liver tumors, typhlitis, intestinal cancer (mice)
Helicobacter pylori 26695	Spiral	Yes	Microaerophilic	Host-associated; stomach	Gastritis, peptic ulcer disease, gastric cancer (humans)
H. pylori J99	Spiral	Yes	Microaerophilic	Host-associated; stomach	Gastritis, peptic ulcer disease, gastric cancer (humans)
Nautilia sp. Am-H	Rod	Yes	Anaerobic	Environmental	Not pathogenic
Thiomicrospira denitrificans	Spiral	No	Anaerobic	Environmental	Not pathogenic
Wolinella succinogenes DSM 1740	"Spiral, curved"	Yes	Microaerophilic	Host-associated; stomach (bovine rumen)	Not pathogenic

[a]Based on prokaryotic genome projects in the NCBI Entrez Genome Project database (http://www.ncbi.nlm.nih.gov/genomes/lproks.cgi).

Table 2
ε-Proteobacteria Genome Sequencing Projects From the NCBI Entrez Genome Project[a]

Organism	Size[b]	Percent GC	ORFs[c]	Virulence determinants						GenBank	Released	Center
				CDT[d]	VacA	CagA	Ure	CiaB	T4SS			
Campylobacter jejuni RM1221	1.78	31%	1838	Yes	No	No	No	Yes	No[f]	CP000025	1/7/05	TIGR
C. jejuni subsp. *jejuni* NCTC 11168	1.64	31%	1654	Yes	No	No	No	Yes	No[f]	AL111168	2/25/00	Sanger Institute
Helicobacter hepaticus ATCC 51449	1.8	36%	1875	Yes	No	No	Yes	Yes	HHG1	AE017125	6/18/03	University of Wuerzburg
Helicobacter pylori J99	1.67	39%	1590	No	Yes	Yes	Yes	No	*cag*	AE000511	8/7/97	TIGR
H. pylori 26695	1.64	39%	1495	No	Yes	Yes	Yes	No	*cag*	AE001439	1/29/99	ASTRA Max Planck Institute
Wolinella succinogenes DSM 1740	2.11	48.5%	2046	No	No	No	No	Yes	Yes	BX571656	9/23/03	
Campylobacter coli RM2228	1.86[e]	31.1%	–	–	–	–	–	–	–	–	In progress	TIGR
Campylobacter fetus	1.5[e]	40%	–	–	–	–	–	–	–	–	In progress	IIB-UNSAM
Campylobacter lari RM2100	1.56[e]	29.6%	–	–	–	–	–	–	–	–	In progress	TIGR
Campylobacter upsaliensis RM3195	1.77[e]	34.3%	–	–	–	–	–	–	–	–	In progress	TIGR
Nautilia sp. Am-H	–	–	–	–	–	–	–	–	–	–	In progress	TIGR
Thiomicrospira denitrificans	–	36%	–	–	–	–	–	–	–	–	In progress	DOE Joint Genome Institute

[a] Based on prokaryotic genome projects in the NCBI Entrez Genome Project database (http://www.ncbi.nlm.nih.gov/genomes/lproks.cgi).
[b] Genome size, MB.
[c] Predicted number of open reading frames.
[d] Cytolethal distending toxin.
[e] Estimated.
[f] Other strains contain a type IV secretion system on the pVir plasmid.

mucosa around the turn of the last century (for review *see* ref. *1*). These early morpho-
logical observations stood in contrast to the widely held belief that the stomach was
essentially sterile by virtue of the low pH of gastric acid. Some 100 yr later, Marshall
and Warren described spiral-shaped bacteria in biopsy specimens of human gastric
mucosa, successfully cultured and isolated the organisms, and speculated (correctly)
that the bacteria caused chronic gastritis and peptic ulcer disease *(2)*. Originally believed
to be *Campylobacter* species, these organisms were subsequently recognized to belong
to a distinct taxon, and were ultimately named *Helicobacter pylori*, the type species for
the new genus *(3)*. It is now generally accepted that *H. pylori* infection is the most
common bacterial infection worldwide, and that infection is typically acquired during
childhood. All infected individuals develop chronic gastritis, and remain persistently
infected despite a robust immune response (for reviews *see* refs. *4* and *5*). The intense
interest in *H. pylori* pathogenesis is owing to the fact that it is the major causative factor in peptic ulcer disease *(6)*, and is strongly associated with the development of gastric adenocarcinoma (leading to its classification as a definite human carcinogen *[7]*),
as well as mucosa-associated lymphoid tissue (MALT) lymphoma *(8)*.

2.2. Other Gastric Helicobacter Species

In addition to *H. pylori*, there are five named, two proposed, and three uncultured
Helicobacter species that have been identified in the stomach of different animal species
(for review *see* ref. *1*). *Helicobacter mustelae* causes gastroduodenal disease in its natural host the ferret *(3)*, and provided the first animal model system that allowed for the
study of host-to-host transmission, chronic gastritis, immunity, eradication, and candidate bacterial virulence determinants (using isogenic mutants) of a gastric *Helicobacter*
species. *Helicobacter felis* naturally infects dogs and cats *(9)*. It is probably not the
most common gastric *Helicobacter* species in these hosts, and it is not clear how important it is as a cause of canine and feline gastroduodenal disease, but it was fortuitously
found to readily infect the stomach of laboratory mice (for review *see* ref. *10*). The
rodent model of *H. felis* has been extensively utilized to investigate chronic gastritis,
gastric atrophy (parietal cell loss), vaccine responses, and immunopathogenesis, as
well as gastric neoplasia, including adenocarcinoma in C57BL/6 and hypergastrinemic
INS-GAS mice, and MALT lymphoma in BALB/c mice (for review *see* ref. *11*). *Helicobacter bizzozeronni (12)* and *Helicobacter salomonis (13)* have been isolated from the
stomach of dogs, and together with *H. felis* form a cluster of three closely related species.
Helicobacter acinonychis (14), formerly known as *H. acinonyx*, was originally isolated
from captive cheetahs, and appears to contribute to clinically significant gastroduodenal
disease in these and other big cats. "*Helicobacter suncus*" isolated from Asian house
(musk) shrews *(15)* and "*Helicobacter cetorum*" isolated from wild and captive dolphins
and a beluga whale *(16)* are species that have been proposed but not yet validated.

Perhaps the most commonly observed bacteria in the stomach of animals with naturally occurring *Helicobacter* species infections are larger than and more tightly coiled
than *H. pylori*. Similar organisms are infrequent causes of human infection. These
organisms are uncultured, morphologically indistinguishable, and have been characterized primarily by sequencing amplified 16S rDNA (genes encoding 16S ribosomal
RDNA) products (for review *see* ref. *1*). Provisional status for such uncultured organisms is designated by the category *Candidatus* (Latin for candidate) followed by a

descriptive epithet *(17)*. *Candidatus* Helicobacter heilmannii *(18)* includes organisms in humans (formerly "*H. heilmanii*" type 2) and domestic and exotic cats, which cannot be distinguished from the cultured species *H. felis, H. bizzozeronni,* and *H. salomonis* by 16S rDNA sequence. *Candidatus* Helicobacter suis *(19)* includes organisms in pigs, as well as in humans and nonhuman primates (formerly "*H. heilmanii*" type 1). *Candidatus* Helicobacter bovis *(20)* has been identified in the abomasum of cattle.

2.3. *Enterohepatic* Helicobacter *Species*

An even greater number of *Helicobacter* species with comparable diversity to that of the gastric organisms have been recognized to inhabit the intestinal tract, liver, and biliary tree of humans and animals, which are collectively referred to as enterohepatic *Helicobacter* species (for reviews *see* refs. *1* and *21*). The 16 named species of enterohepatic *Helicobacter* species include host-restricted organisms that generally have only been isolated from one or a few host species (mostly rodents), as well as organisms with broader host ranges and zoonotic potential. There are also reports of provisionally named and unnamed enterohepatic *Helicobacter* species, which almost guarantees the emergence of additional novel species and expansion of the genus for the foreseeable future. The first enterohepatic *Helicobacter* species to be cultured and isolated was *Helicobacter muridarum (22)*, although it would not be formally named until almost a decade later *(23)*. This organism, isolated from the intestine of rats and mice, is spiral-shaped and possesses periplasmic fibers, in addition to polar flagella, similar to those of *H. felis*. The pathogenic potential of *H. muridarum* is not clear. It apparently causes subclinical infection in immunocompetent rodents, but it can cause gastritis under conditions where it persists in the stomach *(24)*, and triggers typhlocolitis when monoassociated with immunodeficient mice following adoptive transfer with effector T cells *(25)*. Based on 16S rDNA sequence similarity, the closest relative of *H. muridarum* is *Helicobacter hepaticus (26)*.

H. hepaticus is the best characterized enterohepatic *Helicobacter* species (for review *see* ref. *1*). *H. hepaticus* causes subclinical infection in resistant strains of mice, mild to moderate typhlitis in susceptible strains of mice, and severe typhlocolitis that resembles idiopathic inflammatory bowel disease in mutant lines of mice with altered immune function. In some strain backgrounds, this severe inflammatory bowel disease-like condition progresses to adenocarcinoma *(27)*. In susceptible strains, *H. hepaticus* also causes chronic hepatitis and hepatocellular carcinoma, particularly in males (for review *see* ref. *11*). Most recently, *H. hepaticus* has been shown to cause cholesterol gallstone formation in C57L mice fed a cholelithogenic diet *(28)*. *H. hepaticus*, like *H. muridarum*, possesses an active urease enzyme. Unlike *H. muridarum, H. hepaticus* lacks periplasmic fibers. Instead, it is a simple spiral-shaped rod with sheathed polar flagella. With the exception of *Helicobacter aurati (29)*, which has been isolated from the intestine of Syrian hamsters, is associated with gastritis, produces urease, and has periplasmic fibers, the other enterohepatic *Helicobacter* species with limited host range do not produce urease and lack periplasmic fibers. *Helicobacter rodentium (30), Helicobacter typhlonius (31),* and *Helicobacter ganmani (32)* have been isolated from the intestinal tissue of mice. *Helicobacter cholecystus (33)* and *Helicobacter mesocricetorum (34)* have been isolated from Syrian hamsters, whereas *Helicobacter pametensis* has been isolated from birds *(35)*.

2.4. Emergence of Enterohepatic Helicobacter Species as Human Pathogens

The enterohepatic *Helicobacter* species associated with human disease include both urease producing organisms with perplasmic fibers, and organisms that lack urease activity and periplasmic fibers (for reviews *see* refs. *1* and *21*). The urease producing organisms with periplasmic fibers resemble *H. muridarum*, but are fusiform, rather than being spiral shaped. Based on their similar morphology, these organisms have been called the "*Flexispira rappini*" group. 16S rDNA sequence analysis suggests that these organisms can be grouped into at least 10 distinct taxa *(36)*. Two of these taxa are named species, *Helicobacter bilis (37)* and *Helicobacter trogontum (38)*, originally isolated from mice and rats, respectively. Attempts have been made to simplify the classification of these taxa by grouping three of them (taxa 1, 4, and 5) with *H. trogontum (39)*, and three of them (taxa 2, 3, and 8) with *H. bilis (40)*, based on dot-blot hybridization and similarity of partial *ureB* and *hsp60* sequences, respectively. Disparity between *ureB* and *hsp60* results indicate that additional phylogenetic studies are needed; nonetheless, some members of the "*F. rappini*" group, and *H. bilis* and taxon 8 in particular, have been documented in cases of diarrhea *(41,42)*, bacteremia, and related sequelae in compromised patients *(43–48)*. Studies have also implicated these organisms in human biliary tract disease, including chronic cholecystitis *(49)* and/or cholangiocarcinoma *(50–52)*. These associations are based on amplification of specific polymerase chain reaction (PCR) products, but so far culture and isolation from human gallbladder or common bile duct have not been successful.

Of the urease-negative enterohepatic *Helicobacter* species without periplasmic fibers associated with human disease (for reviews *see* refs. *1* and *21*), *Helicobacter cinaedi* and *Helicobacter fennelliae (53,54)* have been isolated from the rectum of homosexual men with or without proctitis, from individuals with diarrhea or bacteremia, and from a variety of animal species, including monkeys and cats (*H. cinaedi*) and hamsters and dogs (both *H. cinaedi* and *H. fennelliae*). *Helicobacter canis (55)*, *Helicobacter pullorum (56)*, and *Helicobacter canadensis (57)* have all been isolated from humans with diarrhea and from animals, including dogs and cats (*H. canis*), chickens (*H. pullorum*), and geese (*H. canadensis*). *H. pullorum* has also been identified by PCR in cases of human biliary tract disease.

3. Genome Analysis and Functional Genomics

3.1. H. pylori Genome Structure

H. pylori was the first bacterial species to have the genome of two independent isolates sequenced *(58,59)*. Analysis of the genomes and how they contribute to our understanding of *H. pylori* biology have been reviewed *(60,61)*. *H. pylori* 26695 was isolated in the United Kingdom from a patient with gastritis, and was sequenced by The Institute for Genomic Research (TIGR), whereas *H. pylori* J99 was isolated in the United States from a patient with duodenal ulcer and was sequenced as a collaboration between Genome Therapeutics and Astra AB (now AstraZeneca PLC). The 26695 genome is 24 kb larger, but both genomes contain a total guanine and cytosine content of 39% (Table 2). Both genomes contain two copies of the 16S and 23S-5S rRNA genes in the same relative location, but 26695 has an additional third copy of the 5S rRNA gene. *H. pylori* 26695 has 1552 putative coding genes (1590 in the primary annotation), 57 more than

H. pylori J99. In both strains, approximately two-thirds of the genes can be assigned a predicted function, whereas one-third are either hypothetical conserved, or have no known orthologs. Interestingly, most of the genes in the genomes of the two strains occur at the same relative position, with only 10 discrete segments ranging in size from 1 to 83 kb being transposed or inverted. This was somewhat unexpected because it had been observed that virtually all *H. pylori* strains appear unique by high resolution geno-typing methods, such as pulsed field gel electrophoresis (PFGE) of chromosomal endonu-clease fragments, random amplified polymorphic DNA (RAPD), and repetitive element PCR (Rep-PCR). Much of this genome diversity can be attributed a high degree of nucleotide variation within individual genes of different *H. pylori* strains, particularly in the third "wobble" position of codon triplets, which produces silent (equivalent amino acid) changes.

3.2. H. pylori *Population Genetics*

The nucleotide sequence diversity in *H. pylori* is among the highest seen in any bac-terial species. This high allelic diversity could result from a high mutation rate, a high rate of recombination, or both. Mosaicism in virulence and housekeeping genes indicates that *H. pylori* has a high rate of recombination. If recombination occurs between two strains during a mixed infection, phylogenetic relationships will become scrambled, meaning that alleles at different loci will be in linkage equilibrium, which is character-istic of species with panmictic population structures. In fact, a comparison between J99 and isolates collected 7 yr subsequent to the patient's original endoscopy revealed both deletions and insertions in the genome compared with the original J99 strain *(62)*. As *H. pylori* is transmitted vertically within families, it has spread slowly as a result of human migration between geographic areas *(63)*. This biogeographical diversity is con-sistent with historical human migration, such as the appearance of West African strains in African Americans because of the slave trade, as well as with accepted prehistoric human migration, such as East Asian strains in Native Americans as a result of theorized Ice Age migrations across the Bering land bridge more than 12,000 yr ago *(64)*. Other *Helicobacter* species, including *H. mustelae* in ferrets *(65)* and *H. hepaticus* in mice *(66, 67)*, do not exhibit such genome diversity. It is not yet clear if these species are clonal because they have a lower rate of recombination than *H. pylori*, or if they appear clonal because they have spread relatively rapidly within their host population since their first appearance, or since the most recent selective sweep, and there has not yet been suffi-cient time for divergence to accumulate.

3.3. H. pylori *Functional Genomics*

Features of the genome have yielded insight into the biology of *H. pylori*. In regard to gene regulation, both of the sequenced strains possess four orthologs of histadine kinase sensor proteins, seven orthologs of DNA-binding response regulators, and only three σ factors: RpoD (σ^{70}), RpoN (σ^{54}), and FliA (σ^{28}). No stationary phase σ factor RpoS or heat shock α factor RpoH are evident. Subsequent to the primary annotation of the genomes, the anti-α factor FlgM was identified and confirmed to regulate FliA-depen-dent gene expression *(68,69)*. Remarkably, over 20 DNA restriction and modification systems are present in each strain (>4% of all genes), many of which are strain-specific. In some cases, the restriction component is not active, although the cognate methylase is, suggesting that they play a role in gene regulation. Another striking feature of the genome sequences is the large number of low-abundance outer membrane proteins, rather than a limited number of predominant outer membrane proteins seen in other bacteria. There are at least five paralogous families of outer membrane proteins containing 3–33 members each, some of which are strain-specific. Some of these genes, as well as the genes that encode glycosyltransferases involved in lipopolysaccharide biosynthesis, contain homopolymeric or dinucleotide repeats that can mediate metastable regulation (phase variation) by a slipped-strand mechanism. Some of these repeats are in open reading frames (ORFs) where changes in repeat number (owing to slippage of DNA polymerase) will lead to frameshifts and translation of truncated proteins, whereas others are in 5' intergenic regions where changes in repeat length affect promoter activity. Because many of these so-called contingency genes are predicted to affect the bacterial surface, it seems likely that they play a role in modulating antigenicity, perhaps to avoid immune recognition.

Information about *H. pylori* energy metabolism and biosynthetic pathways was also gained from the genome sequence, including identification of the membrane-embedded F_0 and catalytic F_1 components of ATP synthase, dehydrogenases, menaquinone, cytochromes, and a terminal oxidase for respiration-coupled oxidative phosphorylation. Fumarate reductase, which may play a role in anaerobic and/or aerobic respiration, as well as superoxide dismutase, catalase, and two peroxidases, presumably to detoxify oxygen species, are also present. Nonetheless, the genetic basis of microaerophily remains incompletely understood. Genes encoding enzymes for nitrogen assimilation, the tricarboxylic acid cycle, and nucleotide metabolism have also been identified, and a metabolic model has been constructed from genomic information that provides an in silico network of almost 400 *H. pylori* enzymatic and transport reactions *(70)*. More recently, a global transposon mutagenesis approach has been used to identify essential genes *(71)*.

3.4. H. pylori *Genomics and Virulence*

Many of the defined *H. pylori* virulence determinants such as flagella, motility, and urease activity, are found in all strains *(4,5)*. Conversely, the immunodominant cytotoxin-associated gene antigen, CagA, and the vacuolating cytotoxin, VacA, are present in some strains and absent in others. Although CagA and VacA are now known to be unlinked, strains that possess the *cagA* gene frequently express a functional *vacA* gene, and these strains are more frequently associated with peptic ulcer disease or cancer than strains without *cagA* that don't express *vacA (72)*. CagA is now known to be part of, and the only identified substrate of, a type IV secretion system *(72,73)*. Type IV secretion systems are ancestrally related to Gram-negative bacterial conjugation systems, and are used by pathogens to deliver protein substrates into eukaryotic host cells *(74)*. No type III secretion system, which are used by *Salmonella, Shigella, Yersinia*, and enteropathogenic and enterohemorrhagic *Escherichia coli* to deliver protein substrates into host cells and are evolutionarily related to flagellar systems, is present in *H. pylori*. The best-studied type IV secretion systems are in *Agrobacterium tumefaciens (virB* system), *Bordetella pertussis (ptl* system), and *Legionella pneumophila (dot/icm* system) (for review *see* ref. *75)*. Both of the sequenced strains have type IV secretion systems, but 26695 was subsequently found to have lost the ability to secrete CagA, presumably as a consequence of laboratory passage *(76)*. The *cag* type IV secretion system is encoded on a 40-kb pathogenicity island (PAI). Prominent Vir orthologs in the *cag* PAI include the inner membrane localized ATPases VirB4 (CagE or HP0544), VirB11 (HP0525), and the coupling protein VirD4 (HP0524) (for reviews *see* refs. *75* and *77*). Systematic mutagenesis of genes in the *cag* PAI has shown 18 of the 27 to be necessary for CagA translocation, and 14 to be necessary for induction of secretion of the chemokine interleukin (IL)-8 *(76,78)*.

Mutants lacking CagA are still able to induce IL-8 secretion, whereas all of the type IV secretion systems components, with the exception VirD4, are required for the induction of IL-8 secretion. Furthermore, VirD4 is required for CagA translocation, confirming that IL-8 induction is CagA independent. Type IV secretion system-dependent IL-8 induction involves nuclear factor κB (NF-κB) and AP-1 activation (for reviews *see* refs. *77* and *79*). On the other hand, CagA-dependent signaling begins when CagA localizes to the cell membrane at the site of *H. pylori* attachment. Here, CagA colocalizes with the tight-junction scaffolding protein, ZO-1, and the transmembrane protein junctional adhesion molecule, leading to disruption of epithelial barrier function *(80)*. CagA is also a substrate for host cell Src kinases, and tyrosine phosphorylation is required for other cellular changes, including interactions with the tyrosine phosphatase, SHP-2, MAP kinase signaling (such as the extracellular signal-regulated protein kinase [ERK]), cell scattering, and cell elongation *(81,82)*. These cellular phenotypes are believed to correlate with increased proliferation and migration of gastric epithelial cell in vivo. It turns out that J99 has an allele of *cagA* that encodes a protein that cannot be phosphorylated, limiting the utility of this strain for studying some of these cellular effects *(77)*.

4. Comparative Genomics of *Helicobacter* and *Campylobacter* Species

4.1. Genome Sequencing of ε-Proteobacteria

In the 10 yr since the first complete bacterial genome sequence, that of *Haemophilus influenzae* Rd, was published *(83)*, 230 microbial genome sequences have been completed, including Archea and members of the Bacteroidetes, Chlamydiae, Cyanobacteria, Firmicutes, Proteobacteria, and Spirochaetes groups of Bacteria (National Center for Biotechnology Information Entrez Genome Project database prokaryotic projects database; http://www.ncbi.nlm.nih.gov/genomes/lproks.cgi). An additional 388 genome-sequencing projects are in progress. Comparative genomics will provide the blueprints for understanding microbial survival strategies, host- and niche-specific adaptations, pathogenic mechanisms, as well as evolutionary relationships among microbes. The complete genome sequences of three *Helicobacter* strains, including two *H. pylori* isolates *(58,59)* and the enterohepatic *Helicobacter* species *H. hepaticus* ATCC 51449 *(67)*, have been determined. The genomes of two *C. jejuni* strains, NCTC 11168 and RM1221, and *Wolinella succinogenes* DSM 1740, have also been completed. Comparative analysis of these ε-Proteobacteria genome sequences has been detailed elsewhere *(84)*. Here, we emphasize evolutionary relationships and the distribution of virulence determinants among the different host-adapted ε-Proteobacteria.

4.2. Comparative Genomics and Virulence

In spite of differences in host specificity and sites of colonization (Table 1), *H. pylori*, *H. hepaticus*, and *C. jejuni* share approx 50% of their genes, with similarities ranging

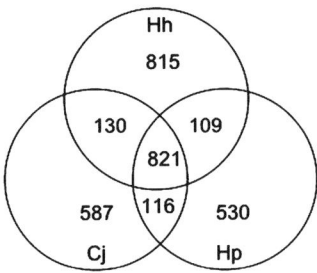

Fig. 1. Comparative gene content of *H. pylori* 26695 (Hp), *H. hepaticus* ATCC 51449 (Hh), and *C. jejuni* NCTC 11168 (Cj). Shown in the Venn diagram are the numbers of shared and species-specific open reading frames.

from 45 to 50% (Fig. 1 *[67,84]*). The majority of these orthologs encode proteins involved in housekeeping functions, including transcription, translation, energy metabolism, and biosynthetic processes. All three organisms possess a small genome, similar coding capability, and a relatively low number of copies of ribosomal RNA genes. The high degree of homology among their genome contents is consistent with their similar morphology and growth requirements (Table 1). It has been proposed that these three organisms, along with *W. succinogenes*, are derived from a common ancestor with a larger genome, and have been subject to reductive evolution leading to reduced genome sizes *(84)*. Subsequent to diverging from the common ancestor, each organism has acquired distinct genome features that contribute to their individual lifestyle and physiology. There are 544, 810, and 585 species-specific genes for *H. pylori*, *H. hepaticus*, and *C. jejuni*, respectively *(84)*. In addition, a total of 109 *H. hepaticus* ORFs have orthologs in the *H. pylori* genome, but not in the *C. jejuni* genome, whereas 130 *H. hepaticus* ORFs have orthologs in the *C. jejuni* genome, but not in the *H. pylori* genome. The species-specific genes are likely to play important roles in pathogen–host interactions. Although some have known function, such as the CagA protein of *H. pylori*, many are uncharacterized genes that are clustered and linked to known and candidate virulence genes. Functional characterization of these species-specific genes should provide insight into the adaptations that permit each microbe to exploit their individual niche.

Genome diversity also exists between strains of a given species. This is particularly true for isolates of *H. pylori*. Like *H. pylori*, the two sequenced strains of *C. jejuni* are also syntenic *(85)*. The major difference between the NCTC 11168 and RM1221 genomes is the presence of four large integrated elements in the strain RM1221 genome. These include a *Campylobacter* mu-like phage (CMLP1), two prophage elements containing ORFs predicted to encode phage-related endonucleases, methylases, or repressors, but no phage structural proteins, and an element without phage-related ORFs that appears to be an integrated plasmid *(85)*. Interestingly, this putative integrated plasmid contains many proteins that are similar to those encoded within the 71-kb *H. hepaticus* genomic island HHGI1, suggesting this genomic island could also be plasmid-derived. There is 98.4% amino acid identity between the 1468 orthologs present in *C. jejuni* 11168 and RM1221 *(85)*, considerably greater than the 93.4% identity between *H. pylori* 26695 and J99 orthologs *(61)*. In sharp contrast to *H. pylori*, *C. jejuni* and *H. hepaticus* have a paucity of restriction-modification loci. There is also no evidence of a chromosomally encoded type IV secretion system in either of the sequenced *C. jejuni* strains, although the pVir plasmid of *C. jejuni* 81-176 has been shown to encode a functional type IV secretion system *(86)*. Neither sequenced *C. jejuni* strain contains plasmids, but the *C. coli*, *C. lari*, and *C. upsaliensis* strains that are currently being sequenced do (Table 2) *(85)*. Although the large plasmids from the non-*C. jejuni Campylobacter* species also contain type IV secretion system orthologs, these are more likely to be DNA conjugation systems than protein-delivery virulence determinants *(85)*. It appears that *H. hepaticus* may be similar to *C. jejuni* in the sense that strains exhibit less allelic variation than *H. pylori*, and major differences consist of the presence or absence of discrete genetic regions. Five of twelve strains examined for gene content by microarray hybridization were found to be missing all or part of the HHGI1 *(67)*. The role the *H. hepaticus* HHGI1 and the putative type IV secretion system it encodes in virulence have not yet been definitively established.

4.3. Sources of Genome Diversity

Progressive accumulation of point mutations over time, horizontal gene transfer, and transposon-mediated gene mobility and recombination have been proposed to explain significant genetic variation among *H. pylori* strains *(63,87–89)*. The ε-Proteobacteria tend to lose genetic information, and it has been suggested that the sequenced *Helicobacter* and *Campylobacter* genomes document a trend to reduce the fraction of noncoding sequence in the drive to smaller genomes *(84)*. This ongoing process is reflected in the large number of pseudogenes in *H. pylori*, and the general trend for genome plasticity resulting from high rates of mutation and recombination. In addition to deletion of genetic information, there is ample evidence for gain of genetic information in the sequenced genomes. Horizontal transfer may account for as much as 5–6% of the total genetic information in the *Helicobacter* species *(84)*. The presence of orthologous genes in *H. hepaticus* and *C. jejuni* that are not present in *H. pylori* raises the intriguing possibility of horizontal transfer in the intestine between enterohepatic *Helicobacter* species and *Campylobacter* species. One obvious example is cytolethal dystending toxin (CDT), which is required for persistent infection by *H. hepaticus (90)* and *C. jejuni (91)*, which is not present in *H. pylori* (Table 2). Although it is possible that gastric *Helicobacter* species have lost the genes encoding CDT, its conservation in *Campylobacter* species and presence in some enterohepatic *Helicobacter* species but not others *(92–94)*, suggests that this virulence determinant was probably acquired by enterohepatic *Helicobacter* species after splitting off from a last common ancestor. Whether gastric *Helicobacter* species are specialized descendents of an ancestral intestinal *Helicobacter* species, or enterohepatic *Helicobacter* species evolved from gastric species by acquiring genes for intestinal colonization/infection, may become more apparent with the sequencing of additional enterohepatic *Helicobacter* species genomes.

4.4. Perspectives on Genomics of the Genera Helicobacter *and* Campylobacter

The availability of complete genome sequences for additional *Helicobacter* and *Campylobacter* species will establish a minimum set of genes common to bacteria in the order Campylobacterales. This will facilitate the identification of common and unique virulence determinants, and permit a better understanding of the evolution of this important group of emerging human and animal pathogens. The long-term goal for such studies is to develop better preventive and therapeutic strategies against infection and disease in humans, which includes elimination of food-borne pathogens from animals and animal-derived products. Advances in molecular biology and biotechnology continue to make genome sequencing ever more rapid and inexpensive. Indeed, genome sequencing of four non-*C. jejuni Campylobacter* species are in progress (Table 2). Given the recent recognition of diverse *Helicobacter* species in human diseases, the availability of the complete genomes for additional *Helicobacter* species is clearly warranted. It may require a significant technological breakthrough to permit genome sequencing of the uncultured gastric *Helicobacter* species, but genome sequences for many of the cultured gastric and enterohepatic *Helicobacter* species is desirable and may, in turn, provide insight into methods for successful cultivation of the *Candidatus* species. The first non-*H. pylori* gastric *Helicobacter* species to be sequenced is *H. mustelae*, which at the time of writing is in the finishing/gap closure phase (Wellcome Trust Sanger Institute; http://www.sanger.ac.uk/Projects/H_mustelae). It seems likely that additional enterohepatic *Helicobacter* species genome projects will soon be underway.

Genome comparisons alone do not provide all the details necessary to understand *Helicobacter* biology and pathogenesis. Genome annotations are frequently based on similarity to genes in disparate organisms, and may not accurately or completely reflect the true role of these genes in vivo. Consequently, predicted functions of genes must be experimentally validated. In addition, a large number of predicted ORFs have no detectable orthologs in the database. These unique genes are good candidates for the most interesting and important biological and pathogen–host interaction activities. Nevertheless, comparative genome analysis has provided a successful means to begin to decipher the lifestyles of, and the relationships between, these bacteria.

References

1. Solnick, J. V. and Schauer, D. B. (2001) Emergence of diverse *Helicobacter* species in the pathogenesis of gastric and enterohepatic diseases. *Clin. Microbiol. Rev.* **14,** 59–97.
2. Marshall, B. J. and Warren, J. R. (1984) Unidentified curved bacilli in the stomach of patients with gastritis and peptic ulceration. *Lancet* **1,** 1311–1315.
3. Goodwin, C. S., Armstrong, J. A., Chilvers, T., et al. (1989) Transfer of *Campylobacter pylori* and *Campylobacter mustelae* to *Helicobacter* gen. nov. as *Helicobacter pylori* comb. nov. and *Helicobacter mustelae* comb. nov., respectively. *Int. J. Syst. Bacteriol.* **39,** 397–405.
4. Dunn, B. E., Cohen, H., and Blaser, M. J. (1997) *Helicobacter pylori. Clin. Microbiol. Rev.* **10,** 720–741.
5. Suerbaum, S. and Michetti, P. (2002) *Helicobacter pylori* infection. *N. Engl. J. Med.* **347,** 1175–1186.
6. Anonymous. (1994) NIH Consensus Conference. *Helicobacter pylori* in peptic ulcer disease. NIH Consensus Development Panel on *Helicobacter pylori* in Peptic Ulcer Disease. *JAMA* **272,** 65–69.
7. Anonymous. (1994) Schistosomes, liver flukes and *Helicobacter pylori*. IARC Working Group on the Evaluation of Carcinogenic Risks to Humans. Lyon, 7–14 June 1994. *IARC Monogr. Eval. Carcinog. Risks Hum.* **61,** 1–241.
8. Parsonnet, J., Hansen, S., Rodriguez, L., et al. (1994) *Helicobacter pylori* infection and gastric lymphoma. *N. Engl. J. Med.* **330,** 1267–1271.
9. Paster, B. J., Lee, A., Fox, J. G., et al. (1991) Phylogeny of *Helicobacter felis* sp. nov., *Helicobacter mustelae*, and related bacteria. *Int. J. Syst. Bacteriol.* **41,** 31–38.

10. Fox, J. G. and Lee, A. (1997) The role of *Helicobacter* species in newly recognized gastrointestinal tract diseases of animals. *Lab. Anim. Sci.* **47,** 222–255.

11. Rogers, A. B. and Fox, J. G. (2004) Inflammation and Cancer. I. Rodent models of infectious gastrointestinal and liver cancer. *Am. J. Physiol. Gastrointest. Liver Physiol.* **286,** G361–G366.

12. Hanninen, M. L., Happonen, I., Saari, S., and Jalava, K. (1996) Culture and characteristics of *Helicobacter bizzozeronii,* a new canine gastric Helicobacter sp. *Int. J. Syst. Bacteriol.* **46,** 160–166.

13. Jalava, K., Kaartinen, M., Utriainen, M., Happonen, I., and Hanninen, M. L. (1997) *Helicobacter salomonis* sp. nov., a canine gastric *Helicobacter* sp. related to *Helicobacter felis* and *Helicobacter bizzozeronii. Int. J. Syst. Bacteriol.* **47,** 975–982.

14. Eaton, K. A., Dewhirst, F. E., Radin, M. J., et al. (1993) *Helicobacter acinonyx* sp. nov., isolated from cheetahs with gastritis. *Int. J. Syst. Bacteriol.* **43,** 99–106.

15. Goto, K., Ohashi, H., Ebukuro, S., et al. (1998) Isolation and characterization of *Helicobacter* species from the stomach of the house musk shrew (*Suncus murinus*) with chronic gastritis. *Curr. Microbiol.* **37,** 44–51.

16. Harper, C. G., Feng, Y., Xu, S., et al. (2002) *Helicobacter cetorum* sp. nov., a ureasepositive *Helicobacter* species isolated from dolphins and whales. *J. Clin. Microbiol.* **40,** 4536–4543.

17. Murray, R. G. and Stackebrandt, E. (1995) Taxonomic note: implementation of the provisional status Candidatus for incompletely described procaryotes. *Int. J. Syst. Bacteriol.* **45,** 186–187.

18. O'Rourke, J. L., Solnick, J. V., Neilan, B. A., et al. (2004) Description of 'Candidatus Helicobacter heilmannii' based on DNA sequence analysis of 16S rRNA and urease genes. *Int. J. Syst. Evol. Microbiol.* **54,** 2203–2211.

19. De Groote, D., van Doorn, L. J., Ducatelle, R., et al. (1999) 'Candidatus Helicobacter suis', a gastric helicobacter from pigs, and its phylogenetic relatedness to other gastrospirilla. *Int. J. Syst. Bacteriol.* **49,** 1769–1777.

20. De Groote, D., van Doorn, L. J., Ducatelle, R., et al. (1999) Phylogenetic characterization of 'Candidatus Helicobacter bovis', a new gastric helicobacter in cattle. *Int. J. Syst. Bacteriol.* **49,** 1707–1715.

21. Fox, J. G. (2002) The non-*H. pylori* helicobacters: their expanding role in gastrointestinal and systemic diseases. *Gut* **50,** 273–283.

22. Phillips, M. W. and Lee, A. (1983) Isolation and characterization of a spiral bacterium from the crypts of rodent gastrointestinal tracts. *Appl. Environ. Microbiol.* **45,** 675–683.

23. Lee, A., Phillips, M. W., O'Rourke, J. L., et al. (1992) *Helicobacter muridarum* sp. nov., a microaerophilic helical bacterium with a novel ultrastructure isolated from the intestinal mucosa of rodents. *Int. J. Syst. Bacteriol.* **42,** 27–36.

24. Lee, A., Chen, M., Coltro, N., et al. (1993) Long term infection of the gastric mucosa with *Helicobacter* species does induce atrophic gastritis in an animal model of *Helicobacter pylori* infection. *Zentralbl. Bakteriol.* **280,** 38–50.

25. Jiang, H. Q., Kushnir, N., Thurnheer, M. C., Bos, N. A., and Cebra, J. J. (2002) Monoassociation of SCID mice with *Helicobacter muridarum,* but not four other enterics, provokes IBD upon receipt of T cells. *Gastroenterology* **122,** 1346–1354.

26. Fox, J. G., Dewhirst, F. E., Tully, J. G., et al. (1994) *Helicobacter hepaticus* sp. nov., a microaerophilic bacterium isolated from livers and intestinal mucosal scrapings from mice. *J. Clin. Microbiol.* **32,** 1238–1245.

27. Erdman, S. E., Poutahidis, T., Tomczak, M., et al. (2003) CD4+ CD25+ regulatory T lymphocytes inhibit microbially induced colon cancer in Rag2-deficient mice. *Am. J. Pathol.* **162,** 691–702.

28. Maurer, K. J., Ihrig, M. M., Rogers, A. B., et al. (2005) Identification of cholelithogenic enterohepatic helicobacter species and their role in murine cholesterol gallstone formation. *Gastroenterology* **128,** 1023–1033.

29. Patterson, M. M., Schrenzel, M. D., Feng, Y., et al. (2000) *Helicobacter aurati* sp. nov., a urease-positive *Helicobacter* species cultured from gastrointestinal tissues of Syrian hamsters. *J. Clin. Microbiol.* **38,** 3722–3728.

30. Shen, Z., Fox, J. G., Dewhirst, F. E., et al. (1997) *Helicobacter rodentium* sp. nov., a urease-negative *Helicobacter* species isolated from laboratory mice. *Int. J. Syst. Bacteriol.* **47,** 627–634.

31. Franklin, C. L., Gorelick, P. L., Riley, L. K., et al. (2001) *Helicobacter typhlonius* sp. nov., a novel murine urease-negative *Helicobacter* species. *J. Clin. Microbiol.* **39,** 3920–3926.

32. Robertson, B. R., O'Rourke, J. L., Vandamme, P., On, S. L., and Lee, A. (2001) *Helicobacter ganmani* sp. nov., a urease-negative anaerobe isolated from the intestines of laboratory mice. *Int. J. Syst. Evol. Microbiol.* **51,** 1881–1889.

33. Franklin, C. L., Beckwith, C. S., Livingston, R. S., et al. (1996) Isolation of a novel *Helicobacter* species, *Helicobacter cholecystus* sp. nov., from the gallbladders of Syrian hamsters with cholangiofibrosis and centrilobular pancreatitis. *J. Clin. Microbiol.* **34,** 2952–2958.

34. Simmons, J. H., Riley, L. K., Besch-Williford, C. L., and Franklin, C. L. (2000) *Helicobacter mesocricetorum* sp. nov., A novel *Helicobacter* isolated from the feces of Syrian hamsters. *J. Clin. Microbiol.* **38,** 1811–1817.

35. Dewhirst, F. E., Seymour, C., Fraser, G. J., Paster, B. J., and Fox, J. G. (1994) Phylogeny of *Helicobacter* isolates from bird and swine feces and description of *Helicobacter pametensis* sp. nov. *Int. J. Syst. Bacteriol.* **44,** 553–560.

36. Dewhirst, F. E., Fox, J. G., Mendes, E. N., et al. (2000) '*Flexispira rappini*' strains represent at least 10 *Helicobacter* taxa. *Int. J. Syst. Evol. Microbiol.* **50,** 1781–1787.

37. Fox, J. G., Yan, L. L., Dewhirst, F. E., et al. (1995) *Helicobacter bilis* sp. nov., a novel *Helicobacter* species isolated from bile, livers, and intestines of aged, inbred mice. *J. Clin. Microbiol.* **33,** 445–454.

38. Mendes, E. N., Queiroz, D. M., Dewhirst, F. E., Paster, B. J., Moura, S. B., and Fox, J. G. (1996) *Helicobacter trogontum* sp. nov., isolated from the rat intestine. *Int. J. Syst. Bacteriol.* **46,** 916–921.

39. Hanninen, M. L., Utriainen, M., Happonen, I., and Dewhirst, F. E. (2003) *Helicobacter* sp. flexispira 16S rDNA taxa 1, 4 and 5 and Finnish porcine *Helicobacter* isolates are members of the species *Helicobacter trogontum* (taxon 6). *Int. J. Syst. Evol. Microbiol.* **53,** 425–433.

40. Hanninen, M. L., Karenlampi, R. I., Koort, J. M., Mikkonen, T., and Bjorkroth, K. J. (2005) Extension of the species *Helicobacter bilis* to include the reference strains of *Helicobacter* sp. flexispira taxa 2, 3 and 8 and Finnish canine and feline flexispira strains. *Int. J. Syst. Evol. Microbiol.* **55,** 891–898.

41. Romero, S., Archer, J. R., Hamacher, M. E., Bologna, S. M., and Schell, R. F. (1988) Case report of an unclassified microaerophilic bacterium associated with gastroenteritis. *J. Clin. Microbiol.* **26,** 142–143.

42. Archer, J. R., Romero, S., Ritchie, A. E., et al. (1988) Characterization of an unclassified microaerophilic bacterium associated with gastroenteritis. *J. Clin. Microbiol.* **26,** 101–105.

43. Gerrard, J., Alfredson, D., and Smith, I. (2001) Recurrent bacteremia and multifocal lower limb cellulitis due to *Helicobacter*-like organisms in a patient with X-linked hypogammaglobulinemia. *Clin. Infect. Dis.* **33,** E116–E118.

44. Iten, A., Graf, S., Egger, M., Tauber, M., and Graf, J. (2001) *Helicobacter* sp. flexispira bacteremia in an immunocompetent young adult. *J. Clin. Microbiol.* **39,** 1716–1720.

45. Cuccherini, B., Chua, K., Gill, V., et al. (2000) Bacteremia and skin/bone infections in two patients with X-linked agammaglobulinemia caused by an unusual organism related to *Flexispira/Helicobacter* species. *Clin. Immunol.* **97**, 121–129.

46. Weir, S., Cuccherini, B., Whitney, A. M., et al. (1999) Recurrent bacteremia caused by a "*Flexispira*"-like organism in a patient with X-linked (Bruton's) agammaglobulinemia. *J. Clin. Microbiol.* **37**, 2439–2445.

47. Tee, W., Leder, K., Karroum, E., and Dyall-Smith, M. (1998) "*Flexispira rappini*" bacteremia in a child with pneumonia. *J. Clin. Microbiol.* **36**, 1679–1682.

48. Sorlin, P., Vandamme, P., Nortier, J., et al. (1999) Recurrent "*Flexispira rappini*" bacteremia in an adult patient undergoing hemodialysis: case report. *J. Clin. Microbiol.* **37**, 1319–1323.

49. Fox, J. G., Dewhirst, F. E., Shen, Z., et al. (1998) Hepatic *Helicobacter* species identified in bile and gallbladder tissue from Chileans with chronic cholecystitis. *Gastroenterology* **114**, 755–763.

50. Matsukura, N., Yokomuro, S., Yamada, S., et al. (2002) Association between *Helicobacter bilis* in bile and biliary tract malignancies: *H. bilis* in bile from Japanese and Thai patients with benign and malignant diseases in the biliary tract. *Jpn. J. Cancer Res.* **93**, 842–847.

51. Murata, H., Tsuji, S., Tsujii, M., et al. (2004) *Helicobacter bilis* infection in biliary tract cancer. *Aliment. Pharmacol. Ther.* **20**, 90–94.

52. Kobayashi, T., Harada, K., Miwa, K., and Nakanuma, Y. (2005) *Helicobacter* genus DNA fragments are commonly detectable in bile from patients with extrahepatic biliary diseases and associated with their pathogenesis. *Dig. Dis. Sci.* **50**, 862–867.

53. Vandamme, P., Falsen, E., Rossau, R., et al. (1991) Revision of *Campylobacter*, *Helicobacter*, and *Wolinella* taxonomy: emendation of generic descriptions and proposal of *Arcobacter* gen. nov. *Int. J. Syst. Bacteriol.* **41**, 88–103.

54. Totten, P. A., Fennell, C. L., Tenover, F. C., et al. (1985) *Campylobacter cinaedi* (sp. nov.) and *Campylobacter fennelliae* (sp. nov.): two new *Campylobacter* species associated with enteric disease in homosexual men. *J. Infect. Dis.* **151**, 131–139.

55. Stanley, J., Linton, D., Burnens, A. P., et al. (1993) *Helicobacter canis* sp. nov., a new species from dogs: an integrated study of phenotype and genotype. *J. Gen. Microbiol.* **139**, 2495–2504.

56. Stanley, J., Linton, D., Burnens, A. P., et al. (1994) *Helicobacter pullorum* sp. nov.-genotype and phenotype of a new species isolated from poultry and from human patients with gastroenteritis. *Microbiology* **140**, 3441–3449.

57. Fox, J. G., Chien, C. C., Dewhirst, F. E., et al. (2000) *Helicobacter canadensis* sp. nov. isolated from humans with diarrhea as an example of an emerging pathogen. *J. Clin. Microbiol.* **38**, 2546–2549.

58. Tomb, J. F., White, O., Kerlavage, A. R., et al. (1997) The complete genome sequence of the gastric pathogen *Helicobacter pylori*. *Nature* **388**, 539–547.

59. Alm, R. A., Ling, L. S., Moir, D. T., et al. (1999) Genomic-sequence comparison of two unrelated isolates of the human gastric pathogen *Helicobacter pylori*. *Nature* **397**, 176–180.

60. Ge, Z. and Taylor, D. E. (1999) Contributions of genome sequencing to understanding the biology of *Helicobacter pylori*. *Annu. Rev. Microbiol.* **53**, 353–387.

61. Alm, R. A. and Trust, T. J. (1999) Analysis of the genetic diversity of *Helicobacter pylori*: the tale of two genomes. *J. Mol. Med.* **77**, 834–846.

62. Israel, D. A., Salama, N., Krishna, U., et al. (2001) *Helicobacter pylori* genetic diversity within the gastric niche of a single human host. *Proc. Natl. Acad. Sci. USA* **98**, 14,625–14,630.

63. Falush, D., Kraft, C., Taylor, N. S., et al. (2001) Recombination and mutation during long-term gastric colonization by *Helicobacter pylori*: estimates of clock rates, recombination size, and minimal age. *Proc. Natl. Acad. Sci. USA* **98**, 15,056–15,061.

64. Falush, D., Wirth, T., Linz, B., et al. (2003) Traces of human migrations in *Helicobacter pylori* populations. *Science* **299,** 1582–1585.

65. Taylor, D. E., Chang, N., Taylor, N. S., and Fox, J. G. (1994) Genome conservation in *Helicobacter mustelae* as determined by pulsed-field gel electrophoresis. *FEMS Microbiol. Lett.* **118,** 31–36.

66. Saunders, K. E., McGovern, K. J., and Fox, J. G. (1997) Use of pulsed-field gel electro-phoresis to determine genomic diversity in strains of *Helicobacter hepaticus* from geo-graphically distant locations. *J. Clin. Microbiol.* **35,** 2859–2863.

67. Suerbaum, S., Josenhans, C., Sterzenbach, T., et al. (2003) The complete genome sequence of the carcinogenic bacterium *Helicobacter hepaticus*. *Proc. Natl. Acad. Sci. USA* **100,** 7901–7906.

68. Colland, F., Rain, J. C., Gounon, P., Labigne, A., Legrain, P., and De Reuse, H. (2001) Identification of the *Helicobacter pylori* anti-σ^{28} factor. *Mol. Microbiol.* **41,** 477–487.

69. Josenhans, C., Niehus, E., Amersbach, S., et al. (2002) Functional characterization of the antagonistic flagellar late regulators FliA and FlgM of *Helicobacter pylori* and their effects on the *H. pylori* transcriptome. *Mol. Microbiol.* **43,** 307–322.

70. Schilling, C. H., Covert, M. W., Famili, I., Church, G. M., Edwards, J. S., and Palsson, B. O. (2002) Genome-scale metabolic model of *Helicobacter pylori* 26695. *J. Bacteriol.* **184,** 4582–4593.

71. Salama, N. R., Shepherd, B., and Falkow, S. (2004) Global transposon mutagenesis and essential gene analysis of *Helicobacter pylori*. *J. Bacteriol.* **186,** 7926–7935.

72. Censini, S., Lange, C., Xiang, Z., et al. (1996) *cag*, a pathogenicity island of *Helicobacter pylori*, encodes type I-specific and disease-associated virulence factors. *Proc. Natl. Acad. Sci. USA* **93,** 14,648–14,653.

73. Akopyants, N. S., Clifton, S. W., Kersulyte, D., et al. (1998) Analyses of the *cag* pathoge-nicity island of *Helicobacter pylori*. *Mol. Microbiol.* **28,** 37–53.

74. Christie, P. J. (2001) Type IV secretion: intercellular transfer of macromolecules by sys-tems ancestrally related to conjugation machines. *Mol. Microbiol.* **40,** 294–305.

75. Nagai, H. and Roy, C. R. (2003) Show me the substrates: modulation of host cell function by type IV secretion systems. *Cell Microbiol.* **5,** 373–383.

76. Fischer, W., Puls, J., Buhrdorf, R., Gebert, B., Odenbreit, S., and Haas, R. (2001) Systematic mutagenesis of the *Helicobacter pylori cag* pathogenicity island: essential genes for CagA translocation in host cells and induction of interleukin-8. *Mol. Microbiol.* **42,** 1337–1348.

77. Bourzac, K. M. and Guillemin, K. (2005) *Helicobacter pylori*-host cell interactions medi-ated by type IV secretion. *Cell Microbiol.* **7,** 911–919.

78. Selbach, M., Moese, S., Meyer, T. F., and Backert, S. (2002) Functional analysis of the *Helicobacter pylori cag* pathogenicity island reveals both VirD4-CagA-dependent and VirD4-CagA-independent mechanisms. *Infect. Immun.* **70,** 665–671.

79. Peek, R. M. Jr. (2001) IV. *Helicobacter pylori* strain-specific activation of signal transduc-tion cascades related to gastric inflammation. *Am. J. Physiol. Gastrointest. Liver Physiol.* **280,** G525–G530.

80. Amieva, M. R., Vogelmann, R., Covacci, A., Tompkins, L. S., Nelson, W. J., and Falkow, S. (2003) Disruption of the epithelial apical-junctional complex by *Helicobacter pylori* CagA. *Science* **300,** 1430–1434.

81. Higashi, H., Tsutsumi, R., Muto, S., et al. (2002) SHP-2 tyrosine phosphatase as an intra-cellular target of *Helicobacter pylori* CagA protein. *Science* **295,** 683–686.

82. Higashi, H., Nakaya, A., Tsutsumi, R., et al. (2004) *Helicobacter pylori* CagA induces Ras-independent morphogenetic response through SHP-2 recruitment and activation. *J. Biol. Chem.* **279,** 17,205–17,216.

83. Fleischmann, R. D., Adams, M. D., White, O., et al. (1995) Whole-genome random sequencing and assembly of *Haemophilus influenzae* Rd. *Science* **269,** 496–512.

84. Eppinger, M., Baar, C., Raddatz, G., Huson, D. H., and Schuster, S. C. (2004) Comparative analysis of four Campylobacterales. *Nat. Rev. Microbiol.* **2,** 872–885.

85. Fouts, D. E., Mongodin, E. F., Mandrell, R. E., et al. (2005) Major structural differences and novel potential virulence mechanisms from the genomes of multiple campylobacter species. *PLoS Biol.* **3,** e15.

86. Bacon, D. J., Alm, R. A., Hu, L., et al. (2002) DNA sequence and mutational analyses of the pVir plasmid of *Campylobacter jejuni* 81-176. *Infect. Immun.* **70,** 6242–6250.

87. Jiang, Q., Hiratsuka, K., and Taylor, D. E. (1996) Variability of gene order in different *Helicobacter pylori* strains contributes to genome diversity. *Mol. Microbiol.* **20,** 833–842.

88. Covacci, A., Falkow, S., Berg, D. E., and Rappuoli, R. (1997) Did the inheritance of a pathogenicity island modify the virulence of *Helicobacter pylori*? *Trends Microbiol.* **5,** 205–208.

89. Suerbaum, S., Smith, J. M., Bapumia, K., et al. (1998) Free recombination within *Helicobacter pylori. Proc. Natl. Acad. Sci. USA* **95,** 12,619–12,624.

90. Ge, Z., Feng, Y., Whary, M. T., et al. (2005) Cytolethal distending toxin is essential for *Helicobacter hepaticus* colonization in outbred Swiss Webster mice. *Infect. Immun.* **73,** 3559–3567.

91. Fox, J. G., Rogers, A. B., Whary, M. T., et al. (2004) Gastroenteritis in NF-κB-deficient mice is produced with wild-type *Camplyobacter jejuni* but not with *C. jejuni* lacking cytolethal distending toxin despite persistent colonization with both strains. *Infect. Immun.* **72,** 1116–1125.

92. Chien, C. C., Taylor, N. S., Ge, Z., Schauer, D. B., Young, V. B., and Fox, J. G. (2000) Identification of *cdtB* homologues and cytolethal distending toxin activity in enterohepatic *Helicobacter* spp. *J. Med. Microbiol.* **49,** 525–534.

93. Taylor, N. S., Ge, Z., Shen, Z., Dewhirst, F. E., and Fox, J. G. (2003) Cytolethal distending toxin: a potential virulence factor for *Helicobacter cinaedi. J. Infect. Dis.* **188,** 1892–1897.

94. Young, V. B., Knox, K. A., and Schauer, D. B. (2000) Cytolethal distending toxin sequence and activity in the enterohepatic pathogen *Helicobacter hepaticus. Infect. Immun.* **68,** 184–191.

7

The Organization of *Leptospira* at a Genomic Level

Dieter M. Bulach, Torsten Seemann, Richard L. Zuerner, and Ben Adler

Summary

The complete nucleotide sequences of three and, in the not too distant future, six strains from the genus *Leptospira* will be available. Managing and maintaining these data will be a perpetual problem unless a system is devised to address this issue. We propose a central role for the International Union of Microbiological Societies Subcommittee on the Taxonomy of *Leptospira* in maintaining and updating the annotated leptospiral genome sequences. The first step in this process is provided as part of this publication, namely a revision of the annotation of the three published genomes, and an internet location for the current versions of these genomes.

Key Words: *Leptospira;* comparative genomics; spirochetes; genome annotation.

1. Introduction

Leptospirosis is a zoonosis of worldwide distribution. Significant efforts aimed at raising the standard of diagnostics *(1)* and the collection and collation of disease data (via LeptoNet; www.leptonet.net) will correct what is likely to be a significant underestimation of the contribution of this disease to human morbidity and mortality. Leptospirosis is caused by pathogenic serovars from the genus *Leptospira* and in the current species classification schema these pathogenic serovars are usually classified as *Leptospira interrogans*, *Leptospira borgpetersenii*, *Leptospira kirschneri*, *Leptospira noguchii*, *Leptospira meyeri*, *Leptospira weilii*, or *Leptospira santarosai*. Analysis based on 16S ribosomal RNA (rRNA) gene analysis shows that these serovars are found in one of the three phylogenic clades into which the genus can be divided *(2)*. The severity of leptospirosis varies from a mild febrile flu-like illness that is rarely fatal, to a disease causing multi-organ failure and mortality rates of up to 20% where inadequate patient support is available. There is an underlying association between severity of disease and serovar and to some extent species, with severe disease being often associated with serovars from *L. interrogans* and *L. kirschneri (3)*.

The relative abundance of sequenced leptospiral genomes, and for that matter spirochetal genomes, has been stimulated because of the difficulties these bacteria have posed for genetic manipulation. Only in the last few years has it been possible to transform *Leptospira*, and this is only in one strain of the saprophyte *L. biflexa*. Studies have compared the genetic layout of strains of *L. interrogans* and demonstrated that gene layout is variable between strains; speculatively, this genetic shuffling may be related to the multitude of insertion sequence (IS) elements present. This genetic fluidity is paradoxical, given the apparent stability of leptospiral strains in long-term serial passage,

From: *Bacterial Genomes and Infectious Diseases*
Edited by: V. L. Chan, P. M. Sherman, and B. Bourke © Humana Press Inc., Totowa, NJ

although this long-term stability is measured in the context of the absence of change to the lipopolysaccharide phenotype. Clearly, pathogenic *Leptospira* undergoes considerable phenotypical change in the transition from in vivo growth to in vitro culture, as indicated by strains that have undergone long term in vitro culture having low infectivity/virulence when returned to in vivo conditions.

In the absence of reliable systems for the genetic manipulation of *Leptospira*, comparative genomics provides a means by which we can begin to understand the genetic basis for differences in virulence and the effects of long term passage. The background and history of sequenced strains is therefore critical. Although two *L. interrogans* strains have been sequenced, the serovar Lai strain has undergone long term in vitro passage *(4)*, whereas the serovar Copenhageni strain was isolated from a patient with severe leptospirosis and has undergone minimal in vitro passage *(5)*. Likewise, the *L. borgpetersenii* serovar, Hardjobovis, is a low-passage, human isolate. Pulsed-field gel electrophoresis analysis of the genomic DNA showed it to be a type A strain (unpublished data). These strains have a wide geographical distribution and are the most common Hardjobovis isolate.

As previously mentioned, the genus *Leptospira* is divided into three broad phylogenic clades. *L. interrogans* and *L. borgpetersenii* are representative of the diversity of one of the clades containing the pathogenic serovars, with *L. interrogans* serovars Lai and Copenhageni associated with severe leptospirosis with high mortality, and Hardjobovis with a much milder disease that is almost never fatal *(2)*. Moreover, the maintenance hosts for Copenhageni and Lai are rodent species, whereas bovine species are the usual maintenance hosts for Hardjobovis. The sequenced strains, therefore, provide an excellent insight into the genetic diversity of a key subgroup of the genus *Leptospira* and perhaps even a means to investigate a genetic basis for disease severity.

Even in the short time since the release of the serovar Lai sequence, it has been necessary to review and update the annotation of this sequence. In light of the imminent release of several additional leptospiral genome sequences, it is essential to develop a strategy that continues to revise the annotation of existing genomes as further applicable data are published, thereby ensuring the maximum benefit is derived from the annotated sequences. Ultimately, the system should ensure that the annotations are up-to-date, consistent, and error-free.

This chapter will attempt to provide an overview of the genomic similarities and differences between the sequenced strains, and propose a system to make orthologous genes recognizable by using a unique identifier for orthologous, leptospiral genes. This chapter will also examine the relationship between *Leptospira* and the other sequenced spirochetes (*Treponema pallidum [6]*, *Treponema denticola [7]*, *Borrelia burgdorferi [8]*, and *Borrelia garinii [7]*).

2. Similarities

2.1. Comparing the Leptospiral Genomes

The primary difficulty in comparing the leptospiral genomes is the identification of orthologous genes. In particular, the different naming schemes used by each of the annotation groups make it difficult to recognize orthologous genes. Another, perhaps less obvious, problem can arise where there are differences in the process used to decide whether or not a reading frame is a protein coding sequence (CDS). Ussery and Hallin

Table 1
Start Codon Frequencies for all Protein Coding Sequence Features in *Leptospira*

Start codon	Serovar		
	Copenhageni	Lai	Hardjobovis
ATG	2733	2807	2572
TTG	479	468	402
GTG	221	220	175
CTG	17	19	19
ATT	0	0	1

have noted that the number of CDS features has been overestimated in the Lai and Copenhageni genomes *(9)*, thereby interfering with comparisons to genomes where a more conservative approach to the annotation of CDS features has been made. Given that there was no relationship between the groups that annotated the three leptospiral genomes, it is no real surprise that the annotations of the Lai, Copenhageni, and Hardjobovis genomes differ significantly. Adopting the method used to annotate the Hardjobovis genome, we have revised the Lai and Copenhageni annotations. A significant benefit of the reannotation process has been that each set of orthologous genes from the sequenced genomes has been assigned a unique identifier, and the start of the coding region has been revised and made consistent across orthologous genes. Accurate estimation of the start of the coding regions is critical to postgenomic studies where the prediction of the subcellular location of the encoded protein is required. Moreover, this type of revision has led to the identification of potential pseudogenes, where there is a frameshift early in the coding region, and the identification of genes that have been interrupted by the incorporation of IS elements into the genome. It is worth noting the frequency of the different start codons that are predicted to be used by *Leptospira* (Table 1). Although ATG is the most common start codon, TTG, GTG, and CTG occur in diminishing frequencies. Based on similarity analysis, it is clear there is at least one Hardjobovis CDS that has ATT as its start codon.

2.2. Overview of the Genome Sequences

The genome of *L. borgpetersenii* serovar Hardjobovis strain L550 comprises two circular chromosomes of 3,614,446 bases and 317,336 bases, with an overall guanine and cytosine (G+C) content of 41.3%. The density of CDS sequences across the genome is 80.3%, with an average gene size of 931 bases. In total 3111 and 292 CDS features were annotated on chromosome 1 and chromosome 2, respectively. The *L. borgpetersenii* genome is smaller than the *L. interrogans* genomes (Table 2) and codes for proportionally fewer genes. The G+C content is higher than that found for *L. interrogans,* consistent with previous estimates of G+C content *(3)*. The relatively lower density of coding regions found in *Leptospira* compared with *Escherichia coli* may be related to the fluidity of the arrangement of genes on the genome. Consistent with this viewpoint, the gene layout found in the sequenced *Borrelia* genomes is conserved and the density of coding regions is around 95% *(10)*. Also likely to contribute to genome fluidity is the

Table 2
Genome Features in *Leptospira*

Feature	Copenhageni CI	Copenhageni CII	Lai CI	Lai CII	Hardjobovis CI	Hardjobovis CII
Size (bp)	4,277,185	350,181	4,332,241	358,943	3,614,446	317,336
G+C content (%)	35.1	35.0	36.0	36.1	41.0	41.2
Protein-coding percentage	73	73	73	73	80	80
Protein-coding (CDS)						
With assigned function	1811	161	1901	159	2000	177
Conserved and hypothetical	1643	113	2459	208	1121	115
Total	3454	274	4360	367	3121	292
Revised annotation	3375	279	3327	300		
Difference (%)	2	0	31	22		
Transfer RNA genes	37	0	37	0	37	0
Ribosomal RNA genes						
23S*rrl*	2	0	1	0	2	0
16S*rrs*	2	0	2	0	2	0
5S*rrf*	1	0	1	0	1	0

CI, chromosome 1; CII, chromosome 2.

number of ISs. The IS elements are more numerous in the Hardjobovis genome than in the sequenced *L. interrogans* strains. Fifteen different elements were identified in the sequenced leptospiral strains, the most numerous of which was IS*1533* with 95 complete copies spread across both chromosomes of the Hardjobovis genome (Table 3). Remarkably, none of the other sequenced spirochetes contains recognizable IS elements.

Like serovar Copenhageni, the Hardjobovis strain has one *rrf*, two *rrs*, and two *rrl* genes, with each of the rRNA genes distributed on chromosome 1. Unlike the *L. interrogans* strains, *L. borgpetersenii* has a characteristic IVS sequence that interrupts the *rrl* gene, resulting in an rRNA profile which has no 23S rRNA band *(11)*. Each of the three sequenced genomes contains 37 tRNA genes; in all cases, these genes are encoded on chromosome 1. The origin of replication for chromosome 1 is located upstream of *gidA* and the gene layout in this region of chromosome 1 is conserved among the three sequenced leptospiral strains. The origin of replication for chromosome 2 was identified by a G+C skew. The differences between the origins of replication indicate that there are different mechanisms for replication of each of the chromosomes.

2.3. Comparison of Gene Layout

The Copenhageni and Lai genomes are colinear except for the inversion of a 1000-kb fragment that proportionally flanks the origin of replication on chromosome 1 and several indels. A notable indel has occurred at the second copy of the 23S rRNA gene, where a probable deletion event in serovar Lai has left only a gene remnant. The difference is assumed to be a deletion from the Lai genome, because in all other leptospiral strains, there are two copies of this gene (Table 2).

Table 3
Summary of Insertion Sequence Elements in *Leptospira*

Insertion sequence (IS)[a]		Serovar[b]					
		Copenhageni		Lai		Hardjobovis	
		CI	CII	CI	CII	CI	CII
IS*1533*	(SPT0001)	–	–	–	–	88 (6)	7 (0)
IS*1500*	(SPT0002)	6 (2)	–	7 (1)	1 (1)	–	–
IS*1501*	(SPT0003)	3 (1)	–	1 (1)	2 (0)	–	–
IS*1501a*	(SPT0004)	–	–	–	–	7 (1)	1 (0)
IS*Lin2*	(SPT0005)	1 (2)	–	0 (1)	–	1 (0)	–
IS*Lin3*	(SPT0006)	1 (0)	–	2 (1)	–	–	–
IS*Lin1*	(SPT0007)	12 (2)	–	41 (2)	3 (0)	0 (1)	1 (1)
IS*1502*	(SPT0008)	1 (0)	–	1 (0)	–	–	–
YhgA-like	(SPT0011)	–	–	4 (1)	–	5 (2)	–
IS*Lbp1*	(SPT0015)	–	–	0 (1)	–	15 (2)	1 (0)
IS*Lbp2*	(SPT0016)	–	–	–	–	1 (2)	1 (0)
IS*Lbp3*	(SPT0017)	–	–	–	–	1 (1)	1 (0)
IS*1477*-like	(SPT0018)	–	–	–	–	2 (0)	–
IS*Lbp4*	(SPT0019)	–	–	–	–	1 (0)	–
IS*Lbp5*	(SPT0020)	–	–	–	–	9 (2)	–

[a]Name of IS element. IS*Lin2*, IS*Lin3*, IS*Lbp1*, IS*Lbp2*, IS*Lbp3*, IS*Lbp4*, and IS*Lbp5* are proposed element names based on the current nomenclature scheme (http://www-is.biotoul.fr/is.html). The number in parentheses is the leptospiral ortholog number.

[b]The number of complete IS elements and chromosome distribution of the IS elements is tabulated. In each instance, the number in parenthesis is the count of incomplete elements.

CI, chromosome I; CII, chromosome 2.

Interspecies comparison of gene layout reveals extensive rearrangement of genes. Differences in gene layout appear to be the result of rearrangements that have occurred frequently enough for it to be rare for synteny to extend beyond 20 kb (Fig. 1). This high frequency of gene rearrangement is the underlying cause of the scattering across the genome of sets of genes that are normally arranged in operons, for example the *lpx* genes for lipid-A biosynthesis. This adds a level of complexity to the genetic analysis of this pathway, especially given the unique structure of leptospiral lipid-A, such that all the biosynthetic genes will not necessarily be identifiable by similarity *(12)*. The process of genetic rearrangement does not seem to be limited to intrachromosomal events, with the occurrence of a small number of genes on chromosome 1 in one species and chromosome 2 in the other. The limited exchange may be the result of a lower frequency of rearrangement between chromosomes or some unknown limitations on the genes that can be exchanged onto chromosome 2, as may be indicated by the absence of tRNA and rRNA genes on chromosome 2.

2.4. The Annotated Genomes

Our revised annotation of the Lai genome has removed more than 1100 CDS features resulting in a final count of 3613 CDS features. The revision included updating the predicted translation initiation sequence for more than 10% of the Lai and Copenhageni

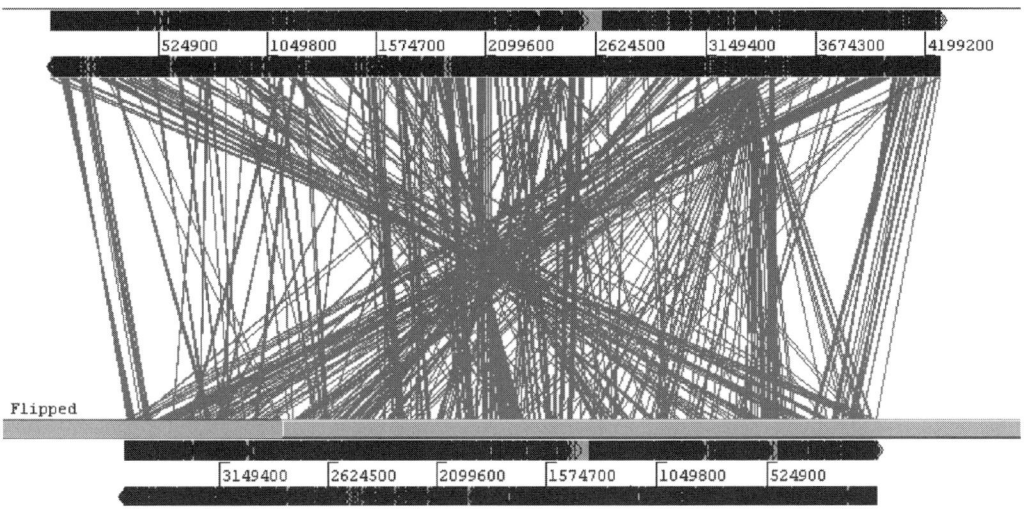

Fig. 1. Visual overview of the regions of similarity between *Leptospira interrogans* serovar Copenhageni and *L. borgpetersenii* serovar Hardjobovis. The Artemis comparison tool was used to display a TBlastn search result (subject: Copenhageni chromosome 1; query: Hardjobovis chromosome 1). The Copenhageni chromosome (bases 1 through 4,277,185) is represented by the top bar and the Hardjobovis chromosome 1 (bases 3,614,446 through 1) is represented by the bottom bar. Regions of similarity are indicated by the bars between the chromosomes.

CDS features. The majority of the deleted features encoded predicted proteins of less than 50 amino acids. Generally, these small CDS features had no orthologous reading frame in *L. borgpetersenii* L550, no similarity to other sequences in GenBank, lacked credible translation initiation sequences, and were not found by the GeneMarkS gene-finder *(13)*. Similarly, our revised annotation of the Copenhageni genome contains almost 200 fewer CDS features, resulting in 3529 CDS features. Although the Hardjobovis genome is around 10% smaller than the *L. interrogans* genomes, it contains 3409 CDS features. However, the total number of CDS features is a misleading figure to use to compare the genomes, because more than 6% of the *L. borgpetersenii* genome comprises IS element sequences. This is significantly higher than either the Lai or Copenhageni genomes, both around 2% of the genome. In addition, the number of CDS features includes the coding regions of pseudogenes and there is more than three times the number of pseudogenes in *L. borgpetersenii* strain L550 (161 pseudogenes) compared with either of the *L. interrogans* genomes. By excluding both the IS element-related CDS features and the CDS features associated with pseudogenes, a more rational comparison of the coding capacity of the genomes can be made. These figures show the similarity in the coding capacity of the *L. interrogans* genomes, with 3390 CDS features for Lai and 3388 CDS features for Copenhageni, and the substantially lower coding

Table 4
Overview of Total Number of Protein Coding Sequence Features in *Leptospira*

	Serovar					
	Copenhageni		Lai		Hardjobovis	
	CI	CII	CI	CII	CI	CII
Total CDS	3250	279	3313	300	3119	290
Excluding pseudogenes and IS elements	3113	275	3118	272	2603	237
Unique hypothetical proteins[a]	805	82	823	82	583	70
Conserved hypothetical proteins[b]	501	35	496	34	378	32

CI, chromosome 1; CII, chromosome 2.
[a]Unique hypothetical protein means hypothetical proteins unique to the genus *Leptospira* (thus far).
[b]Conserved hypothetical proteins refers to hypothetical proteins found in other genera.

capacity of the *L. borgpetersenii* L550 strain with 2840 CDS features. Remarkably, between 23 and 26% of the CDS features encode proteins that are unique to the genus *Leptospira* (Table 4). When the hypothetical proteins for which there are homologs in other genera are included, the total proportion of hypothetical proteins is between 37 and 42%.

The similarities between the leptospiral genomes are considerable, with just over 2700 CDS features (approx 80%) with orthologs in each of the three genomes. Although a number of approaches can be used to compare the coding capacities of related genomes, we have chosen to include only intact CDS features, because it is our intention to compare the predicted protein functions encoded by the genomes. We have excluded features encoded on IS elements because the high copy number of these features will artificially inflate particular functions. 612 CDS features are present in both Lai and Copenhageni but not in Hardjobovis, whereas 238 CDS features are found exclusively in Hardjobovis. These figures exclude CDS features associated with pseudogenes. It is interesting to note that there are CDS features that are unique to either the Copenhageni genome (82 features) or Lai genome (42 features). In addition, there are four CDS features that are found on the Lai and Hardjobovis genomes and not the Copenhageni genome, and 13 CDS features found on both the Copenhageni and Hardjobovis genomes, and not the Lai genome. This final group of features is consistent with lateral transfer of genetic material between different species of *Leptospira*.

The Clusters of Orthologous Groups of proteins (COGs) classification scheme *(14)* can be used to provide a general overview of the functions of the reduced set of CDS features that excludes the IS element-related CDS features and pseudogenes associated CDS features. COG classification depends on proteins being similar to a protein with a characterized function. The high proportion of unique and conserved hypothetical proteins means that COGs were assigned to 1967 out of the 2840 Hardjobovis CDS features, and 2176 of the 3388 Copenhageni CDS features. The breakdown of the numbers of CDS features assigned to each COG is shown in Table 5. The most notable differences between the Copenhageni and Hardjobovis genomes is that Hardjobovis has fewer "signal transduction" proteins (cluster T) and fewer "metabolism" proteins (spread across clusters G, E, H, I, and P), indicating that Hardjobovis may exist in a less varied

Table 5
Overview of Clusters of Orthologous Groups Classification
of Predicted Leptospiral Proteins

COG category	COG description	Serovar	
		Hardjobovis	Copenhageni
Information, storage, and processing			
[J]	Translation, ribosomal structure, and biogenesis	168	143
[A]	RNA processing and modification	–	–
[K]	Transcription	77	79
[L]	Replication, recombination, and repair	87	101
[B]	Chromatin structure and dynamics	2	2
Cellular processes and signaling			
[D]	Cell cycle control, cell division, chromosome partitioning	22	22
[Y]	Nuclear structure	–	–
[V]	Defense mechanisms	33	37
[T]	Signal transduction mechanisms	149	190
[M]	Cell wall/membrane/envelope biogenesis	203	218
[N]	Cell motility	85	90
[Z]	Cytoskeleton	–	–
[W]	Extracellular structures	–	–
[U]	Intracellular trafficking, secretion, and vesicular transport	34	30
[O]	Posttranslational modification, protein turnover, chaperones	93	96
Metabolism			
[C]	Energy production and conversion	112	115
[G]	Carbohydrate transport and metabolism	62	74
[E]	Amino acid transport and metabolism	128	143
[F]	Nucleotide transport and metabolism	52	50
[H]	Coenzyme transport and metabolism	104	111
[I]	Lipid transport and metabolism	84	99
[P]	Inorganic ion transport and metabolism	64	80
[Q]	Secondary metabolites biosynthesis, transport, and catabolism	13	15
Poorly characterized			
[R]	General function prediction only	227	279
[S]	Function unknown	168	202

environment (fewer cluster T proteins) and the apparent reduction in "metabolism" proteins may indicate an increased host dependency.

3. Overview of Core Leptospiral Features

3.1. Metabolism

Studies of the metabolism and enzyme activities of *Leptospira*, most of which were performed more than 30 yr ago, provide a framework for predicting the sets of meta-

Table 6
Overview of the Predicted Metabolic Capabilities of the Sequenced Spirochetes

	Leptospira	Treponema pallidum	Treponema denticola	Borrelia burgdorferi
Oxidative phosphorylation	Yes	No	No	No
Electron transfer system	Yes	No	No	No
Glycolysis	No	Yes	Yes	Yes
Gluconeogenesis	Yes	No	Yes	No
Pentose phosphate pathway	Yes	Yes	Yes	Yes
Lipopolysaccharide biosynthesis	Yes	No	No	No
Fatty acid metabolism (β-oxidation)	Yes[a]	No	Yes	No
Fatty acid biosynthesis	Yes[b]	No	Yes	No
Glycerol metabolism	Yes	No	Yes	No
Nucleotide biosynthesis	Yes	No	Yes	No
Amino acid degradation and interconversion	Yes[c]	No	Yes	No

[a]Long chain fatty acids (C15 or longer) or shorter chain fatty acids can be utilized if there is a mixture of long and short chain fatty acids (3).

[b]Biochemical studies indicate that *Leptospira* can modify long chain fatty acids, but is unable to synthesize them from pyruvate or acetate (3).

[c]One study on a serovar Pomona strain indicates that there is no interconversion of arginine, leucine, valine, and phenylalanine. Glutamate and aspartate, and alanine and glycine, are interconverted (3).

bolic/biosynthetic enzymes that are likely to be encoded by the sequenced leptospiral genomes (3). Among others, these studies anticipate the presence of enzymes involved in oxidative phosphorylation, the electron transfer chain, β-oxidation of long chain fatty acids, gluconeogenesis, and the pentose phosphate pathway. Many of the enzymes in these pathways have been predicted with a high degree of certainty based on similarity and, therefore, provide supporting evidence for the earlier biochemical studies. Although it would be possible to fill many of these gaps in biochemical pathways with likely candidates, it is our opinion that this should be left to future functional studies.

In summary, leptospires use phospholipids and fatty acids as their major carbon and energy sources. Phospholipids are digested extracellularly to long chain fatty acids and glycerol, both of which are then transported into the cell and metabolized. Speculatively, the primary role of the leptospiral sphingomyelinases is in nutrient acquisition via the preliminary digestion of phospholipids prior to transport into the cell; their role in virulence is incidental.

All sugars are synthesized *de novo* and there is no biochemical evidence or bioinformatic inference for a glycolytic pathway in *Leptospira*; the gene identified by Nascimento et al. (5) as a glucokinase (EC 2.7.1.2, LA1437, LIC12312, SPN1953) is, by similarity, more likely to be a regulatory protein. The components of the electron transport system include an NADH dehydrogenase complex, a fumerate reductase complex, a cytochrome O ubiquinol oxidase-like cytochrome c oxidase complex, as well as an F_0F_1-type ATPase complex (EC 3.6.3.14). There is a marked difference between the predicted metabolic capabilities of *Leptospira* and the other sequenced spirochetes. A general comparative overview is provided in Table 6.

3.2. Regulation of Gene Expression

Leptospira appears to use a number of known strategies to tailor expression patterns to the diverse environments normally encountered by this bacterium. Like the other spirochetes, *Leptospira* appears to have both Sigma-54 (LA2404, LIC11545, SPN1246) and Sigma-70 (LA2232, LIC11701, and SPN1385) transcription initiation factors. Sigma-70 is essential for the regulation of transcription during exponential growth *(15)*. Specialized sigma factors, such as the extracytoplasmic function (ECF) sigma factors, which respond to specific environmental signals, are not found in the *Borrelia* genomes, whereas there are one and two examples in the *T. pallidum* and *T. denticola* genomes, respectively. By contrast, each of the leptospiral genomes has 11 ECF sigma factors. Notably, these factors are not encoded exclusively on chromosome 1. In other bacteria these factors are normally cotranscribed with an antisigma regulatory factor that is normally located in the inner membrane *(16)*. It is not surprising, perhaps, that none of the 11 leptospiral ECF sigma factor genes is located near the genes encoding the 11 anti-sigma regulatory factors. About 20 anti-sigma factor antagonists are encoded on each of the leptospiral genomes. The second main class of response regulators found in *Leptospira* are the two component response regulators. Each of the leptospiral genomes contains approx 80 genes that encode either sensor histidine kinases, response regulators, or both. This is in contrast to the *T. pallidum* and *B. burgdorferi* genomes, each of which has less than 20 of these genes. Two general classes of response regulators can be identified, those with DNA binding domains, and those with diguanylate cyclase activity *(17)*. Diguanylate cyclase domain regulators are found on each of the spirochete genomes, with seven found on the *L. borgpetersenii* genome, and 14 on the *L. interrogans* genomes. The difference is because of an apparent gene duplication event in *L. interrogans* resulting in seven paralogous copies of one of the response regulator genes (LIC11125-LIC11131 and LA2926-LA2933, excluding LA2928).

Core components of the heat shock response are conserved in all the sequenced spirochetes, including DnaK, GroES, GroEL, and GrpE. However, the regulation of the expression of these genes in the *T. pallidum*, *T. denticola*, *B. burgdorferi*, and *B. garinii* differs from *Leptospira* because of the absence of an HrcA ortholog *(18)*.

3.3. Endoflagella

Each of the two endoflagella in *Leptospira* located in the periplasmic space arise from the end of the bacterium. After cell division, each daughter cell retains one parental endoflagellum and a new endoflagellum is formed at the end where cell division occurred. There are between 47 and 49 endoflagellar-related genes in each of the sequenced, leptospiral genomes; cumulatively, these genes cover more than 45 kb. Comparison of the leptospiral, endoflagellar-related proteins with those encoded on the other sequenced spirochetal genomes reveals that the assembly process and structure of the endoflagellum is conserved across the spirochetes. Predominantly, the differences relate to the number of paralogous copies of key genes encoding endoflagellar structural proteins.

3.4. Lipopolysaccharide

Lipopolysaccharide (LPS) is a major component of the outer membrane of *Leptospira* and a key interface between the bacterium and the surrounding environment. None of the other sequenced spirochetes is genetically capable of producing an attached, surface-exposed polysaccharide. An insight into the complexity of the leptospiral LPS structure is indicated by the more than 200 recognized serovars, each of which synthesizes a unique LPS structure. Even more remarkable is the fact that metabolically, all sugars incorporated into the LPS must be synthesized *de novo (3)*. This structure thus comes at a considerable energy cost to the cell, and therefore, LPS must confer an enormous, but as yet poorly understood, benefit to *Leptospira*.

LPS is normally comprised of three parts, lipid-A, core polysaccharide, and a variable *O*-antigen side chain polysaccharide. In *Leptospira*, the genes encoding the enzymes for the biosynthesis of the polysaccharide component of LPS are predominantly found on a single 100-kb region of the genome. This region is remarkable because almost exclusively, the genes are encoded on the one strand. Toward the 3' end of the region there is a cluster of approx 17 genes for which there are orthologs in each of the sequenced leptospiral genomes; included in this region are the thymidine dTDP-rhamnose biosynthesis genes. Immediately upstream, there is synteny between the *L. interrogans* serovars, but there is no such conservation of gene layout between species. Toward the 5' end of the region, there are some genes that appear to encode proteins that have no role in LPS biosynthesis.

3.5. Protein Secretion Systems

Based on similarity, *Leptospira* has a Sec translocase that resembles that of *E. coli*. This complex is responsible for the translocation of both inner membrane and extracytoplasmic proteins *(19)*. The processing of the extracytoplasmic proteins occurs via the signal peptidases. Orthologs of Lgt, Lnt, and LspA (signal peptidase II) are present, indicating that *Leptospira* processes lipoproteins using a mechanism similar to other Gram-negative bacteria, although the specificity of the signal recognition sequence is different *(20)*. In addition, proteins related to LolA, LolC, and LolD are present and are probably involved in the transport and incorporation of lipoproteins into the outer membrane. The signal directing, newly translocated lipoproteins to the outer membrane is encoded by the second amino acid of the mature lipoprotein; again, this signal differs from that found for Gram-negative bacteria *(20)*. Two LepB-related proteins are encoded on each of the sequenced genomes indicating a possible redundancy for signal peptidase I. The Sec translocase and related signal peptidases are found in the other sequenced spirochetes, although the processing of lipoproteins in *Borrelia* may differ because of the absence of an Lgt ortholog. Sec-independent secretion of proteins occurs via the twin-arginine translocase *(21)*. The sequenced leptospiral genomes each encode orthologs for two of the three proteins normally found in the translocase. Although no leptospiral proteins have been shown to be secreted by a twin-arginine translocase, it appears to be a likely mechanism for secretion in *Leptospira*. Thus far, no other spirochete encodes proteins for an equivalent translocase (Table 7).

Leptospira appears to have the genetic capability to secrete proteins via both type I and type II secretion systems, but there appears to be no type III secretion system. Proteins related to components of this secretion system are likely to be involved in endoflagella biosynthesis rather than a secretion system. In the absence of specific functional data, candidate outer membrane proteins and ABC transporter proteins are present and, therefore, it is possible that type I secretion complexes exist for *Leptospira*. Genes encoding proteins for the terminal branch of the general secretory pathway (type II secretion)

Table 7
Leptospiral Proteins Involved in Translocation and Extracellular Protein Maturation

Protein	Lai number	Copenhageni number	Leptospiral ortholog number	Function
Sec translocase				
SecA	LA1960	LIC11944	SPN1609	Sec protein translocation
SecY	LA0759	LIC12853	SPN2440	Sec protein translocation
SecE	LA3426	LIC10747	SPN0540	Sec protein translocation
SecG	LA1695	LIC12095	SPN1742	Sec protein translocation
YajC	LA1141	LIC12540	SPN2160	Sec protein translocation
SecD	LA1142	LIC12538	SPN2158	Sec protein translocation
SecF	LA1143	LIC12537	SPN2157	Sec protein translocation
YidC	LA0178	LIC10157	SPN0004	Sec protein translocation
SecB	–	–	–	No related protein
Signal peptidases and associated proteins				
LepB	LA2788, LA3754	LIC11233, LIC10478	SPN0965, SPN0297	Signal peptidase I
Lnt	LA1124, LA4078	LIC12556, LIC13250	SPN2175[a], SPN2800	Lipoprotein biosynthesis (Apolipoprotein N-acyltransferase)
LspA	LA1336	LIC12389	SPN2022	Signal peptidase II
Lgt	LA3004	LIC11063	SPN0813[b]	Lipoprotein biosynthesis (Prolipoprotein diacylglyceryltransferase)
Lipoprotein, outer membrane targeting				
LolA	LA0410, LA1136	LIC10359, LIC12545	SPN0193, SPN2165[b]	outer membrane incorporation
LolE	LA0273, LA1284, LA2983	LIC10232, LIC12429, LIC11080	SPN0071, SPN2059, SPN0829	outer membrane incorporation (permease)
LolD	LA0274, LA2982	LIC10233, LIC11081	SPN0072, SPN0830	outer membrane incorporation (ATP-binding protein)
Twin arginine translocase				
TatA	LA1926	LIC11980	SPN1638	Sec-independent protein secretion pathway component
TatC	LA1925	LIC11981	SPN1639	Sec-independent protein secretion pathway translocase

[a]No Hardjobovis ortholog.
[b]T. denticola and T. pallidum ortholog, no Borrelia ortholog.

are located in an operon in each of the sequenced leptospiral genomes. Orthologous genes are not found in the other sequenced spirochetal genomes (Table 8).

3.6. Lipoproteins

Although key physical properties, such as membrane localization and expression profiles, of more than 10 leptospiral lipoproteins have been determined, there have been

Table 8
Leptospiral Proteins Involved in Type II Secretion

Protein	Lai number	Copenhageni number	Leptospiral ortholog number
GspK	LA2368	LIC11577	SPN1278
GspG	LA2372	LIC11573	SPN1274
GspF	LA2373	LIC11572	SPN1273
GspE	LA2374	LIC11571	SPN1272
GspD	LA2375	LIC11570	SPN1271
GspC	LA2376	LIC11569	SPN1270

few functional studies, with the exception of the Lig family of lipoproteins *(22)*. The primary interest in this class of proteins has been as a source of cross-protective antigens, in that these proteins are highly conserved in a wide range of serovars *(20)*. Using LipoP to predict potential lipoproteins *(23)*, we estimate, excluding pseudogenes, that there are 96 lipoproteins encoded by the Hardjobovis genome, 136 by the Lai genome, and 147 by the Copenhageni genome. There are 96 lipoproteins for which there is an ortholog in each of the genomes, and of these 14 are pseudogenes in Hardjobovis, and five in Lai. There are 44 lipoproteins that are found in the *L. interrogans* genomes, but not in Hardjobovis, and 13 lipoproteins that are unique to Hardjobovis. There are two and five lipoproteins that are unique to serovar Lai and Copenhageni, respectively. In each instance, more than 75% of the lipoproteins are unique to the genus. In the case of the conserved hypothetical lipoproteins, there are only one or two examples of orthologous proteins being encoded by any other spirochete genome.

4. Concluding Remarks

The remarkable feature of the relationship between *Leptospira* and the other spirochetes is the near absence of a relationship. Although there are fundamental features that have been conserved among the sequenced spirochetes, such as the endoflagella and the systems for protein translocation, the systems used for nutrient acquisition and metabolism have virtually no similarities. However, these differences are probably related to the apparent extensive genome reduction that *T. pallidum* and *Borrelia* have undergone.

The annotation of any genome is an ongoing operation. Studies will reveal that improved prediction methods will provide better indications of function and mistakes will have to be corrected. In addition, leptospiral genomics is by no means a static field, with at least three other genome sequences to be published over the next few years. Given our experience with the comparative analysis of the published leptospiral genomes, where a substantial review of both the Copenhageni and Lai annotation was necessary before any meaningful comparisons could be made, it is necessary to examine systems whereby the leptospiral genome data can be updated and revised. It is clear that this will not be done by the main genetic databases (GenBank, EMBL) and there seems to be no incentive for the groups that initially publish annotated genome sequences to update the annotations. For instance, the *T. pallidum* strain Nichols annotation has not been updated since its initial publication in 1998. There are examples of organisms for which the annotations are reviewed and updated and include the *Pseudomonas aeruginosa* genome (http://www.pseudomonas.com) and the *E. coli* genome (http://ecocyc.org). We consider a key step to be the establishment of a group to take responsibility for maintaining the genome data for the genus *Leptospira*, as well as a website from which the current versions of the annotated genomes can be obtained. Our current version of annotated leptospiral genomes can be obtained from http://vbc.med.monash.edu.au/genomes. The IUMS Subcommittee on the Taxonomy of *Leptospira* seems to be an appropriate body to assume responsibility for maintaining and updating the annotated leptospiral genome sequences.

References

1. Chappel, R. J., Goris, M., Palmer, M. F., and Hartskeerl, R. A. (2004) Impact of proficiency testing on results of the microscopic agglutination test for diagnosis of leptospirosis. *J. Clin. Microbiol.* **42,** 5484–5488.
2. Perolat, P., Chappel, R. J., Adler, B., et al. (1998) *Leptospira fainei sp. nov.*, isolated from pigs in Australia. *Int. J. Syst. Bacteriol.* **48,** 851–858.
3. Faine, S., Adler, B., Bolin, C., and Perolat, P. (1999) *Leptospira and Leptospirosis.* 2nd ed. MediSci, Melbourne, Australia.
4. Ren, S. X., Fu, G., Jiang, X. G., et al. (2003) Unique physiological and pathogenic features of *Leptospira interrogans* revealed by whole-genome sequencing. *Nature* **422,** 888–893.
5. Nascimento, A. L., Ko, A. I., Martins, E. A., et al. (2004) Comparative genomics of two *Leptospira interrogans* serovars reveals novel insights into physiology and pathogenesis. *J. Bacteriol.* **186,** 2164–2172.
6. Fraser, C. M., Norris, S. J., Weinstock, G. M., et al. (1998) Complete genome sequence of *Treponema pallidum*, the syphilis spirochete. *Science* **281,** 375–388.
7. Seshadri, R., Myers, G. S., Tettelin, H., et al. (2004) Comparison of the genome of the oral pathogen *Treponema denticola* with other spirochete genomes. *Proc. Natl. Acad. Sci. USA* **101,** 5646–5651.
8. Fraser, C. M., Casjens, S., Huang, W. M., et al. (1997) Genomic sequence of a Lyme disease spirochaete, *Borrelia burgdorferi. Nature* **390,** 580–586.
9. Ussery, D. W. and Hallin, P. F. (2004) Genome update: annotation quality in sequenced microbial genomes. *Microbiology* **150,** 2015–2017.
10. Glockner, G., Lehmann, R., Romualdi, A., et al. (2004) Comparative analysis of the *Borrelia garinii* genome. *Nucleic Acids Res.* **32,** 6038–6046.
11. Ralph, D. and McClelland, M. (1993) Intervening sequence with conserved open reading frame in eubacterial 23S rRNA genes. *Proc. Natl. Acad. Sci. USA* **90,** 6864–6868.
12. Que-Gewirth, N. L., Ribeiro, A. A., Kalb, S. R., et al. (2004) A methylated phosphate group and four amide-linked acyl chains in *Leptospira interrogans* Lipid A. The membrane anchor of an unusual lipopolysaccharide that activates TLR2. *J. Biol. Chem.* **279,** 25,420–25,429.
13. Besemer, J., Lomsadze, A., and Borodovsky, M. (2001) GeneMarkS: a self-training method for prediction of gene starts in microbial genomes. Implications for finding sequence motifs in regulatory regions. *Nucleic Acids Res.* **29,** 2607–2618.
14. Tatusov, R. L., Fedorova, N. D., Jackson, J. D., et al. (2003) The COG database: an updated version includes eukaryotes. *BMC Bioinformatics* **4,** 41.
15. Paget, M. S. and Helmann, J. D. (2003) The sigma-70 family of sigma factors. *Genome Biol.* **4,** 203.
16. Helmann, J. D. (2002) The extracytoplasmic function (ECF) sigma factors. *Adv. Microb. Physiol.* **46,** 47–110.
17. Galperin, M. Y., Nikolskaya, A. N., and Koonin, E. V. (2001) Novel domains of the prokaryotic two-component signal transduction systems. *FEMS Microbiol. Lett.* **203,** 11–21.

18. Ballard, S. A., Go, M., Segers, R. P., and Adler, B. (1998) Molecular analysis of the *dnaK* locus of *Leptospira interrogans* serovar Copenhageni. *Gene* **216,** 21–29.

19. Nakatogawa, H., Murakami, A., and Ito, K. (2004) Control of SecA and SecM translation by protein secretion. *Curr. Opinion. Microbiol.* **7,** 145–150.

20. Haake, D. A. (2000) Spirochaetal lipoproteins and pathogenesis. *Microbiology* **146,** 1491–1504.

21. Mangels, D., Mathers, J., Bolhuis, A., and Robinson, C. (2005) The core TatABC complex of the twin-arginine translocase in *Escherichia coli*: TatC drives assembly whereas TatA is essential for stability. *J. Mol. Biol.* **345,** 415–423.

22. Matsunaga, J., Barocchi, M. A., Croda, J., et al. (2003) Pathogenic *Leptospira* species express surface-exposed proteins belonging to the bacterial immunoglobulin superfamily. *Mol. Microbiol.* **49,** 929–945.

23. Juncker, A. S., Willenbrock, H., Von Heijne, G., Brunak, S., Nielsen, H., and Krogh, A. (2003) Prediction of lipoprotein signal peptides in Gram-negative bacteria. *Prot. Sci.* **12,** 1652–1662.

8

Listeria monocytogenes

Keith Ireton

Summary

Listeria monocytogenes is a Gram-positive, intracellular bacterial pathogen responsible for severe food-borne illnesses resulting in central nervous system infection or abortion. *L. monocytogenes* induces its own internalization into mammalian cells, escapes from the host cell phagosome, replicates extensively in the cytosol, and spreads from one host cell to another through an F-actin-dependent motility process. Previously, classical genetic approaches were used to identify bacterial virulence factors critical for the intracellular life cycle of *L. monocytogenes*. The recent availability of the nucleotide sequence of *L. monocytogenes* has provided the potential for global analysis of bacterial proteins that affect pathogenesis. In this chapter, ways in which the *L. monocytogenes* genome has been used to probe the functions of bacterial proteins in virulence is discussed. At the end of the chapter, future genomic- or proteomic-based approaches that might improve or expand on current work are highlighted.

Key Words: Genomics; internalins; leucine-rich repeat; *Listeria monocytogenes*; listeriosis; PrfA; lipoprotein; LPXTG; sortase; surface protein.

1. Introduction

Listeria monocytogenes is a Gram-positive, food-borne pathogen capable of causing gastroenteritis and severe systemic infections culminating in meningitis or abortion *(1)*. The incidence of listeriosis in the general population is low (~2500 cases per year in the United States), representing less than 0.1% of all food-borne illnesses *(2)*. However, the high mortality rate of listeriosis (~20%) makes *L. monocytogenes* responsible for almost 30% of deaths caused by food-borne pathogens.

L. monocytogenes is a facultative, intracellular pathogen that occupies several environmental niches *(1)*. This bacterium infects a wide variety of animal hosts in addition to humans, including other mammals (sheep, cattle, goats, pigs, rabbits, mice), birds, and fish. Moreover, *L. monocytogenes* replicates outside of host animals, and is thought to live in the soil as a saprophyte utilizing decaying vegetation as a food source. Thus, in contrast to obligate bacterial pathogens of humans, such as *Neisseria* spp. or *Chlamydia* spp., *L. monocytogenes* exercises multiple lifestyles that are adapted for different environmental situations. Consistent with this greater breadth of habitat, the genome size of *L. monocytogenes* (~3 Mb *[3]*) is larger than those of the aforementioned obligate pathogens *(4–6)*.

Much of our current knowledge of the functions of individual genes in *L. monocytogenes* has focused on those involved in virulence in animal models, and probably humans

From: *Bacterial Genomes and Infectious Diseases*
Edited by: V. L. Chan, P. M. Sherman, and B. Bourke © Humana Press Inc., Totowa, NJ

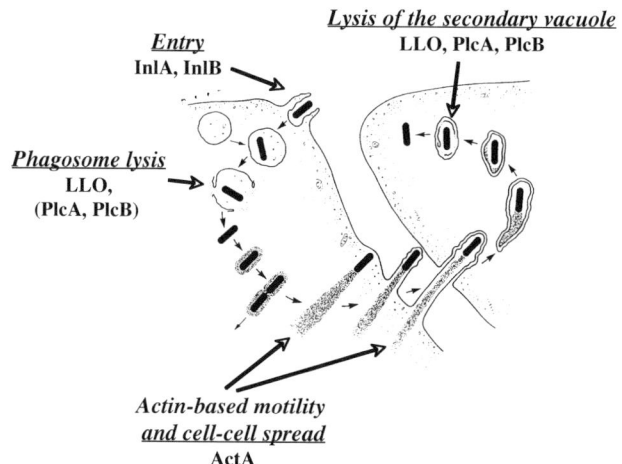

Fig. 1. The intracellular life cycle of *Listeria monocytogenes*. Bacterial virulence factors are indicated next to the stage of infection that they control. (Adapted from ref. *9*, with permission from The Rockefeller University Press.)

(1,7). Originally, classic "phenotype-driven" genetic screens were performed to identify bacterial virulence genes *(1)*, and genetic, biochemical, or immunological approaches were employed to define mouse or human proteins that influence host susceptibility to infection *(8)*. The recent availability of the nucleotide sequences of the genomes of *L. monocytogenes* and its human or animal hosts provides the potential for global analysis of bacterial and host factors that influence pathogenesis. This chapter begins with a brief introduction to the intracellular life cycle of *L. monocytogenes*, and the functions of virulence factors identified through classic genetic screens (**Subheading 8.2.**). Then, the general characteristics of the *L. monocytogenes* genome are described (**Subheading 8.3.**), followed by a discussion on how DNA sequence information has prompted genetic analysis of the roles of particular bacterial genes in virulence (**Subheading 8.4.**). Recent genomic-based studies of bacterial gene expression responses are covered in **Subheading 8.5.** Finally, at the end of the chapter, future genomic- or proteomic-based approaches that might improve or expand on current work are highlighted (**Subheading 8.6.**).

2. Intracellular Life Cycle of *L. monocytogenes*

L. monocytogenes has the ability to penetrate host mammalian cells, to replicate within cells, and to spread from one cell to another while always remaining in the host cytosol. The various stages of the *L. monocytogenes* intracellular life cycle were originally characterized through transmission electron microscopy analysis of infected cultured mammalian cells *(9)*. Genetic approaches led to the identification of bacterial factors controlling the different steps in the life cycle *(1)*. A cartoon based on the seminal electron microscopy work by Tilney and Portnoy is presented in Fig. 1, with virulence proteins indicated next to the step(s) that they regulate. The life cycle is briefly described in **Subheading 2.1.** For a detailed discussion of molecular mechanisms of *Listeria* virulence proteins, the reader is referred to an excellent comprehensive review by Vazquez-Boland and coauthors *(1)*.

2.1. Bacterial Entry, Phagosome Dissolution, and Cell–Cell Spread

On interaction with host mammalian cells, *L. monocytogenes* is rapidly internalized into membranous structures called phagosomes. *Listeria* is engulfed by professional phagocytes, such as macrophages or neutrophils, and also by cells that are generally thought of as nonphagocytic, including epithelial and endothelial cells *(1)*. Internalization ("entry") into nonphagocytic cells requires active participation of both the bacterium and the host cell. The two major bacterial factors that mediate *Listeria* entry into mammalian cells are the surface proteins Internalin A (InlA) and InlB *(10)*. Each of these microbial proteins contains an amino-terminal leucine-rich repeat (LRR) domain that folds into a horseshoe-like structure, followed by an internal immunoglobulin-like inter-repeat (IR) region, and a carboxyl-terminal region that mediates cell surface anchoring *(10,11)*. The LRR regions in InlA and InlB interact with distinct mammalian surface receptors, the cell–cell adhesion molecule, E-cadherin *(12)*, or the receptor tyrosine kinase, Met *(13)*, respectively. Bacterial engagement of E-cadherin or Met promote entry through modulation of mammalian signaling pathways that lead to F-actin-driven remodeling of the host cell surface *(10,14,15)*.

Within approx 30 min after internalization, *L. monocytogenes* lyses the phagosome through the action of the secreted pore-forming hemolysin, listeriolysin O (LLO) *(16)*. In some mammalian cell lines, the secreted phospholipases, PlcA and PlcB, aid LLO in mediating phagosomal escape *(17)*. Once free in the cytosol, pathogens replicate extensively and use the bacterial surface protein, ActA, to recruit host actin into filaments that become organized into tail-like structures *(1,16)*. These "comet tails" propel pathogens through the cytosol, and allow cell–cell spread through formation of a protrusion that is engulfed by an adjacent cell. The engulfed bacterium is found in a double-membranous, "secondary vacuole," which it rapidly dissolves through the concerted action of LLO, PlcA, and PlcB *(18,19)*. Having again accessed the cytosol, *L. monocytogenes* repeats its cycle of replication, motility, and cell–cell spread. With the exception of *inlA* and *inlB*, expression of the genes encoding each of the previously described virulence proteins requires the bacterial transcription factor PrfA *(20,21)*. PrfA directly activates transcription of promoters of these genes by binding to an upstream sequence called a PrfA box. PrfA activity is upregulated in the phagosome and cytosol of host cells, ensuring that maximal synthesis of virulence proteins occurs in the intracellular environment in which they exert their functions.

2.2. Role of Bacterial Virulence Genes in Listeriosis

The intracellular life cycle of *L. monocytogenes* observed in cultured cells is thought to play an essential role in colonization of host tissues during human listeriosis *(1,10, 22)*. Gastroenteritis likely arises from damage resulting from *L. monocytogenes* internalization and proliferation within cells lining the lumen of the small intestine, with a subsequent inflammatory response. In systemic infections, bacterial traversal of the intestinal cell barrier is probably followed by replication in macrophages and possibly hepatocytes, with blood-borne spread to the central nervous systems (CNS) or placenta *(1)*.

Work with animal models indicate that the bacterial genes depicted in Fig. 1 play key roles in virulence. InlA/E–cadherin interaction appears to mediate translocation across the intestinal cell barrier, based on experiments with guinea pigs or transgenic mice expressing human E-cadherin *(23)*. InlB is needed for infection of hepatocytes and efficient colonization of the liver in mice (24,25). Bacterial propagation in the mouse liver, spleen, and other organs also requires LLO, ActA, and PrfA (1), indicating critical roles for intracellular replication and cell–cell spread. In contrast to these virulence factors, PlcA and PlcB have redundant functions in colonization, which reflect the overlapping roles for these phospholipases in mediating escape from primary and secondary phagosomes (18).

In addition to promoting infection of the liver, InlB might also contribute to CNS infections by inducing internalization into brain microvascular endothelial cells (1,22). Alternatively, or additionally, infections of the CNS could be initiated by ActA-mediated spread of bacteria inside phagocytes to endothelium, or by transmigration of infected leukocytes across the endothelial barrier (22,26). Most abortions are thought to arise from L. monocytogenes infection of placental trophoblast cells, followed by entry into the fetal blood stream. Although InlA and InlB promote internalization into primary human trophoblast cultures or placental villous explants (27,28), neither of these bacterial surface proteins play a detectable role in traversal of the fetal–placental barrier in pregnant guinea pigs (28). These findings suggest that infection of trophoblasts in vivo may occur primarily through ActA-mediated cell–cell spread from maternal macrophages.

3. Listeria Genomes

3.1. L. monocytogenes Strain, EGDe and the Nonpathogenic L. innocua Strain, CLIP11262

Although phenotype-driven genetic screens were instrumental in the identification and characterization of the L. monocytogenes virulence factors depicted in Fig. 1 (1,7), the determination of the complete sequence of the L. monocytogenes genome in 2001 ([3] http://genolist.pasteur.fr/ListiList/genome.cgi) has significantly impacted the way in which gene function is being studied. Many laboratories have used DNA sequence information to provide clues as to the potential role of particular genes in pathogenesis. One trend has been to identify genes encoding proteins with sequence similarity to known virulence factors or proteins that are predicted to regulate the localization or activity of previously identified virulence factors. Another strategy, termed "comparative genomics" takes advantage of the availability of the genome sequence of the nonpathogenic Listeria species, L. innocua (3). The logic behind this strategy is that genes present in L. monocytogenes, but absent in L. innocua, might be involved in pathogenesis.

The genomes of the L. monocytogenes strain EGDe (serovar 1/2a) and the L. innocua strain CLIP11262 (serovar 6A) are similar in size (~3 Mb), overall guanine and cytosine content (~40%), gene identity, and gene organization. About 90% of the genes present in L. monocytogenes EGDe are also found in L. innocua, and approx 95% of the genes found in L. innocua are conserved in L. monocytogenes. These findings indicate that the difference in pathogenicity between these two Listeria species is owing to a relatively small number of genes representing a minor proportion of the genome.

A subset of the 270 EGDe-specific genes are likely to contribute to pathogenesis (3,7). Among these are the known L. monocytogenes virulence genes depicted in Fig. 1. In addition to InlA, InlB, and ActA, 26 of the EGDe-specific genes encode putative cell surface proteins. Strikingly, 9 of the 26 surface-anchored proteins contain LRR and IR domains that have sequence similarity to those in InlA and InlB (3,29,30). In addition to these surface proteins, three putative secreted proteins with LRR and IR domains are present in EGDe and absent from *L. innocua (31,32)*. Together with InlA and InlB, these surface and secreted LRR proteins comprise the internalin family *(11,33)*. LRR proteins are widespread in nature, being found in prokaryotic and eukaryotic proteins of diverse function *(34)*. A feature common to virtually all LRR domains is that they mediate protein–protein interactions. However, without exception, different LRR proteins have distinct ligands, indicating that the "horseshoe" structure common to these proteins serves as a scaffold on which specific amino acid clusters are presented for interaction. The multitude of EGDe-specific LRR proteins suggests that internalins, in addition to InlA and InlB, likely play important roles in pathogenesis. This notion has already been confirmed for a small proportion of these *L. monocytogenes* LRR proteins *(30,31,35)*. Apart from internalins and other cell surface or secreted factors, other classes of genes present in EGDe and absent in *L. innocua* include those mediating transport or metabolism of carbohydrates or other nutrients, and transcription factors.

3.2. *Other* L. monocytogenes *Strains*

The genome of EGDe was sequenced because it is arguably the most commonly used strain for molecular virulence and immunological studies. However, EGDe was isolated from an outbreak in rabbits *(1)* and it is not clear if this strain would be highly virulent for humans. *L. monocytogenes* strains are classified into 13 serotypes, which fall into distinct evolutionary lineages *(36,37)*. Serotypes 1/2a, 1/2c, 3a, and 3c are in lineage I; 4b, 4d, 4e, 1/2b, and 3b serotypes occupy lineage II; and 4a and 4c strains comprise lineage III. Interestingly, certain serotypes appear to have greater pathogenic potential than others *(38)*. The majority of epidemic and sporadic cases are associated with 4b strains, despite the fact that 1/2a and 1/2c strains are commonly found in food and the environment *(36)*. Based on this data, the 1/2a strain EGDe probably does not have the full pathogenic potential manifested by 4b strains isolated from food outbreaks.

In an effort to identify bacterial proteins that might contribute to human epidemics, two different groups determined the nucleotide sequences of the genomes of three serotype 4b strains associated with listeriosis outbreaks, and one 1/2a strain responsible for a sporadic illness *(36,37)*. Comparison of genes present in the newly sequenced 1/2a strain and EGDe, and the three 4b strains revealed that individual strains of *L. monocytogenes* are surprisingly genetically diverse, with 2–5% (50–140) of genes being strain-specific. This is approximately the same level of divergence observed between EGDe and *L. innocua* CLIP11262. In one of the studies, DNA microarrays containing EGDe genes were used to screen 93 different *L. monocytogenes* strains of various serotypes, including 25 1/2a and 22 4b strains *(36)*. As expected, genes encoding the major virulence factors characterized in EGDe (InlA, InlB, LLO, ActA, PlcA, PlcB) were conserved among all of the 1/2a or 4b strains analyzed. These findings support the idea that these proteins constitute "core" virulence factors critical for pathogenesis in humans.

The study also led to a more precise determination of genes specific to 1/2a or 4b strains *(36)*. Nineteen genes present in all 1/2a strains and absent in all 4b strains were found. Six of these genes encode potential surface proteins, four of which also contain LRR domains likely to have structures similar to those of InlA and InlB. Transport proteins represent another class of 1/2a-specific genes, with a putative oligopeptide permease ABC transporter and four phosphoenol pyruvate-dependent transport system enzyme II components being absent in 4b strains. Eight genes found in 4b strains, but absent in

1/2a strains, were identified in the microarray study *(36)*. Similar to the situation with 1/2a-specific factors, three of the 4b-specific proteins are surface-anchored LRR proteins. Other proteins present specifically in 4b strains include putative transcription factors and proteins whose function cannot be easily predicted. Based on the microarray data, it would appear that some of the differences in pathogenicity thought to exist between 1/2a and 4b strains is because of a distinct repertoire of LRR and other surface proteins.

3.3. Identification of Virulence Genes Through Comparative Genomics

Given that elucidation of the genomes of 4b strains is very recent, the roles of 4b-specific genes in pathogenesis have yet to be addressed. In contrast, use of the EGDe and *L. innocua* CLIP11262 genomes has already led to the identification of several bacterial genes that affect virulence in animal models. In **Subheading 8.4.**, examples in which EGDe-specific proteins belonging to some of these categories were found to play important roles in pathogenesis are discussed. It should be noted, however, that the presence of a particular gene in both *Listeria* species does not exclude the possibility that it might encode a virulence factor, particularly if the corresponding protein participates in fundamental processes, such as protein secretion or anchoring. Examples of genes conserved among *Listeria* species that contribute to *L. monocytogenes* pathogenesis are also presented.

4. Use of Genomics to Identify *L. monocytogenes* Virulence Proteins

4.1. Surface or Secreted Proteins

4.1.1. Translocation Across the Plasma Membrane

With the exception of the transcription factor PrfA, all of the well-studied *Listeria* virulence factors presented in Fig. 1 are anchored to the bacterial surface or secreted into the external environment, where they interact with host cell membranes or proteins. Each of these *Listeria* proteins contains an amino-terminal signal peptide (SP) sequence typical of substrates of the classical Sec-dependent secretion pathway *(39)*. The genome of *L. monocytogenes* EGDe encodes 219 SP-containing proteins *(3)*, indicating that at least 8% of all proteins are likely to be exported through a Sec apparatus. In *Escherichia coli*, transport through the Sec pathway is driven by the SecA ATPase, which provides energy for translocation and also presents SP-containing proteins to an integral membrane channel composed, in part, of SecY and SecE. *E. coli* and most other bacteria have single copies of these *sec* genes. Interestingly, *L. monocytogenes* and *L. innocua* encode two SecA homologs, SecA1 and SecA2 *(40)*. SecA1 is probably essential *(40)*, and is thought to be needed for export of the majority of SP-containing proteins, including InlA, InlB, LLO, PlcA, PlcB, and ActA. SecA2 is nonessential, required for virulence in mice, and appears to mediate secretion of a smaller subset of proteins, including the cell wall hydrolases p60 and *N*-acetylmuramidase, the fibronectin-binding protein/adhesin, FpbA, and several lipoproteins *(41,42)*. Surprisingly, in contrast to SecA1 substrates, some proteins that are exported in a SecA2-dependent manner lack recognizable SP sequences *(41,42)*. It is not clear how these secreted proteins are recognized by the SecA2 apparatus.

4.1.2. Anchoring of Surface Proteins

After export by the Sec machinery, proteins are either localized to the cell surface or released into the extracellular milieu. Release (secretion) appears to be a default pathway that occurs when a particular protein lacks an anchoring region. There are four different mechanisms, defined by the nature of the anchoring motif, that mediate surface association of proteins in *Listeria* species *(33)*.

4.2. Hydrophobic Membrane Anchor

Some surface proteins have a carboxyl-terminal hydrophobic region that is inserted in the cytoplasmic membrane, followed by a positively charged, cytosolic "stop transfer" sequence. The protein's amino-terminus projects through the peptidoglycan *(33)*. ActA is anchored to the bacterial surface through such a hydrophobic tail-based mechanism, and examination of the EGDe genome identified 10 additional proteins that are likely to have hydrophobic sequences embedded in the cytoplasmic membrane *(33)*. The roles of these other surface proteins in virulence remain to be determined.

4.3. Linkage to the Cell Wall

A second general anchoring mechanism occurs through covalent linkage to the peptidoglycan. Many surface proteins of various Gram-positive bacterial species contain a carboxyl-terminal LPXTG sequence (where X is any amino acid) that undergoes enzymatic cleavage between the threonine and glycine residues, followed by linkage of the carboxyl group of the threonine to meso-diaminopimelic acid in the cell wall *(43)*. The first protein to be demonstrated to be anchored by such a mechanism was *Staphyloccus aureus* protein A. There are 17–21 additional LPXTG-containing proteins in *S. aureus* strains. The staphylococcal enzyme that promotes anchoring of LPXTG sequences is called Sortase A *(43)*. *S. aureus* also encodes a related enzyme, Sortase B, which mediates cell wall linkage of the carboxyl-terminal motif, NPQTN. In contrast to Sortase A, staphylococcal Sortase B may have only one substrate, IsdC *(43,44)*.

Genes encoding proteins similar to Sortases A and B were recently identified in the EGDe and *L. innocua* genomes *(45–47)*. Like *S. aureus*, *Listeria* species exhibit a disproportionality in Sortase A and Sortase B substrates. *L. monocytogenes* InlA contains an LPXTG motif that mediates its cell surface localization *(48)*. In addition to InlA, an astounding 40 other LPXTG-containing proteins are encoded in the EGDe genome *(3,33)*. This number represents about 30% of all surface proteins in this strain. Nineteen of these LPXTG proteins contain LRR domains, and 11 of the LPXTG/LRR proteins are absent from *L. innocua (3,29,30)*. *L. monocytogenes* Sortase B is thought to recognize a carboxyl-terminal NXXTN sequence, and might have only two substrates, SvpA and Lmo2186 *(47,49)*. Both of these *Listeria* proteins exhibit significant amino acid similarity to staphylococcal IsdC *(47)*.

In *S. aureus*, Sortase A plays an important role in virulence *(50)*, underscoring the importance of LPXTG proteins in colonization of host tissues by this pathogen. In *L. monocytogenes*, inactivation of *srtA* caused the expected elimination in cell surface anchoring of InlA and several other LPXTG-containing proteins, and abolished InlA-dependent entry into host cells *(45)*. The *srtA* deletion also resulted in a virulence defect in intravenously or orally inoculated mice. It is noteworthy that the *srtA* mutant was

more attenuated than an *inlA* deletion mutant, which does not exhibit a virulence phenotype unless mice are genetically modified to express human E-cadherin *(23,24)*. Taken
together, these results indicate that, in addition to InlA, other LPXTG-containing proteins influence pathogenesis by acting at a step subsequent to E-cadherin-mediated traversal of the intestinal cell barrier.

In *S. aureus*, Sortase B is dispensable for initial host colonization, and is instead
needed for persistence in host tissues *(44)*. Sortase B and its substrate IsdC promote
heme–iron acquisition, and it is thought that the virulence defect of *srtB* mutants may
be because of impaired scavenging of iron in the host *(43,51)*. Despite the fact that the
putative *L. monocytogenes* substrates, Lmo2186 and SvpA, have sequence similarity
to IsdC *(47)*, these proteins do not have a detectable role in heme–iron utilization in *L.
monocytogenes (49)*. Moreover, deletion of *svpA* or *srtB* does not substantially alter the
median lethal dose (LD50) in intravenously inoculated mice *(47,49)*, suggesting that
surface anchoring by SrtB does not play an essential role in *Listeria* pathogenesis.

4.4. Anchoring to Lipids

A third mechanism of cell surface association is mediated by covalent linkage to lipids
in the bacterial cell membrane *(33)*. Prolipoproteins contain signal peptide sequences
that differ from those in other secreted proteins, chiefly in the presence of a cysteine residue immediately following the site of cleavage. The thiol group in this cysteine is linked
to an *N*-acyl diglyceride group of a glycerophospholipid by the action of an enzyme
called prolipoprotein diacylglyceryl transferase (Lgt). After linkage of the cysteine, the
signal pepide is removed by the type II signal peptidase, SPase II.

The EGDe genome encodes 68 predicted lipoproteins, representing approx 2.5% of
all proteins, and approx 50% of all surface proteins *(3,33)*. Twenty-four (~35%) of the
lipoproteins are substrate-binding components of ABC transporter systems predicted
to promote uptake of sugars, metals, amino acids, or peptides *(3,52)*. OppA, a lipoprotein
that mediates import of oligopeptides, plays a minor role in the growth of *L. monocytogenes* in host mammalian cells *(53)*, suggesting that cytosolic bacteria might use
peptides as a nutrient source. However, bacteria deleted for *oppA* are unaffected for virulence in mice, possibly because of the presence of multiple ABC transporters capable
of importing di- or oligopeptides, or because host sugars can also be utilized as a food
source (*see* **Subheading 4.6.**). The roles of other lipoprotein components of ABC transporters in pathogenesis have not been addressed. LpeA, a lipoprotein with homology
to the PsaA adhesin of *Streptococcus pneumoniae (54)* was found to be required for
internalization of *L. monocytogenes* into epithelial cell lines, but not for virulence in
mice *(55)*. In an effort to determine if lipoprotein maturation is essential for virulence,
a gene (*lsp*) encoding a protein with homology to bacterial SPase II enzymes was identified in the genomes of *L. monocytogenes* EGDe and *L. innocua*. Inactivation of *lsp* in
L. monocytogenes resulted in a partial inhibition in maturation of LpeA, and a minor
virulence defect in mice, perhaps because of an impairment in phagosomal escape *(52)*.
Given the incomplete effect of the *lsp* mutation on SvpA processing, it is difficult to
conclude from these studies whether lipoprotein maturation is critical for *L. monocytogenes* pathogenesis. Importantly, a gene (*lmo1101*) encoding a second putative lipoprotein signal peptidase is present in the *L. monocytogenes* genome, but absent in *L. innocua*
(3). In future work, it would appear worthwhile to determine the effect of inactivation

of this gene, alone and in combination with *lsp*, on lipoprotein maturation and bacterial virulence. Another consideration is that removal of signal peptides might not be important for pathogenic properties of lipoproteins; perhaps it is only crucial that these proteins be anchored to the cytoplasmic membrane. Deletion of the *L. monocytogenes lgt* gene, encoding prolipoprotein diacylglyceryl transferase, might help determine the role of lipoprotein anchoring in virulence.

4.5. Electrostatic Anchoring of Proteins to Lipotechoic Acid

The entry-promoting protein, InlB, is bound to the bacterial surface through interaction of its positively charged carboxyl-terminal domain with negatively charged lipotechoic acid present in the outer leaflet of the cell membrane *(10,33)*. The InlB anchoring domain is comprised of three conserved, approx 80 amino acid "GW" modules containing the dipeptide glycine–tryptophan. Apart from InlB, EGDe contains seven other proteins with GW modules, with six of these also being present in *L. innocua (3,33)*. One of the species-conserved GW proteins, Ami, promotes adhesion to mammalian cells through its carboxyl-terminal GW modules *(56)*. Ami contains an amino-terminal autolysin domain, which is also present in the related GW protein, Auto *(57)*. Auto is absent from *L. innocua*, appears to aid InlA- and InlB-dependent entry into host cells, and is needed for efficient colonization of the intestine and liver of orally inoculated guinea pigs. Based on the findings with Ami and entry, it would be interesting to determine if the GW domain of Auto is sufficient for its role in mammalian cell invasion.

4.5.1. Export of Proteins to the External Environment (Secretion)

At least 105 proteins encoded in the EGDe genome are likely to be fully secreted, as judged by the presence of an SP sequence and the absence of a recognizable anchoring motif. This number represents about 4% of the *L. monocytogenes* proteome. The best characterized of the signal peptide-containing secreted proteins are the virulence factors LLO, PlcA, PlcB, and a metalloprotease (Mpl) that mediates PlcB processing *(1)*. Another secreted protein demonstrated to have a role in virulence in mice is the LRR protein, InlC *(31)*. The *inlC* gene is transcribed at low levels outside of host cells, and its expression is strongly induced on internalization of bacteria *(31,58)*. *inlC* expression is directly activated by the transcription factor PrfA *(31,59)*.

Despite the intriguing data on InlC expression, the mechanism by which this protein affects pathogenesis is not clear. InlC is dispensable for phagosome escape, intracytosolic replication, or cell-to-cell spread in all cell lines tested thus far *(31,60)*. A recent report suggests that InlC might aid InlA-dependent entry into host cells *(61)*. However, the fact that *inlC* expression is strongly upregulated inside host cells suggests that this LRR protein probably also influences postinternalization events. It is tempting to speculate that, like InlA and InlB, InlC may also promote virulence by interacting with one or more mammalian binding partners. In the case of InlC, the expectation is that at least one of its ligands is intracellular. The identification of InlC binding partners will likely provide important clues as to the mechanism(s) by which this protein affects pathogenesis.

In addition to InlC, LLO, PlcA, PlcB, and Mpl, the genome of EGDe encodes 18 other secreted proteins that are absent from *L. innocua (3)* Presently, the role of these secreted proteins in virulence is unknown. Interestingly, two of the uncharacterized secreted

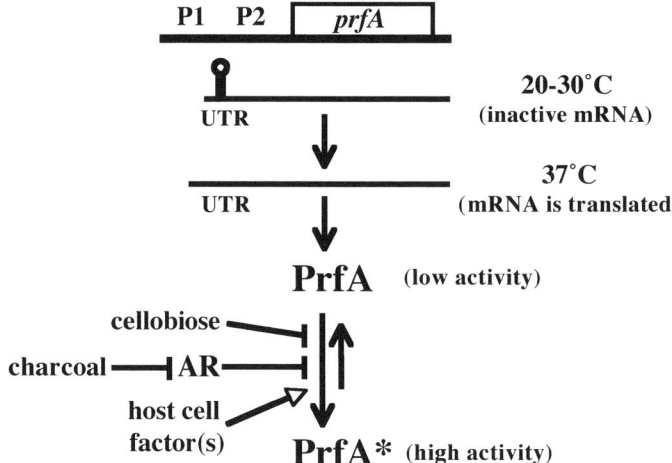

Fig. 2. Multiple levels of regulation of PrfA activity. At low temperatures (20–30°C) typically encountered in the soil, PrfA translation is inhibited owing to occlusion of the ribosome binding site (RBS) in a stem-loop structure formed by the 5' untranslated region (UTR) of the mRNA transcript originating from the P1 promoter or the upstream *plcA* promoter (not shown) *(74)*. At 37°C, the secondary structure in the UTR is melted out, rendering the RBS accessible for translation. Transcripts directed from the P2 promoter lack the stem-loop structure; however, expression from this promoter is low when PrfA is active *(20)*. PrfA protein is regulated at the post-translational level by at least three different conditions. First, readily metabolizable, nonphosphorylated sugars, including the plant-derived dissacharride, cellobiose, inhibit PrfA activity through an undefined mechanism *(20)*. Second, the presence of activated charcoal in the bacterial growth medium enhances PrfA activity through sequestration of an unidentified bacterial "autorepressor" (AR) *(76)*. Finally, one or more host factors stimulate PrfA activity *(78)*.

proteins are members of the LRR family, raising the possibility that they could interact with host factors.

4.6. Proteins Involved in Nutrient Acquisition or Metabolism

A genomic approach was recently used to elucidate mechanisms by which *L. monocytogenes* obtains nutrients for replication in the mammalian cell cytosol *(60,62)*. The intriguing observation that catabolism of hexose phosphates during growth in broth is PrfA-dependent *(62,63)* prompted a search for PrfA-regulated genes that might mediate sugar transport. A scan of the *L. monocytogenes* genome for open reading frames containing an upstream PrfA box led to the identification of a gene encoding a putative hexose phosphate transporter (Hpt) with sequence similarity to a known mammalian glucose-6-phosphate translocase *(62)*. Hpt is absent from *L. innocua*. Subsequent experiments demonstrated that Hpt is needed for utilization of hexose phosphates (but not unphosphorylated sugars) in broth, efficient bacterial replication within host cells, and for virulence in mice. Taken together, these results provide compelling evidence that one or more host hexose phosphate(s) fuels cytosolic growth of *L. monocytogenes*. Glucose-1-phosphate, the precursor and breakdown product of glycogen, is a plausible candidate for the growth-promoting sugar. Glycogen is abundant in the cytosol of hepatocytes *(62,63)*, and *hpt* deletion mutants are defective in colonization of the liver of infected mice *(62)*.

4.7. Transcriptional Regulators

The EGDe genome encodes 24 putative or known transcription factors that are absent from *L. innocua* CLIP11262 *(3)*. Based on amino acid sequence, these transcriptional regulators can be placed into one of several families, including GntR, BglG, Xre, AraC, TetR, LysR, Fur, the CAP/Fnr family, and the response regulator (RR) proteins of two component-systems. Of all these proteins, only functions of the CAP/Fnr family member PrfA and the RRs CesR, LisR, and DegU have been determined. CesR, LisR, and DegU are all needed for efficient colonization of the spleen in intragastrically inoculated mice *(64–66)*. CesR and its cognate histidine protein kinase (HPK), CesK, promote resistance to β-lactam antibiotics, and the virulence defect of the CesR/CesK mutants may reflect a role for this two-component system in controlling cell wall integrity during infection. LisR and its partner HPK, LisK, might also affect the cell wall, because these proteins influence resistance to cephalosporins and the lantabiotic, nisin *(67)*. DegU is a member of a two-component system that promotes the development of genetic competence, degradative enzyme synthesis, and motility in the Gram-positive bacterium *Bacillus subtilis (68)*. *L. monocytogenes* DegU is required for flagellin gene expression and motility *(66)*. However, the fact that flagella are not required for virulence in mice *(66)* suggests that DegU promotes bacterial infections through a process distinct from motility. The transcription factor PrfA has been extensively studied since its discovery in 1991, and has been found to play a critical role in regulation of virulence gene expression *(20)*. **Subheading 4.8.** describes what is known about how the activity of PrfA is controlled, and **Subheading 5.** discusses results from a recent transcriptional profiling study that led to the identification of several classes of PrfA-controlled genes.

4.8. Regulation of PrfA Activity

The virulence genes *hlyA* (encoding LLO), *plcA*, *mpl*, *actA*, *plcB*, and *inlC* are all highly dependent on PrfA for their transcription *(20,21)*. PrfA activity is upregulated on bacterial entry into host cells, resulting in a large induction in expression of these genes *(20)*. Unlike the above genes, which have functions inside mammalian cells, the *inlA* and *inlB* genes are only partly dependent on PrfA for expression. *inlA* and *inlB* comprise an operon with multiple promoters, only one of which is controlled by PrfA *(69,70)*. PrfA-independent expression of InlA and InlB makes intuitive sense, given that their role in internalization into host cells precedes PrfA activation in the cytosol.

PrfA has structural and functional similarity to the Catabolite Activator Protein (CAP)/Fnr family of transcription factors *(20,71,72)*. PrfA boxes located upstream of target promoters conform to the consensus palindromic sequence TTAACANNTGTTAA, where N is any nucleotide *(20)*. The *hly* and *plcA* promoters have "perfect" consensus PrfA boxes, whereas promoters for the remaining virulence genes have PrfA binding sites containing one or two base pair mismatches relative to the consensus. Promoters with perfect consensus PrfA binding sites appear to have higher affinity for PrfA than those with imperfect palindromes *(71)*. In all PrfA-regulated target genes studied to date, the PrfA box is centered approx 41 nucleotides upstream of the transcriptional start site, a situation that mirrors the position of CAP binding sites in so-called class II CAP-dependent promoters *(72)*. CAP-mediated transcriptional activation of class II promoters is known to involve interactions between the transcription factor and the α and σ^{70} subunits of

RNA polymerase (RNAP) holoenzyme *(72,73)*. It is possible that PrfA controls transcription through a similar mechanism.

PrfA is regulated by multiple mechanisms that allow increased transcriptional activity in the infected host (Fig. 2). One mechanism is controlled by temperature *(74)*. At low temperatures typical of growth in the soil (less than 30°C), translation of *prfA* mRNA is inhibited. This inhibition is probably because of occlusion of the ribosome binding site in a folded structure formed by the 5' untranslated region (UTR) of the *prfA* mRNA. Higher temperatures encountered in host tissues (37°C) cause unfolding of the 5' UTR, resulting in access to the ribosome binding site and derepression of translation.

PrfA is also regulated at the post-translational level by several environmental, bacterial, and host-derived factors. Experiments with constitutively activated mutants of PrfA suggest that these factors control conformational changes that regulate the DNA binding activity of this transcriptional regulator *(20,71,75)*. Readily metabolized, nonphosphorylated sugars including glucose, fructose, mannose, and β-glucosides, such as cellobiose, all strongly inhibit PrfA activity *(20)*. In contrast, hexose phosphates, including the Hpt substrate glucose-1-phosphate, fail to impair PrfA activity *(62,63)*. Interestingly, cellobiose is derived from plants and this disaccharide may serve the dual function of sustaining *Listeria* growth on decaying vegetation, while simultaneously preventing wasteful and inappropriate expression of PrfA-dependent virulence genes. Another mode of PrfA inhibition involves a diffusible molecule that is made by *L. monocytogenes* and secreted into culture supernatants *(76)*. This proteinaceous autorepressor (AR) can be removed by addition of activated charcoal to the growth medium. AR might regulate PrfA by direct interaction with the transcription factor or, alternatively, could antagonize the function of a putative PrfA-activating factor *(77)*. Like cellobiose, AR could act to restrain PrfA activity during the saprophytic lifestyle of *L. monocytogenes*. On internalization of individual bacteria into mammalian cell phagosomes and release into the cytosol, AR would likely be diluted, resulting in increased PrfA activity. Finally, the activation state of PrfA also appears to be controlled by one or more mammalian cell protein(s) *(78)*. Host surface proteins could be involved, because upregulation of PrfA activity was found to require bacterial adhesion, but not internalization. Together, the multiple modes of regulation of PrfA ensure that the transcription factor is fully active only when *L. monocyogenes* infects host cells.

5. Use of Genomics or Proteomics to Characterize Changes in Bacterial Gene or Protein Expression

5.1. Identification of PrfA-Controlled Target Genes

Availability of the *L. monocytogenes* genome sequence has made it possible to analyze transcription of all known or predicted open reading frames in response to specific genetic and/or environmental conditions. Buchrieser and colleagues used DNA arrays containing 99% of all predicted open reading frames in the EGDe genome to identify genes that are differentially expressed in wild-type and *prfA*-deletion mutant strains *(21)*. Gene expression was monitored in the presence or absence of charcoal or cellobiose in order to assess the effect of conditions that enhance or inhibit PrfA activity. Seventy PrfA-regulated genes organized in 47 predicted transcription units were identified. These genes fell into three distinct groups, based on whether PrfA augmented or impaired transcription, and on the influence of charcoal or cellobiose on regulation.

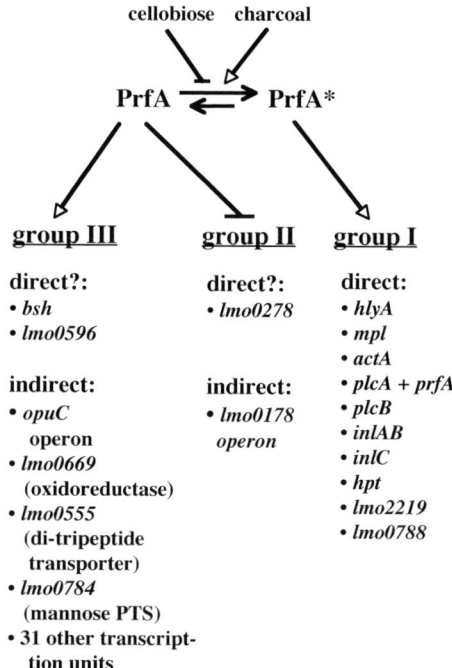

Fig. 3. Three classes of PrfA-regulated genes. Three categories of PrfA-controlled genes were identified in DNA microarray study performed by Buchrieser and colleagues *(21)*. Group I genes include nine well-characterized virulence genes whose expression was previously known to be promoted by PrfA, as well as two novel genes (*lmo2219* and *lmo0788*) of undefined function. Expression of these genes is enhanced by charcoal and impaired by cellobiose, suggesting that a high activity form of PrfA (PrfA*) directs their transcription. Surprisingly, cellobiose did not alter expression of the two group II and 35 group III transcription units identified in the study. These results raise the possibility that a low activity form of PrfA might direct transcription of these genes. Expression of group II genes was repressed by PrfA, whereas transcription of group III genes was enhanced. "Direct?" indicates the presence of an upstream PrfA box, and the possibility the PrfA directly regulates gene expression. "Indirect" indicates that absence of a predicted PrfA-binding site making it improbable that the transcription unit is directly controlled by PrfA.

Altogether, the results suggest the existence of at least two different activation states of PrfA, each of which regulates expression of distinct sets of target genes (Fig. 3).

Group I genes were directly upregulated by PrfA, in a manner enhanced by charcoal and diminished by cellobiose. This group contains key virulence genes already known to be under direct PrfA control, including *hlyA*, *mpl*, *actA*, *plcA*, *plcB*, *inlA*, *inlB*, *inlC*, *hpt*, and *prfA* itself. Two genes of unknown function, *lmo2219* and *lmo0788*, also fall into group I. Based on the presence of upstream PrfA boxes and the distance between these binding sites and the predicted –10 promoter elements, it seems likely that *lmo2219* and *lmo0788* are also directly regulated by PrfA.

PrfA negatively regulates expression of group II genes (Fig. 3). Interestingly, PrfA-mediated repression was insensitive to charcoal or cellobiose, suggesting that regulation of these genes may not require fully active PrfA. Eight group II genes comprising two putative transcription units were identified. One transcription unit contains seven genes (*lmo0178–0184*), some of which encode substrate-binding and permease components of an ABC transport system predicted to import disaccharides, oligosaccharides, and polyols. The second transcription unit, *lmo0278*, is monocistronic and codes for the ATPase component of a sugar transporter. It is possible that the products of the two group II transcription units act together to mediate sugar import. *lmo0278* has a potential PrfA box located only about four nucleotides upstream of its –10 region. The proximity of these binding sites to the –10 might inhibit transcription by preventing binding of RNAP. In contrast to the situation with *lmo0278*, *lmo0178* lacks an upstream PrfA box near predicted –10 or –35 regions of the promoter, and the mechanism by which PrfA downregulates expression of this gene is unclear. It is also not obvious why it would be advantageous for *L. monocytogenes* to impair expression of this transport system through PrfA. One possible explanation might be that the group II genes encode a transporter of sugars that are readily available in the host cell cytosol and capable of inhibiting PrfA activity. In this case, repression of these genes would be crucial for the ability of PrfA to efficiently activate virulence gene expression in mammalian cells. This idea could be tested by identification of the specific carbohydrates transported by the group II gene products and determining the effects of these sugars on PrfA activity.

Group III genes were positively regulated by PrfA, in a manner that was insensitive to cellobiose (Fig. 3). Again, regulation of these genes might not require full PrfA activity. Fifty-three class III genes organized in 37 transcription units were found. Two of these genes, *bsh* and *lmo0596*, contain potential PrfA binding sites located 18–24 nucleotides upstream of the –10 regions of their promoters *(3,79)*. Thus, these genes are likely to be direct targets of PrfA. *bsh* codes for a bile salt hydrolase that plays an important role in virulence, presumably by protecting *L. monocytogenes* from toxic effects of bile salts in the duodenum and liver *(79)*. The recent and unexpected finding that *L. monocytogenes* colonizes the gall bladder in mice *(80)* suggests that *bsh* also might promote bacterial survival in this bile-producing organ. The function of *lmo0596* is unknown.

In contrast to the situation with *bsh* and *lmo0596*, the vast majority of group III genes lack upstream PrfA boxes. It seems likely that PrfA induces expression of one or more intermediary factors that then exert more direct effects on transcription (*see* **Subheading 5.2.**). The functions of about 70% of the indirectly regulated group III genes have been characterized or predicted based on sequence (Fig. 3). One group of proteins that have been extensively studied is the OpuC carnitine/betaine ABC transport system, which promotes growth in high salt conditions (osmotolerance) and is also needed for colonization of the small intestine and livers of infected mice *(81)*. OpuC might promote growth and survival of *L. monocytogenes* in the high salt/low water environment of many food products and also in the duodenum, which has an osmolarity equivalent to approx $0.3\ M$ NaCl *(81)*. The predicted functions of the remainder of the group III proteins include regulation of the cellular redox environment (oxidoreductase- Lmo0669), di- and/or tripeptide transport (Lmo0555), pyruvate-dependent transport system-mediated import of mannose, uptake of glucose or other sugars (Lmo169), and carbon catabolism (succinate semi-aldehyde dehydrogenase- Lmo0913; dihydroxyacetone kinase subunits- Lmo2695-2697). Although not identified in the microarray study, another transcription unit that may be classified as belonging to group III is the *bilE* operon *(82)*. This operon appears to be indirectly regulated by PrfA, and encodes a two-component bile exporter that promotes resistance to bile in vitro and intestinal colonization in mice. Taken together, the findings suggest that PrfA-mediated expression of group III genes may help

L. monocytogenes cope with toxicity associated with osmotic shock, oxidative stress, and bile salts, and perhaps also provide energy for growth in the mammalian cytosol.

5.2. Interaction Between the PrfA and σ^B Regulons

The DNA array study of PrfA-regulated genes revealed an unexpected interaction between regulons controlled by PrfA and the alternative sigma factor, σ^B. In *L. monocytogenes*, *Bacillus subtilis*, and other Gram-positive bacteria, σ^B is activated and promotes survival during multiple stress conditions, including starvation, acid, osmotic, heat, or oxidative shock *(83–86)*. The group I *inlAB* operon, all of the group II transcription units, and about 60% (22/37) of the group III transcription units, including the *opuC* and *bilE* operons, are predicted to contain promoters transcribed by RNAP containing σ^B *(21)*. Moreover, a DNA microarray study comparing gene expression in wild-type and *sigB* deletion strains of *L. monocytogenes* demonstrated that *inlAB*, *bsh*, *opuC*, *bilE*, and at least five of the other group III transcription units indicated in Fig. 3, are indeed controlled by σ^B *(82,87)*. Experiments with a *sigB* null mutant strain support the idea that σ^B-dependent expression of *inlA*, *bsh*, and *opuC* is critical for the function of these genes *(79,83,85,88–91)*. Finally, one of the three promoters (P2) known to drive PrfA expression has consensus σ^B –10 and –35 sites (Fig. 4A), and is transcribed by σ^B-containing RNAP in vitro *(92)*. σ^B-mediated transcription of PrfA from P2 could, at least in part, be responsible for the stimulatory effects of oxidative stress, high osmolarity, and severe heat shock (42–48°C) on PrfA activity *(20)*.

For genes with upstream PrfA boxes, PrfA and σ^B are likely to regulate transcription by acting at different promoters. Both *inlAB* and *bsh* contain multiple promoters *(69,79)*, and sequence information suggests that at least one promoter for each of these transcription units is probably transcribed by RNAP containing the major sigma factor, σ^A (Fig. 4B). Based on spacing between –10 sites and PrfA boxes, it seems likely that PrfA activates expression of the σ^A promoters for both *inlAB* and *bsh*, and that the σ^B promoters are not direct PrfA targets (Fig. 4B). The fact that PrfA has no influence on σ^B-directed transcription of *bsh* in vitro is consistent with this idea *(92)*. The existence of separate promoters for σ^B and PrfA could provide a mechanism for distinct environmental signals to regulate gene expression at different steps of infection. In the mildly acidic conditions (pH 4.5–6.5) and elevated osmolarity (~0.3 M NaCl) of the small intestine, σ^B-dependent transcription of *bsh* and/or *inlAB* could play an important role in acquisition of resistance to bile salts and/or enhancement of bacterial invasion. σ^B is probably uniquely suited to boost gene expression at this early step in infection, as PrfA is inhibited by low pH *(20)*. The contribution of PrfA to *inlAB* and *bsh* transcription might become more important during infection of the liver, where the pH of fluids is closer to neutral. It would be interesting to determine the relative contributions of the different promoters upstream of *inlAB* and *bsh* to intestinal and hepatic colonization using orally inoculated guinea pigs and/or transgenic mice expressing human E-cadherin *(23)*.

A central unresolved question is how PrfA promotes expression of the majority of group III genes, which lack identifiable upstream PrfA-binding sites. Three different models are presented in Fig. 4C. In two of the models, PrfA activates expression of a gene whose product (X) directly stimulates transcription at either σ^B- and/or σ^A-dependent promoters. In the third model, factor X stimulates σ^B activity. A particularly attractive

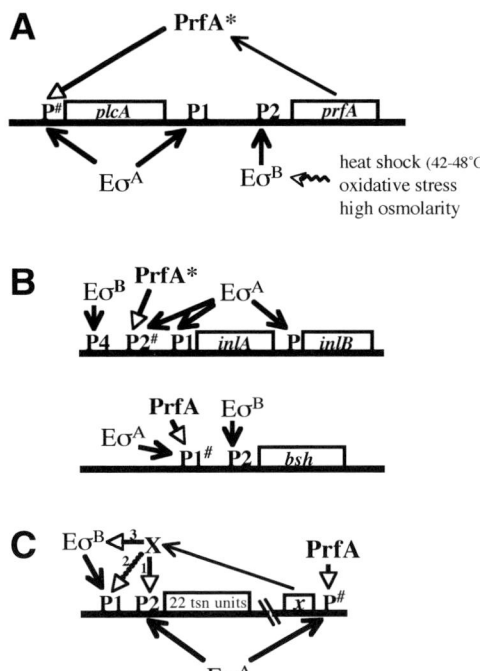

Fig. 4. Interaction between the PrfA and σB regulons (**A**) PrfA transcription. The three promoters directing PrfA expression are depicted. The # symbol indicates that PrfA directly activates transcription of the *plcA* promoter by binding to an upstream PrfA box. Activated PrfA (PrfA*) and RNA polymerase (RNAP) holoenzyme containing σ^B (EσB) control expression of distinct promoters. EσA indicates that transcription of the *plcA* and P1 promoters is directed by RNAP containing the major sigma factor, σ^A. It is worth noting that data in a recent report suggests that P2 is actually comprised of two overlapping promoters, one of which is transcribed by σ^B-containing RNAP and the other of which is recognized by σ^A *(92)*. For the sake of simplicity, only σ^B-dependent transcription is depicted. (**B**) Transcription of *inlAB* and *bsh*. The roles of PrfA and σ^B in controlling gene expression is illustrated. Similar to the situation with the *plcA–prfA* operon, PrfA and σ^B appear to regulate gene expression by acting at distinct promoters. The fact that *bsh* expression is insensitive to cellobiose *(21)* suggests that PrfA might not need to be fully active in order promote transcription of the gene. (**C**) Group III genes indirectly regulated by PrfA. PrfA is predicted to stimulate expression of one or more genes (*x*) that have more direct effects on group III gene expression. *x* could represent a single gene that controls transcription f all group III genes or, alternatively, several *x* genes could exist, each of which regulates expression of distinct subsets of group III transcription units. Three potential modes by which X might indirectly control gene expression are presented. X could be a transcription factor that activates transcription from σ^A- (1) or σ^B- (2) dependent promoters. Alternatively, X might enhance σ^B activity *(3)*.

candidate for "X" in model 3 is RsbV, an anti-antisigma factor that positively regulates σ^B activity in a variety of Gram-positive bacteria *(86)*. *rsbV* was identified as a group III gene in EGDe but, surprisingly, PrfA did not enhance *rsbV* expression in a different *L. monocytogenes* strain that has an almost identical profile of group III genes *(21)*. Hence, *rsbV* transcription alone cannot account for the PrfA-dependency of group III genes. Moreover, *rsbV* lacks upstream PrfA boxes, indicating that it is unlikely to be directly controlled by PrfA. Clearly, understanding the mechanism(s) by which PrfA affects group III gene expression will require identification of factor X. This could be accomplished by genetic screens to identify mutations that simultaneously affect expression of several group III genes. In addition, biochemical studies might lead to the identification of factor(s) that bind group III promoters, providing evidence for models 1 or 2.

6. Future Challenges of Genome-Based Approaches

Although the complete sequence of the *L. monocytogenes* genome has been available for less than 4 yr, it is already clear that genomics has made a large impact on *Listeria* research. Several studies have utilized genome sequence information as an indispensable tool to provide clues about whether a particular gene might be involved in virulence. Approaches include mutating *L. monocytogenes* genes that are absent in *L. innocua* and/or have sequence similarity to genes of known function from other bacteria, and the use of DNA microarrays to identify bacterial genes that might affect *Listeria* pathogenesis. In addition, although not discussed in this review, several recent proteomic studies have led to the identification of bacterial proteins that are secreted or differentially expressed under particular environmental conditions, such as starvation or biofilm formation *(41,90,93–96)*. On the side of the host organism, information on the human and mouse genomes has allowed the study of mammalian genes and proteins that may affect susceptibility to *Listeria (97–100)*.

The work outlined in this chapter represents the first generation of studies utilizing the power of genomics. Future work is likely to expand and improve on the initial studies in several respects. For example, transcriptional profiling experiments that more closely mirror in vivo physiological situations could be performed by analysis of mRNA isolated from infected host cells or mice. The use of bacterial strains deleted for genes encoding PrfA or σ^B could confirm, extend, or modify regulons that had been identified using artificial or simplified in vitro activating or repressing conditions, like growth in charcoal or cellobiose. Simultaneous analysis of host and bacterial gene expression through DNA microarrays has recently been performed in mice infected with a pathogenic *E. coli* strain *(101)*, and similar experiments could probably be performed with *L. monocytogenes*. Such studies could lead to identification of bacterial or host genes that are differentially expressed in various tissues, such as the intestinal epithelium, the liver, or the spleen.

Another likely future direction in *Listeria* genomics is the analysis of bacterial genes that have been previously neglected because the sequence of their protein products did not suggest a precise molecular function. One category of "neglected" genes is those whose proteins have signature motifs that allow them to be confidently placed into general functional classes, such as transcription factors, two-component HPKs or RRs, or LRR proteins. Although sequence information predicts that the corresponding proteins are likely to regulate transcription or promote protein–protein interactions, their exact responses and molecular targets are difficult if not impossible to divine. It will be important to systematically and exhaustively examine the roles of these proteins using functional genomic approaches that are currently being utilized to great effect with the budding yeast *Saccharomyces cerevisiae (102)*. For example, the functions of uncharacterized HPKs, RRs, or other transcription factors can begin to be probed by transcriptional profiling using DNA microarrays of the *L. monocytogenes* genome. In the case of RRs and other DNA binding proteins, the approach can be modified so as to identify target genes that are under direct control of the transcriptional regulator by performing chromatin immunoprecipitation (ChIP) of the DNA binding protein of interest, followed by hybridization to DNA microarrays (chip). This "ChIP-on-chip" approach has been recently employed to identify direct targets of DNA binding proteins in yeast *(102)* and bacteria *(103,104)*. Finally, the near universal involvement of LRR domains in protein–protein interactions *(34)* suggests that the function of an uncharacterized *Listeria* LRR protein can be effectively investigated through comprehensive analysis of protein complexes. This goal could be accomplished through two-hybrid-based screens of genomic libraries using the LRR protein as bait, or affinity purification of a tagged LRR protein followed by identification of copurifying ligands through mass spectrometry. Two-hybrid or mass spectometry strategies have been successfully used to comprehensively characterize protein complexes in the Gram-negative bacterial pathogen *Helicobacter pylori (105)* and in yeast *(102)*. However, it is vital to consider that the two *L. monocytogenes* LRR proteins that have been most extensively studied, InlA and InlB, bind to mammalian ligands. The eleven EGDe-specific LRR proteins whose binding partners have yet to be identified *(3)* could have bacterial and/or host binding partners. Hence a thorough investigation of a given LRR protein will require attempts to isolate binding partners from both the bacterial and mammalian proteomes.

Another future goal will be to use the genomes of serotype 4b epidemic strains to identify bacterial factors that may enhance pathogenic potential. Such factors might be present in 4b strains, but absent in other *L. monocytogenes* serotypes. One potential complication is that existing mouse or guinea pig animal models may not be appropriate for identification of auxillary virulence factors that augment pathogenicity in humans.

Recent advances in inhibition of mammalian gene expression through RNA interference (RNAi) technology suggest that it may become possible in the not-too-distant future to perform genome-wide screens to identify mouse or human genes that regulate various aspects of *Listeria* pathogenesis. A recent report describes a phenotypic screen using microarrays containing spots of mammalian cells in which expression of various target genes were "knocked down" through transfection with small interfering RNAs (siRNAs) *(106)*. Although only seven genes were targeted in the study, plans to assemble comprehensive RNAi collections for the human genome will likely make large-scale genetic screens a reality *(107)*. siRNA–mammalian cell microarrays could be used to elucidate many important events in *Listeria* pathogenesis. For example, a recent study described an innate immune response induced by cytosolic *L. monocytogenes* that results in induction of the β interferon gene and downstream interferon responsive genes *(100)*. When comprehensive RNAi microarrays are available, these could potentially be employed with a mammalian cell line expressing a luciferase reporter driven by the β interferon promoter to identify host genes that sense or transduce signals elicited by cytosolic bacteria. Similarly, siRNA-mammalian cell arrays could be infected with *L. monocytogenes* harboring a transcriptional reporter of a PrfA-controlled target gene, with the goal of finding host genes that stimulate PrfA activity *(78)*. In addition to cell microarrays containing siRNA, microarrays in which a cell line is subjected to a battery of chemical inhibitors have also been described *(106,107)*. In principle, such arrays could be used to screen natural or synthetic compounds for the ability to impair key events in *Listeria* virulence, such as bacterial internalization or actin-dependent movement.

Although several mammalian genes that affect susceptibility to *L. monocytogenes* infection have been identified through gene targeting *(8)*, transgenics *(23)*, or other approaches *(108)*, the list of host susceptibility factors is far from complete. Recently, genome-wide chemical mutagenesis of mice has been performed with the aim of performing unbiased screens for a variety of novel mutant phenotypes *(109,110)*. Efforts are underway to screen mouse mutants for susceptibility to *L. monocytogenes* and other bacterial pathogens (*[111]* http://www.mikrobio.med.tu-muenchen.de/forschung/enu.html).

Finally, in addition to animals or mammalian cell lines, other model systems might prove useful in genomic-based investigations of host susceptibility factors. For example, *L. monocytogenes* establishes infections in *Drosophila melanogaster* cell lines or whole organisms, in a manner that depends on LLO-mediated escape and ActA-dependent cell–cell spread *(112,113)*. The availability of *Drosophila* mutants and the amenability of *Drosophila* cell lines and animals to RNAi-mediated gene silencing may make this metazoan useful for identification of host proteins that play an evolutionarily conserved role in defense against intracellular pathogens. Clearly, genomic-based strategies will play an increasingly important role in investigations of mechanisms of virulence of *L. monocytogenes* and other bacterial pathogens. Of course, equally important is the use of classic genetic and biochemical approaches to thoroughly probe the function of any candidate virulence factor identified from genomics.

References

1. Vazquez-Boland, J. A., Kuhn, M., Berche, P., et al. (2001) *Listeria* pathogenesis and molecular virulence determinants. *Clin. Microbiol. Rev.* **14,** 584–640.
2. Mead, P. S., Slutsker, L., Dietz, V., et al. (1999) Food-related illness and death in the United States. *Emerg. Infec. Dis.* **5,** 607–625.
3. Glaser, P., Frangeul, L., Buchrieser, C., et al. (2001) Comparative genomics of Listeria species. *Science* **294,** 849–852.
4. Tettelin, H. E., Saunders, N. J., Heidelberg, J., et al. (2000) Complete genome sequence of *Neisseria meningitidis* serogroup B strain MC58. *Science* **287,** 1809–1815.
5. Parkhill, J., Achtman, M., James, D. D., et al. (2000) Complete DNA sequence of a serogroup A strain of *Neisseria meningitidis* Z2491. *Nature* **404,** 502–506.
6. Subtil, A. and Dautry-Varsat, A. (2004) Chlamydia: five years A.G. (after genome). *Curr. Opin. Microbiol.* **7,** 85–92.
7. Dussurget, O., Pizarro-Cerda, J., and Cossart, P. (2004) Molecular determinants of *Listeria monocytogenes* virulence. *Annu. Rev. Microbiol.* **58,** 587–610.
8. Pamer, E. (2004) Immune responses to *Listeria monocytogenes*. *Nat. Rev. Immunol.* **4,** 812–823.
9. Tilney, L. G. and Portnoy, D. A. (1989) Actin filaments and the growth, movement, and spread of the intracellular bacterial parasite *Listeria monocytogenes*. *J. Cell Biol.* **109,** 1597–1608.
10. Cossart, P., Pizarro-Cerda, J., and Lecuit, M. (2003) Invasion of mammalian cells by *Listeria monocytogenes*: functional mimicry to subvert cellular functions. *Trends Cell Biol.* **13,** 23–31.
11. Schubert, W. D. and Heinz, D. W. (2003) Structural aspects of adhesion to and invasion of host cells by the human pathogen *Listeria monocytogenes*. *Chembiochem.* **4,** 1285–1291.
12. Mengaud, J., Ohayon, H., Gounon, P., Mège, R. M., and Cossart, P. (1996) E-cadherin is the receptor for internalin, a surface protein required for entry of *Listeria monocytogenes* into epithelial cells. *Cell* **84,** 923–932.

13. Shen, Y., Naujokas, M., Park, M., and Ireton, K. (2000) InlB-dependent internalization of *Listeria* is mediated by the Met receptor tyrosine kinase. *Cell* **103**, 501–510.

14. Cossart, P. and Sansonetti, P. J. (2004) Bacterial invasion: the paradigms of enteroinvasive pathogens. *Science* **304**, 242–248.

15. Sun, H., Shen, Y., Dokainish, H., Holgado-Madruga, M., Wong, A., and Ireton, K. (2005) Host adaptor proteins Gab1 and CrkII promote InlB-dependent entry of *Listeria monocytogenes*. *Cell Microbiol.* **7**, 443–457.

16. Portnoy, D. A., Auerbuch, V., and Glomski, I. J. (2002) The cell biology of *Listeria monocytogenes* infection: the intersection of bacterial pathogenesis and cell-mediated immunity. *J. Cell Biol.* **158**, 409–414.

17. Marquis, H., Doshi, V., and Portnoy, D. A. (1995) The broad-range phospholipase C and a metalloprotease mediate listeriolysin O-independent escape of *Listeria monocytogenes* from a primary vacuole in human epithelial cells. *Infect. Immun.* **63**, 4531–4534.

18. Smith, G. A., Marquis, H., Jones, S., Johnston, N. C., Portnoy, D. A., and Goldfine, H. (1995) The two distinct phospholipases C of *Listeria monocytogenes* have ovelapping roles in escape from a vacuole and cell-to-cell spread. *Infec. Immun.* **63**, 4231–4237.

19. Gedde, M. M., Higgins, D. E., Tilney, L. G., and Portnoy, D. A. (2000) Role of Listeriolysin O in cell-to-cell spread of *Listeria monocytogenes*. *Infect. Immun.* **68**, 999–1003.

20. Kreft, J. and Vazquez-Boland, J. A. (2001) Regulation of virulence genes in *Listeria*. *Int. J. Med. Microbiol.* **291**, 145–157.

21. Milohanic, E., Glaser, P., Coppee, J.-Y., et al. (2003) Transcriptome analysis of *Listeria monocytogenes* identifies three groups of genes differently regulated by PrfA. *Mol. Microbiol.* **47**, 1613–1625.

22. Drevets, D. A., Leenen, P. J. M., and Greenfield, R. A. (2004) Invasion of the central nervous system by intracellular bacteria. *Clin. Microbiol. Rev.* **17**, 323–347.

23. Lecuit, M., Vandormael-Pournin, S., Lefort, J., et al. (2001) A transgenic model for Listeriosis: role of internalin in crossing the intestinal barrier. *Science* **292**, 1722–1725.

24. Dramsi, S., Biswas, I., Maguin, E., Braun, L., Mastroeni, P., and Cossart, P. (1995) Entry of *L. monocytogenes* into hepatocytes requires expression of InlB, a surface protein of the internalin multigene family. *Mol. Microbiol.* **16**, 251–261.

25. Gaillard, J. L., Jaubert, F., and Berche, P. (1996) The *inlAB* locus mediates the entry of *Listeria monocytogenes* into hepatocytes in vivo. *J. Exp. Med.* **183**, 359–369.

26. Join-Lambert, O. F., Ezine, S., Le Monnier, A., et al. (2005) *Listeria monocytogenes* infected bone marrow myeloid cells promote bacterial invasion of the central nervous system. *Cell Microbiol.* **7**, 167–180.

27. Lecuit, M., Nelson, D. M., Smith, S. D., et al. (2004) Targeting and crossing of the human maternofetal barrier by Listeria monocytogenes: Role of internalin interaction with trophoblast E-cadherin. *Proc. Natl. Acad. Sci. USA* **101**, 6142–6157.

28. Bakardjiev, A. I., Stacy, B. A., Fisher, S. J., and Portnoy, D. A. (2004) Listeriosis in the pregnant guinea pig: a model of vertical transmission. *Infect. Immun.* **72**, 489–497.

29. Dramsi, S., Dehoux, P., Lebrun, M., Goossens, P. L., and Cossart, P. (1997) Identification of four new members of the internalin multigene family of Listeria monocytogenes EGD. *Infect. Immun.* **65**, 1615–1625.

30. Raffelsbauer, D., Bubert, A., Engelbrecht, F., et al. (1998) The gene cluster inlC2DE of *Listeria monocytogenes* contains additional new internalin genes and is important for virulence in mice. *Mol. Gen. Genet.* **260**, 144–158.

31. Engelbrecht, F., Chun, S.-K., Ochs, C., et al. (1996) A new PrfA-regulated gene of *Listeria monocytogenes* encoding a small, secreted protein which belongs to the family of internalins. *Mol. Microbiol.* **21**, 823–837.

32. Glaser, P., Frangeul, L., Buchreiser, C., et al. (2001) Comparative genomics of *Listeria* species. *Science* **294**, 849–852.
33. Cabanes, D., Dehoux, P., Dussurget, O., Frangeul, L., and Cossart, P. (2002) Surface proteins and the pathogenic potential of *Listeria monocytogenes. Trends Microbiol.* **10**, 238–245.
34. Kobe, B. and Kajava, A. V. (2001) The leucine-rich repeat as a protein recognition motif. *Curr. Opin. Struct. Biol.* **11**, 725–732.
35. Schubert, W. D., Gobel, G., Diepholz, M., et al. (2001) Internalins from the human pathogen Listeria monocytogenes combine three distinct folds into a contiguous internalin domain. *J. Mol. Biol.* **312**, 783–794.
36. Doumith, M., Cazalet, C., Simoes, N., et al. (2004) New aspects regarding evolution and virulence of *Listeria monocytogenes* revealed by comparative genomics and DNA arrays. *Infect. Immun.* **72**, 1072–1083.
37. Nelson, K. E., Fouts, D. E., Mongodin, E. F., et al. (2004) Whole genome comparisons of serotype 4b and 1/2a strains of the food-borne pathogen *Listeria monocytogenes* reveal new insights into the core genome components of this species. *Nucl. Acids Res.* **32**, 2386–2395.
38. Kathariou, S. (2002) *Listeria monocytogenes* virulence and pathogenicity, a food safety perspective. *J. Food Prot.* **65**, 1811–1829.
39. Mori, H. and Ito, K. (2001) The Sec protein-translocation pathway. *Trends Microbiol.* **9**, 494–500.
40. Lenz, L. L. and Portnoy, D. A. (2002) Identification of a second *Listeria secA* gene associated with protein secretion and the rough phenotype. *Mol. Microbiol.* **45**, 1043–1056.
41. Lenz, L. L., Mohammadi, S., Geissler, A., and Portnoy, D. A. (2003) SecA2-dependent secretion of autolytic enzymes promotes *Listeria monocytogenes* pathogenesis. *Proc. Natl. Acad. Sci. USA* **100**, 12,432–12,437.
42. Dramsi, S., Bourdichon, F., Cabanes, D., Lecuit, M., Fsihi, H., and Cossart, P. (2004) FbpA, a novel multifunctional *Listeria monocytogenes* virulence factor. *Mol. Microbiol.* **53**, 639–649.
43. Ton-That, H., Marraffini, L. A., and Schneewind, O. (2004) Protein sorting to the cell wall envelope of Gram-positive bacteria. *Biochem. Biophys. Acta* **1694**, 269–278.
44. Mazmanian, S. K., Ton-That, H., Su, K., and Schneewind, O. (2002) An iron-regulated sortase anchors a class of surface protein during S*taphylococcus aureus* pathogenesis. *Proc. Natl. Acad. Sci. USA* **99**, 2293–2298.
45. Bierne, H., Mazmanian, S. K., Trost, M., et al. (2002) Inactivation of the *srtA* gene in *Listeria monocytogenes* inhibits anchoring of surface proteins and affects virulence. *Mol. Microbiol.* **43**, 869–881.
46. Garandeau, C., Reglier-Poupet, H., Dubail, I., Beretti, J.-L., Berche, P., and Charbit, A. (2002) The sortase SrtA of *Listeria monocytogenes* is involved in processing of internalin and in virulence. *Infect. Immun.* **70**, 1382–1390.
47. Bierne, H., Garandeau, C., Pucciarelli, M. G., et al. (2004) Sortase B, a new class of sortase in *Listeria monocytogenes. J. Bacteriol.* **186**, 1972–1982.
48. Lebrun, M., Mengaud, J., Ohayon, H., Nato, F., and Cossart, P. (1996) Internalin must be on the bacterial surface to mediate entry of *Listeria monocytogenes* into epithelial cells. *Mol. Microbiol.* **21**, 579–592.
49. Newton, S. M. C., Klebba, P. E., Raynaud, C., et al. (2005) The *svpA-srtB* locus of *Listeria monocytogenes*: fur-mediated iron regulation and effect on virulence. *Mol. Microbiol.* **55**, 927–940.

50. Mazmanian, S. K., Liu, G., Jensen, E. R., Lenoy, E., and Schneewind, O. (2000) *Staphylococcus aureus* sortase mutants defective in the display of surface proteins and in the pathogenesis of animal infections. *Proc. Natl. Acad. Sci. USA* **97,** 5510–5515.

51. Mazmanian, S. K., Skaar, E. P., Gaspar, A. H., et al. (2003) Passage of heme-iron across the envelope of *Staphylococcus aureus*. *Science* **299,** 906–909.

52. Reglier-Poupet, H., Frehel, C., Dubail, I., et al. (2003) Maturation of lipoproteins by type II signal peptidase is required for phagosomal escape of *Listeria monocytogenes*. *J. Biol. Chem.* **278,** 49,469–49,477.

53. Borezee, E., Pellegrini, E., and Berche, P. (2000) OppA of *Listeria monocytogenes*, an oligopeptide-binding protein required for bacterial growth at low temperature and involved in intracellular survival. *Infect. Immun.* **68,** 7069–7077.

54. Sampson, J. S., O'Connor, S. P., Stinson, A. R., Tharpe, J. A., and Russel, H. (1994) Cloning and nucleotide sequence of psaA, the *Streptococcus pneumoniae* gene encoding a 37-kilodalton protein homologous to previously reported *Streptococcus* sp. adhesins. *Infec. Immun.* **62,** 319–324.

55. Reglier-Poupet, H., Pellegrini, E., Charbit, A., and Berche, P. (2003) Identification of LpeA, a PsaA-like membrane protein that promotes cell entry by *Listeria monocytogenes*. *Infect. Immun.* **71,** 474–482.

56. Milohanic, E., Jonquieres, R., Cossart, P., Berche, P., and Gaillard, J.-L. (2001) The autolysin Ami contributes to the adhesion of *Listeria monocytogenes* to eukaryotic cells via its cell wall anchor. *Mol. Microbiol.* **39,** 1212–1224.

57. Cabanes, D., Dussurget, O., Dehoux, P., and Cossart, P. (2004) Auto, a surface associated autolysin of *Listeria monocytogenes* required for entry into eukaryotic cells and virulence. *Mol. Microbiol.* **51,** 1601–1614.

58. Bubert, A., Sokolovic, Z., Chun, S. K., Papatheodorou, L., Simm, A., and Goebel, W. (1999) Differential expression of *Listeria monocytogenes* virulence genes in mammalian host cells. *Mol. Gen. Genet.* **261,** 323–336.

59. Luo, Q., Rauch, M., Marr, A. K., Muller-Altrock, S., and Goebel, W. (2004) *In vitro* transcription of the *Listeria monocytogenes* virulence genes *inlC* and *mpl* reveals overlapping PrfA-dependent and -independent promoters that are differentially activated by GTP. *Mol. Microbiol.* **52,** 39–52.

60. Goetz, M., Bubert, A., Wang, G., et al. (2001) Microinjection and growth of bacteria in the cytosol of mammalian host cells. *Proc. Natl. Acad. Sci. USA* **98,** 12,221–12,226.

61. Bergmann, B., Raffelsbauer, D., Kuhn, M., Goetz, M., Hom, S., and Goebel, W. (2002) InlA- but not InlB-mediated internalization of *Listeria monocytogenes* by non-phagocytic mammalian cells needs the support of other internalins. *Mol. Microbiol.* **43,** 557–570.

62. Chico-Calero, I., Suarez, M., Gonzalez-Zorn, B., et al; European Listeria Genome Consortium. (2002) Hpt, a bacterial homolog of the microsomal glucose- 6-phosphate translocase, mediates rapid intracellular proliferation in *Listeria*. *Proc. Natl. Acad. Sci. USA* **99,** 431–436.

63. Ripio, M., Brehm, K., Lara, M., Suarez, M., and Vazquez-Boland, J. A. (1997) Glucose-1-phosphate utilization by *Listeria monocytogenes* is PrfA dependent and coordinately expressed with virulence factors. *J. Bact.* **179,** 7174–7180.

64. Kallipolitis, B. H. and Ingmer, H. (2001) Listeria monocytogenes response regulators important for stress tolerance and pathogenesis. *FEMS Microbiol. Lett.* **204,** 111–115.

65. Kallipolitis, B. H., Ingmer, H., Gahan, C. G., Hill, C., and Sogaard-Andersen, L. (2003) CesRK, a two-component signal transduction system in *Listeria monocytogenes*, responds to the presence of cell wall-acting antibiotics and affects beta-lactam resistance. *Antimicrob. Agents Chemother.* **47,** 3421–3429.

66. Knudsen, G. M., Olsen, J. E., and Dons, L. (2004) Characterization of DegU, a response regulator in *Listeria monocytogenes*, involved in regulation of motility and contributes to virulence. *FEMS Microbiol. Lett.* **240,** 171–179.

67. Cotter, P. D., Guinane, C. M., and Hill, C. (2002) The LisRK signal transduction system determines the sensitivity of *Listeria monocytogenes* to nisin and cephalosporins. *Antimicrob. Agents Chemother.* **46,** 2784–2790.

68. Ogura, M. and Tanaka, T. (2002) Recent progress in Bacillus subtilis two-component regulation. *Front. Biosci.* **7,** 1815–1824.

69. Dramsi, S., Kocks, C., Forestier, C., and Cossart, P. (1993) Internalin-mediated invasion of epithelial cells by *Listeria monocytogenes* is regulated by the bacterial growth state, temperature and the pleiotropic activator prfA. *Mol. Microbiol.* **9,** 931–941.

70. Lingnau, A., Domann, E., Hudel, M., et al. (1995) Expression of the Listeria monocytogenes EGD *inlA* and *inlB* genes, whose products mediate bacterial entry into tissue culture cell lines, by PrfA-dependent and -independent mechanisms. *Infect. Immun.* **63,** 3896–3903.

71. Vega, Y., Rauch, M., Banfield, M. J., et al. (2004) New *Listeria monocytogenes* prfA* mutants, transcriptional properties of PrfA* proteins and structure-function of the virulence regulator PrfA. *Mol. Microbiol.* **52,** 1553–1565.

72. Busby, S. and Ebright, R. H. (1999) Transcription activation by catabolite activator protein (CAP). *J. Mol. Biol.* **293,** 199–213.

73. Dove, S. L., Darst, S. A., and Hochschild, A. (2003) Region 4 of sigma as a target for transcription regulation. *Mol. Microbiol.* **48,** 863–874.

74. Johansson, J., Mandin, P., Renzoni, A., Chiaruttini, C., Springer, M., and Cossart, P. (2002) An RNA thermosensor controls expression of virulence genes in *Listeria monocytogenes*. *Cell* **110,** 551–561.

75. Shetron-Rama, L. M., Mueller, K., Bravo, J. M., Bouwer, H. G. A., Way, S. S., and Freitag, N. E. (2003) Isolation of *Listeria monocytogenes* mutants with high-level *in vitro* expression of host cytosol-induced gene products. *Mol. Microbiol.* **48,** 1537–1551.

76. Ermolaeva, S., Novella, S., Vega, Y., Ripio, M.-T., Scortti, M., and Vazquez-Boland, J. A. (2004) Negative control of *Listeria monocytogenes* virulence genes by a diffusible autorepressor. *Mol. Microbiol.* **52,** 601–611.

77. Bockmann, R., Dickneite, C., Goebel, W., and Bohne, J. (2000) PrfA mediates specific binding of RNA polymerase of *Listeria monocytogenes* to PrfA-dependent virulence gene promoters resulting in a transcriptionally active complex. *Mol. Microbiol.* **36,** 487–497.

78. Renzoni, A., Cossart, P., and Dramsi, S. (1999) PrfA, the transcriptional activator of virulence genes, is upregulated during interaction of *Listeria monocytogenes* with mammalian cells and in eukaryotic cell extracts. *Mol. Microbiol.* **34,** 552–561.

79. Dussurget, O., Cabanes, D., Dehoux, P., et al. (2002) *Listeria monocytogenes* bile salt hydrolase is a PrfA-regulated virulence factor involved in the intestinal and hepatic phases of listeriosis. *Mol. Microbiol.* **45,** 1095–1106.

80. Hardy, J., Francis, K. P., DeBoer, M., Chu, P., Gibbs, K., and Contag, C. H. (2004) Extracellular replication of *Listeria monocytogenes* in the murine gall bladder. *Science* **303,** 851–853.

81. Sleator, R. D., Wouters, J., Gahan, C. G. M., Abee, T., and Hill, C. (2001) Analysis of the role of OpuC, an osmolyte transport system, in salt tolerance and virulence potential of *Listeria monocytogenes*. *Appl. Environ. Microbiol.* **67,** 2692–2698.

82. Sleator, R. D., Wemekamp-Kamphuis, H. H., Gahan, C. G. M., Abee, T., and Hill, C. (2005) A PrfA-regulated bile exclusion system (BilE) is a novel virulence factor in *Listeria monocytogenes*. *Mol. Microbiol.* **55,** 1183–1195.

83. Becker, L. A., Cetin, M. S., Hutkins, R. W., and Benson, A. K. (1998) Identification of the gene encoding the alternative sigma factor sigma B from *Listeria monocytogenes* and its role in osmotolerance. *J. Bacteriol.* **180**, 4547–4554.

84. Chaturongakul, S. and Boor, K. J. (2004) RsbT and RsbV contribute to {sigma}B-dependent survival under environmental, energy, and intracellular stress conditions in *Listeria monocytogenes*. *Appl. Environ. Microbiol.* **70**, 5349–5356.

85. Sue, D., Fink, D., Wiedmann, M., and Boor, K. J. (2004) Sigma B-dependent gene induction and expression in *Listeria monocytogenes* during osmotic and acid stress conditions simulating the intestinal environment. *Microbiology* **150**, 3843–3855.

86. Price, C. W. (2000) Protective function and regulation of the general stress response in *Bacillus subtilis* and related Gram-positive bacteria, in *Bacterial Stress Responses* (Storz, B. and Hengge-Aronis, R. eds.). ASM Press, Washington, DC, pp. 179–197.

87. Kazmierczak, M. J., Mithoe, S. C., Boor, K. J., and Wiedmann, M. (2003) *Listeria monocytogenes sigma* B regulates stress response and virulence functions. *J. Bacteriol.* **185**, 5722–5734.

88. Kim, H., Boor, K. J., and Marquis, H. (2004) *Listeria monocytogenes* sigma B contributes to invasion of human intestinal epithelial cells. *Infect. Immun.* **72**, 7374–7378.

89. Fraser, K. R., Sue, D., Wiedmann, M., Boor, K., and O'Byrne, C. P. (2003) Role of sigma B in regulating the compatible solute uptake systems of *Listeria monocytogenes*: osmotic induction of opuC Is sigma B dependent. *Appl. Environ. Microbiol.* **69**, 2015–2022.

90. Sue, D., Boor, K. J., and Wiedmann, M. (2003) Sigma B-dependent expression patterns of compatible solute transporter genes *opuCA* and lmo1421 and the conjugated bile salt hydrolase gene bsh in *Listeria monocytogenes*. *Microbiology* **149**, 3247–3256.

91. Cetin, M. S., Zhang, C., Hutkins, R. W., and Benson, A. K. (2004) Regulation of transcription of compatible solute transporters by the general stress sigma factor, sigma B, in *Listeria monocytogenes*. *J. Bacteriol.* **186**, 794–802.

92. Rauch, M., Luo, Q., Muller-Altrock, S., and Goebel, W. (2005) SigB-dependent in vitro transcription of prfA and some newly identified genes of *Listeria monocytogenes* whose expression is affected by PrfA in vivo. *J. Bacteriol.* **187**, 800–804.

93. Schaumburg, J., Diekmann, O., Hagendorff, P., et al. (2004) The cell wall subproteome of *Listeria monocytogenes*. *Proteomics* **4**, 2991–3006.

94. Weeks, M. E., James, D. C., Robinson, G. K., and Smales, C. M. (2004) Global changes in gene expression observed at the transition from growth to stationary phase in *Listeria monocytogenes* ScottA batch culture. *Proteomics* **4**, 123–135.

95. Tremoulet, F., Duche, O., Namane, A., Martinie, B., and Labadie, J. C.; European Listeria Genome Consortium. (2002) Comparison of protein patterns of *Listeria monocytogenes* grown in biofilm or in plantonic mode by proteomic analysis. *FEMS Microbiol. Lett.* **210**, 25–31.

96. Helloin, D., Jansch, L., and Phan-Thanh, L. (2003) Carbon starvation survival of *Listeria monocytogenes* in plantonic state and in biofilm: a proteomic study. *Proteomics* **3**, 2052–2064.

97. Cohen, P., Bouaboula, M., Bellis, M., et al. (2000) Monitoring cellular responses to *Listeria monocytogenes* with oligonucleotide arrays. *J. Biol. Chem.* **275**, 11,181–11,190.

98. Baldwin, D., Vanchinathan, V., Brown, P., and Theriot, J. (2002) A gene-expression program reflecting the innate immune response of cultured intestinal epithelial cells to infection by *Listeria monocytogenes*. *Genome Biology* **4**, R2.1–R2.14.

99. Pizarro-Cerda, J., Jonquieres, R., Gouin, E., Vandekerckhove, J., Garin, J., and Cossart, P. (2002) Distinct protein patterns associated with *Listeria monocytogenes* InlA- or InlB-phagosomes. *Cell Microbiol.* **4**, 101–115.

100. McCaffrey, R. L., Fawcett, P., O'Riordan, M., et al. (2004) A specific gene expression program triggered by Gram-positive bacteria in the cytosol. *Proc. Natl. Acad. Sci. USA* **101,** 11,386–11,391.

101. Motley, S. T., Morrow, B. J., Liu, X., et al. (2004) Simultaneous analysis of host and pathogen interactions during an *in vivo* infection reveals local induction of host acute phase response proteins, a novel bacterial stress response, and evidence of a host-imposed metal ion limited environment. *Cell Microbiol.* **6,** 849–865.

102. Bader, B. D., Heilbut, A., Andrews, B., Tyers, M., Hughes, T., and Boone, C. (2003) Functional genomics and proteomics: charting a multidimensional map of the yeast cell. *Trends Cell Biol.* **13,** 344–356.

103. Molle, V., Fujita, M., Jensen, S. T., et al. (2003) The Spo0A regulon of *Bacillus subtilis*. *Mol. Microbiol.* **50,** 1683–1701.

104. Molle, V., Nakaura, Y., Shivers, R. P., et al. (2003) Additional targets of the *Bacillus subtilis* global regulator CodY identified by chromatin immunoprecipitation and genome-wide transcript analysis. *J. Bacteriol.* **185,** 1911–1922.

105. Rain, J. C., Selig, L., Reuse, H. D., et al. (2001) The protein-protein interaction map of *Helicobacter pylori*. *Nature* **409,** 211–215.

106. Bailey, S. N., Sabatini, D. M., and Stockwell, B. R. (2004) Microarrays of small molecules embedded in biodegradable polymers for use in mammalian cell-based screens. *Proc. Natl. Acad. Sci. USA* **101,** 16,144–16,149.

107. Carpenter, A. E. and Sabatini, D. M. (2004) Systematic genome-wide screens of gene function. *Nature Rev. Genet.* **5,** 11–22.

108. Boyartchuk, V., Rojas, M., Yan, B.-S., et al. (2004) The host resistance locus sst1 controls innate immunity to *Listeria monocytogenes* infection in immunodeficient mice. *J. Immunol.* **173,** 5112–5120.

109. Nolan, P. M., Peters, J., Strivens, M., et al. (2000) A systematic, genome-wide, phenotype-driven mutagenesis programme for gene function studies in the mouse. *Nat. Genet.* **25,** 440–443.

110. Hrabe de Angelis, M. H., Flaswinkel, H., Fuchs, H., et al. (2000) Genome-wide, large-scale production of mutant mice by ENU mutagenesis. *Nat. Genet.* **25,** 444–447.

111. Walduck, A., Rudel, T., and Meyer, T. F. (2004) Proteomic and gene profiling approaches to study host responses to bacterial infection. *Curr. Opin. Microbiol.* **7,** 33–38.

112. Cheng, L. W. and Portnoy, D. A. (2003) Drosophila S2 cells: an alternative infection model for *Listeria monocytogenes*. *Cell Microbiol.* **5,** 875–885.

113. Mansfield, B. E., Dionne, M. S., Schneider, D. S., and Freitag, N. E. (2003) Exploration of host-pathogen interactions using *Listeria monocytogenes* and *Drosophila melanogaster*. *Cell Microbiol.* **5,** 901–911.

9

Mycobacterial Genomes

David C. Alexander and Jun Liu

Summary

Tuberculosis (TB), caused by *Mycobacterium tuberculosis*, remains a major cause of death around the world. Diseases caused by nontuberculous mycobacteria are increasingly associated with immuno-compromised individuals. The availability of whole-genome sequences of mycobacterial species in the past several years has revolutionized TB research. This chapter provides an overview of the biology of mycobacteria and the diseases that they cause, with emphasis on how recent advances in genomics have improved our knowledge of the lifestyle and phylogeny of these organisms.

Key Words: Mycobacterial genomes; *Mycobacterium tuberculosis*; Mycobacteria.

"If one judges of the importance of a disease according to its distribution, and according to the degree in which it menaces health and induces death prematurely, tuberculosis assumes the first rank in human pathology...it has been known as far back as the memory of man extends, and has unceasingly decimated the race for hundreds and thousands of years," (1)

Prof. Dr. Georg Cornet, 1904 (19th century Bacteriologist)

1. Introduction

One hundred years later, mycobacterial diseases retain their first rank as menaces to human and animal health. Despite global initiatives and five decades of chemotherapeutics, tuberculosis (TB) caused by *Mycobacterium tuberculosis* remains a common bacterial disease. An estimated 2 billion people are infected with *M. tuberculosis* and 2 million succumb to TB each year *(2)*. Leprosy, caused by *Mycobacterium leprae*, inflicts disfigurement and untold human suffering. Effective treatments for leprosy are available, but attempts at eradication have failed, and more than 600,000 new cases are reported each year *(3)*. Buruli ulcer, a deadly necrotizing skin disease caused by *Mycobacterium ulcerans*, is often described as an emerging infection, but in endemic regions it is more common than leprosy. Diseases caused by atypical, nontuberculous mycobacteria, such as members of the *Mycobacterium avium* complex (MAC), used to be rare. However, the epidemic of HIV infection and AIDS has been accompanied by a surge in these opportunistic mycobacterial infections, which are often difficult to treat. Animals also suffer. Vigilant farming practices have reduced the incidence of *Mycobacterium bovis* infection, but outbreaks of bovine TB still occur. Johne's disease, a fatal inflammatory bowel disease of livestock, caused by the *Mycobacterium avium* subspecies *paratuberculosis*, remains endemic in domestic herds *(4)*.

Historically, mycobacterial research has been hampered by the fastidious nutritional requirements, extraordinarily slow growth rates of these organisms and, especially with

From: *Bacterial Genomes and Infectious Diseases*
Edited by: V. L. Chan, P. M. Sherman, and B. Bourke © Humana Press Inc., Totowa, NJ

M. tuberculosis, the high risk of contagion. During the past 15 yr, the development of effective molecular biology tools has rendered these bacteria more amenable to genetic and biochemical studies. The availability of whole-genome sequences, starting with *M. tuberculosis* strain H37Rv in 1998 *(5)*, has revolutionized TB research. The genome sequences of *M. bovis (6)*, *M. leprae (7)*, and *M. tuberculosis* strain CDC1551 *(8)* have now been published. Additional sequencing projects are underway for the genomes of *M. ulcerans*, *Mycobacterium marinum*, *Mycobacterium microti*, *Mycobacterium smegmatis*, two *M. avium* subspecies, several strains of *M. tuberculosis* and the vaccine strain, Bacille Calmette-Guérin (BCG).

This chapter provides an overview of mycobacteria and the diseases that they cause, with an emphasis on how recent advances in genomics have enriched our understanding of both the biology and phylogeny of these organisms. The focus of the first section is TB and *M. tuberculosis*. As illustrated in the second section, the role of other mycobacteria should not be underestimated. The final section focuses on specific gene families and virulence factors that distinguish mycobacteria from other prokaryotes.

2. *Mycobacterium tuberculosis* and the Genus *Mycobacterium*

The genus *Mycobacterium* comprises more than 70 species *(9)*. A few, notably *M. tuberculosis*, *M. leprae*, and *M. ulcerans*, cause significant morbidity and mortality. Others, including *M. kansasii*, *M. fortuitum*, *M. abscessus*, *M. xenopi*, *M. chelonae*, and the *M. avium* complex, are responsible for occasionally lethal opportunistic infections *(10–12)*. However, the vast majority are harmless environmental organisms, common in water and soil *(13)*. Under the microscope, mycobacteria are small, rod-shaped bacteria. They are Gram-positive organisms, but are best distinguished by their characteristic acid-fast staining. This acid-fastness is a property of the mycobacterial cell wall, an unusual, lipid-rich structure that forms a hydrophobic, low permeability barrier and provides innate protection against many antimicrobial agents. Traditionally, mycobacteria have been classified according to growth rate and pigmentation (e.g., the Runyon Groups), and further subdivided on the basis of biochemical reactions (e.g., niacin production, nitrate reduction, drug resistance), serotypes, bacteriophage susceptibility, and cell wall lipid profiles. However, these have been superseded by molecular methods, especially DNA sequencing and polymerase chain reaction-based tests, which are rapid and require little starting material, both of which are important considerations when dealing with slow-growing and hazardous organisms. Sequencing of 16S ribosomal DNA (rDNA) and the 16S–23S rDNA internal transcribed spacer (ITS) region has been used to establish a phylogeny of *Mycobacterium* species. The latter has revealed that the ITS of fast-growing mycobacteria is longer and structurally distinct from the ITS of slow-growers, and supports the traditional distinction based on growth rate *(14)*. Sequencing allows discrimination between isolates that are phenotypically indistinguishable *(15)*, and has uncovered phylogenetic differences (i.e., sequevars) within individual species *(14)*.

3. *Mycobacterium tuberculosis* and the Global Impact of Tuberculosis

M. tuberculosis was first described in 1882 by the eminent microbiologist Robert Koch *(16)*, but it has been with us since antiquity. Known as tuberculosis, consumption, phthisis, and the white plague, evidence of *M. tuberculosis* disease has been found in ancient manuscripts, sculptures, and wall paintings. In recent years, *M. tuberculosis* DNA has been extracted from mummies *(17,18)*. At the end of the 19th century, TB was the leading cause of death in the Western world, killing one in seven. Advances of the 20th century, including antibiotics, succeeded in almost eliminating the disease from Europe and the Americas. Despite a recent resurgence in Eastern Europe, the annual incidence of TB in most of the developed world remains low, with less than 50 cases per 100,000 people *(19)*. Globally, the situation is much worse. With more than 250 cases per 100,000 people, the incidence of TB is highest in Africa, but because of its larger population, the total number of cases is greater in Asia. The World Health Organization estimates that *M. tuberculosis* is responsible for 2 million deaths and 8.8 million new infections each year, 80% of which occur in developing countries *(20)*. One important impediment to global efforts to control TB is the ongoing HIV epidemic *(21,22)*. There is an overlap in the geographical distribution of these infectious diseases and, in the year 2000, an estimated 11% of new adult TB cases were also infected with HIV. At greatest risk for both *M. tuberculosis* and HIV infections are people in their prime working and reproductive years, between 14 and 49 yr of age. Widespread illness in this age group has profound social and economical effects. Without a workforce to maintain and fund the local medical, educational, and business infrastructure, the health of an entire community suffers.

The enormous cost of *M. tuberculosis* is matched by the complex lifestyle of this facultative intracellular pathogen *(23,24)*. It is transmitted between people, most often by an aerosol route. The cough of a tuberculous individual generates tiny nasal droplets, no larger than 5 µm in diameter, which contain live bacteria. New infections occur when these droplets are inhaled and penetrate to the alveoli of the respiratory tract, where *M. tuberculosis* is ingested by alveolar macrophages. Although macrophages typically destroy invading microbes, *M. tuberculosis* has the ability to subvert these phagocytic cells *(25,26)*. During the first 2 wk of infection, the bacteria slowly but continuously replicate inside of the macrophages. Two to eight weeks postinfection, cell-mediated immunity develops. At this stage, activated T-lymphocytes and noninfected macrophages act to control the growth of *M. tuberculosis*. In most individuals (~90%), the infection stops here. The immune response generates a granuloma around the *M. tuberculosis* that prevents the bacteria from spreading. With the infection contained, active disease does not develop. However, in some infected individuals, especially children under 5 yr and immunocompromised adults, the primary *M. tuberculosis* infection cannot be contained. The bacilli continue to replicate, host tissue is destroyed, and active TB develops. Although most often associated with the lungs (pulmonary TB), *M. tuberculosis* can attack anywhere in the body including the bones (Pott's disease), brain (TB meningitis), lymph nodes (scrofula), and intestinal tract.

M. tuberculosis is a tenacious pathogen. Even when the primary infection is contained, the bacteria within the granuloma can survive for decades, persisting in a special dormant state *(27,28)*. When the immune system is compromised, by such factors as malnutrition, HIV infection, diabetes, renal disease, chemotherapy, or extensive corticosteroid therapy, reactivation of the disease can occur *(29)*. The protective granuloma disintegrates, and the long dormant *M. tuberculosis* revives and spreads unchecked.

Most cases of TB will respond to antibiotics. Standard regimens involve daily treatment with four drugs: isoniazid, rifampin, pyrazinamide, and ethambutol, for 2 mo, followed by 4 mo of isoniazid and rifampin. Failure to comply can result in a relapse and the emergence of multi-drug resistant strains, which no longer respond to these agents *(30–32)*. Effective treatment of multi-drug resistant strains requires quarantine of the patient and up to 24 mo of drug therapy. Atypical nontuberculous mycobacterial organisms, responsible for opportunistic infections in AIDS patients, tend to exhibit natural antibiotic resistance. Combined with the immunocompromised state of the host, these infections are, therefore, extraordinarily difficult to eradicate.

4. The Genome of *Mycobacterium tuberculosis*

The complete genome sequence of *M. tuberculosis* H37Rv (a virulent strain isolated in 1905 and then propagated in vitro) was published in 1998 *(5)*. The circular genome comprises of 4,411,532 bp and has a mean guanine and cytosine content of 65.6%. The original annotation identified 3974 genes encoding 3924 proteins and 50 stable RNA. Initially overlooked, an additional 82 protein encoding genes have since been added. Genes have been identified via (1) sequence homology to known proteins in other microorganisms, (2) experiments using two-dimensional electrophoresis and mass spectrometry, and (3) bioinformatic techniques that examine *M. tuberculosis* codon usage. The approx 4000 genes are classified into 11 broad functional groups. Of these, 52% are assigned with precise or putative functions, with the remaining 48% being conserved hypotheticals or functionally unknown genes.

The publication of the H37Rv genome was followed by completion of the *M. tuberculosis* strain CDC1551 genome sequence (isolated in 1995 and responsible for a outbreak in the United States) *(8,33)*, and the partial genome sequencing of *M. tuberculosis* strain 210 (a representative of the W/Beijing strains, responsible for the majority of cases in Asia and the former Soviet Union, as well as outbreaks in the United States) *(34,35)*. Analysis of these sequences reveals single nucleotide polymorphisms (SNPs, including both synonymous and nonsynonymous substitutions), large sequence polymorphisms (LSPs; genetic deletions/insertions), plus variations in the numbers and types of mobile elements (e.g., transposons and prophages) among *M. tuberculosis* isolates. For example, comparison of the whole genome sequences of H37Rv (4.41 Mbp) and CDC1551 (4.40 Mbp) revealed approx 1100 SNPs *(8)*. Approximately 65% of the SNPs are nonsynonymous substitutions, which is unusual because it is generally thought that many nonsynonymous substitutions are lost during purifying selection, as demonstrated in other bacteria, such as *Escherichia coli* and *Salmonella enterica*. LSP analysis indicates that *M. tuberculosis* exhibits less genetic diversity than other bacteria. Only 74 LSPs longer than 10 bp were identified between H37Rv and CDC1551 *(8)*. In a larger study of 100 clinical isolates, a total of 68 distinct deletion events, ranging in size from 105 bp to approx 12 kb, were identified *(36)*. Together, these LSPs represent approx 186 kb (~4.2%) of the H37Rv genome and affect 224 (~5.5%) genes, including genes in all major functional categories. However, among individual isolates no more than 41 kb or 50 genes were deleted. In contrast, differences between the K12 and O157 strains of *E. coli* affect more than 1300 genes *(37)*. Even so, these studies indicate that a degree of polymorphism does exist between different *M. tuberculosis* strains, which is consistent with the phenotypical diversity observed among clinical isolates *(38)*. Numerous methods, most commonly restriction fragment length profiles, spoligotypes, IS*6110* profiles, and SNP analysis are used to characterize clinical isolates *(39–41)*. These differences provide insight into the epidemiology of outbreaks, infectivity, and virulence of individual strains. Notably, half of LSPs between H37Rv and CDC1551 involve genes encoding Pro–Pro–Glu (PPE) and Pro–Glu (PE)-polymorphic GC-rich repetitive sequence (PGRS) family proteins *(8)*, which are considered antigens important for human immunity (*see* **Subheading 11.2.**). Genetic studies also provide insight into the evolution of *M. tuberculosis*. Current evidence indicates that the W/Beijing strains, such as 210, are more ancestral than CDC1551 and H37Rv, and contain genes that are no longer present in the more recently derived strains *(38,42,43)*.

5. The *Mycobacterium tuberculosis* Complex

The *M. tuberculosis* complex refers to a number of genetically related human and animal pathogens that share 99.9% similarity at the nucleotide level and are indistinguishable by 16S rDNA sequencing. These include *M. tuberculosis, M. africanum, M. microti* (voles), *M. caprae* (goats), *M. bovis*, as well as the BCG vaccine strains and a variety of isolates from unusual sources, such as *M. pinnipedii* (from seal lions and fur seals), and the dassie bacillus (from *Procavia capensis*, the hyrax, or dassie). The animal strains are responsible for zoonotic transmission of TB to humans, especially via ingestion of infected meat or milk. Indeed, it was long believed that TB was an animal disease that managed to jump the species barrier. However, genetic interrogation of *M. tuberculosis* complex isolates, together with genome sequencing of *M. microti, M. bovis* AF2122/97 (a virulent strain from Great Britain) *(6)*, and BCG Pasteur (a vaccine strain maintained in Paris since 1923), reveals that the opposite is true. *M. tuberculosis* originated with humans and was transmitted to animals *(44)*. The current phylogeny indicates that the *M. tuberculosis* complex evolved from a human strain via successive, unidirectional deletion events (*see* Fig. 1).

M. africanum refers to human TB isolates from parts of Africa which, on the basis of biochemical tests, were considered distinct from both *M. tuberculosis* and *M. bovis*. However, this classification scheme has proven unreliable as biochemical markers do not correlate with genetic data, which identify at least three groups of *M. africanum* *(42,45)*. Some so-called *M. africanum* strains are genetically indistinguishable from *M. tuberculosis*. A second group contains a single genomic deletion, or region of difference (RD), called RD9, which affects seven genes. In the third group of *M. africanum*, RD9 is deleted along with the LSPs, RD7, RD8, and RD10. These four deletions are conserved across all animal strains. An additional five deletions are found in both *M. caprae* and *M. bovis*. Taken together, these RDs provide a scheme for the reductive evolution of the *M. tuberculosis* complex from human to cattle. Additional deletions define branches within the *M. tuberculosis* complex. For example, a small deletion, called TbD1, is a marker of "modern" TB *(42)*. Multiple genomic deletions distinguish the dassie bacillus from other species *(46)*. Additional variations are unique to specific host–pathogen pairs. The biological roles of individual RDs have yet to be established. Some may be hot spots for genetic rearrangement, but most are believed to influence disease transmission and progression. Consistent with both of these ideas, several overlapping, but nonidentical deletions have occurred in discrete isolates of the *M. tuberculosis* complex. For example, the RD1 deletion, which has occurred independently in *M. microti*, the dassie bacillus, and BCG impairs the secretion of key immunodominant antigens and impacts bacterial virulence. In contrast, the RD5 deletion, which affects five genes in

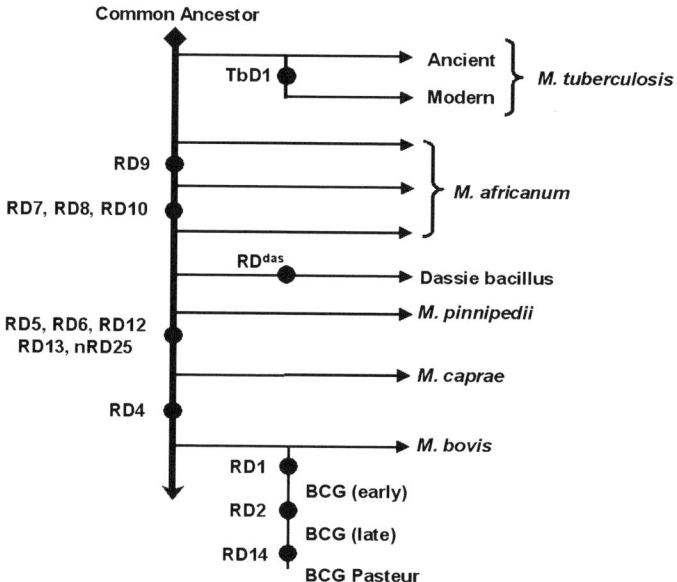

Fig. 1. Phylogenetic tree of the *Mycobacterium tuberculosis* complex. This phylogeny was generated using a variety of molecular markers, but only major genomic deletion events (filled circles) are indicated. Events on the thick vertical axis have accumulated over time and affect successive members of the complex. Events on the thin horizontal axes are only found in specific members of the complex. (Adapted from refs. *42,44–46,54,55,58.*)

M. microti, seven genes in the dassie bacillus, and eight genes in BCG Pasteur, is associated with transposition of the mobile genetic element, insertion sequence (IS)*6110 (44)*.

The role of SNPs in the evolution of the *M. tuberculosis* complex is less clear. The genome of *M. bovis* is 99.95% identical to that of *M. tuberculosis*. Fewer than 2500 SNPs have been identified and at least one-third are synonymous changes that do not alter amino acid sequence *(6)*. However, frameshifts and other variations do occur. A point mutation in *ald* (alanine dehydrogenase) prevents *M. bovis* and BCG strains from catabolizing the amino acid alanine *(47)*. *M. bovis*, although not BCG, also has a mutation in *pykA* (pyruvate kinase), such that glycolytic intermediates are not converted to pyruvate or used in the TCA cycle. Other SNPs alter the antigenic repertoire of *M. bovis*, and are likely to impact host–pathogen interactions *(6)*. Curiously, some genes that are defective in *M. bovis* are also psuedogenes in *M. leprae*.

5.1. Bacille Calmette-Guérin

Live attenuated vaccines have reduced the incidence, and even eliminated, many important bacterial and viral diseases. The original BCG was derived between 1908 and 1921 by 230 in vitro passages of an *M. bovis* strain. This attenuated BCG was found to be nonpathogenic in guinea pigs, yet sufficiently immunogenic to protect against a challenge with virulent *M. tuberculosis (48,49)*. In the preantibiotic era, the promise of an effective vaccine against TB made BCG popular and, starting in 1923, stocks were distributed around the world. BCG vaccination of newborns remains common in many countries with an estimated 100 million doses of BCG administered each year. BCG is safe and probably protects children against TB meningitis. However, in randomized controlled trials the efficacy of BCG against adult pulmonary disease has ranged from 0 to 80%, and remains a matter of debate *(50,51)*. Efficacy may be diminished by prior exposure to environmental mycobacteria, which can alter the immune response to the vaccine *(52,53)*. However, BCG itself is also likely to blame. Between 1923 and the 1960s (when lyophilized stocks were finally established), BCG strains continued to be propagated in vitro. Different laboratories employed different culture media and passaged strains according to different schedules. BCG continued to adapt to laboratory conditions, plus there was selection for strains that produced few side effects, yet still promoted tuberculin conversion (wrongly considered an indicator of immunogenicity). The net result has been the creation of numerous, genetically heterogeneous BCG substrains. Although the original BCG of 1921 has been lost, genetic analysis of BCG substrains indicates that the initial attenuation event was the deletion of RD1 *(54,55)*. Indeed, experimental deletion of RD1 impairs the virulence of *M. tuberculosis* H37Rv in a mouse model of TB *(56,57)*. As with the *M. tuberculosis* complex, evolution of BCG substrains has involved multiple unidirectional deletion events. For example, the RD2 region is deleted from all substrains obtained from the Pasteur Institute after 1931, whereas nRD18 is only deleted in strains derived after 1933 *(54,58)*.

The need for an effective vaccine against *M. tuberculosis* remains. Although some research is directed at the generation of attenuated *M. bovis* or *M. microti* strains, DNA vaccines, and protein preparations, much of research continues to focus on improving the BCG strain *(59–61)*. Despite its shortcomings, BCG has an excellent safety record and, as a live persistent vaccine, it exhibits long lasting and complex immunostimulatory properties. Work with recombinant BCG, modified to produce immunogenic mycobacterial antigens, also shows promise against TB *(62,63)*. Recombinant BCG is also being used to fight other infections *(64,65)*. For example, BCG expressing an antigen from *Schistosoma mansoni* appears to protect mice from this helminthic infection *(66)*. In addition, BCG exhibits antitumor properties and is used as effective treatment for bladder cancer *(67)*.

6. *Mycobacterium leprae*

On February 28, 1873, G. H. Armauer Hansen identified bacilli in nodules removed from a patient with leprosy *(68)*. Previous epidemiological evidence suggested that leprosy was transmissible, but it was Hansen's finding that established leprosy as an infectious bacterial disease. Leprosy, often called Hansen's disease, affects millions and is endemic in India, Vietnam, and the Philippines, with approx 630,000 new infections occurring each year *(3)*. Despite its global importance, leprosy remains difficult to study. Animal models do not accurately emulate human disease, and the mechanisms of transmission are poorly defined. *M. leprae* is a slow growing obligate intracellular bacterium that has never been cultured in vitro. The complete genome sequence of *M. leprae* strain TN revealed that this organism has undergone massive gene decay; that is, an extreme form of the reductive evolution seen in the *M. tuberculosis* complex *(7)*. The genome is 3.3 Mbp, which is 1.1 Mbp smaller than that of *M. tuberculosis*. Of the estimated 3720 *M. leprae* open reading frames, more than 1100 are pseudogenes that no longer encode proteins. Even so, biosynthetic pathways for most molecules (e.g., amino acids, nucleotides) and cell wall components (e.g., peptidoglycan, arabinogalactan, mycolic acids, lipids) remain intact *(7,69)*. Conversely, many catabolic pathways, transcriptional regulators, transport proteins, polyketide synthesis systems, detoxification enzymes, and DNA repair processes are impaired. These defects likely account for the fastidious growth requirements of *M. leprae*. Many putative virulence factors, such as the *mce* genes, PE/PPE genes, and all PE-PRGS genes are also absent or defective. As in some strains of BCG, the *mma3* gene, required for synthesis of methoxymycolates, is defective *(70)*. Similar to *M. avium* subsp. *paratuberculosis*, the mycobactin siderophore biosynthesis (*mbt*) operon is nonfunctional, suggesting that iron metabolism is impaired. At the same time, *M. leprae* has an extra NRAMP-like metal transporter, which may compensate for the *mbt* defect. Other genes not found in *M. tuberculosis*, and possibly beneficial to intracellular survival of *M. leprae*, include a uridine phosphorylase, a eukaryote-like adenylate cyclase, and a putative sugar transport system.

The chronology of gene decay in *M. leprae* has been a matter of some debate. Did the loss of regulatory genes precipitate or follow the loss of metabolic functions? The isolation of an ancestral *M. leprae*, with a more intact genome, could answer this question, but no such organism has been identified. Epidemiological studies of *M. leprae* are hampered by the lack of molecular typing methods. One contribution of genome sequence analysis has been the identification of polymorphic regions suitable for molecular typing *(71,72)*. Such characterization is integral to an improved understanding of disease transmission. Although several natural reservoirs have been suggested, including armadillos, insects, soil, and water, the sources for human infection remain a mystery.

7. The *Mycobacterium avium* Complex

MAC includes a variety of genetically related species with diverse pathogenic potential *(10)*. *M. avium* subsp. *avium* (*Maa*) is common in the environment. It causes avian tuberculosis and sporadic infections of wild mammals (e.g., deer), as well as opportunistic infections in immunocompromised humans. *M. avium* subsp. *silvaticum* (*Mas*), the so-called wood pigeon bacillus, is primarily a bird pathogen. *M. avium* subsp. *paratuberculosis* (*Map*) causes Johne's disease and, although the hypothesis remains controversial, has been implicated as a cause of Crohn's Disease, a chronic inflammatory bowel disease in humans *(73–75)*. MAC organisms exhibit greater heterogeneity than members of the *M. tuberculosis* complex. Multiple sequevars have been revealed by rDNA analysis and unidirectional deletion events cannot account for relationships between all isolates *(9,76)*. Different branches appear to have acquired new genetic material via horizontal transfer *(77)*. Genome comparison of *Maa* strain 104 (a human pulmonary isolate) and *Map* strain K10 (a Johne's disease isolate from a cow) has emphasized variations both within MAC, and between MAC and the *M. tuberculosis* complex.

Maa strain 104 (5.4 Mbp) and *Map* strain K10 (4.5 Mbp) have larger genomes than *M. tuberculosis* and encode several hundred more genes. Orthologs to many *M. tuberculosis* genes exist, but there are some notable differences. For example, *Maa* encodes one-third as many PE/PPE genes as *M. tuberculosis*. Although the function of these repetitive proteins is unknown, they are thought to contribute to the antigenic diversity of mycobacteria. The RD1 region, considered important for virulence, is missing from MAC. At the same time, *Maa* possesses more transcriptional regulatory genes and dedicates a larger portion of its genome to lipid metabolism. Extra genes, especially those involved in transcriptional regulation or associated with cell wall functions, likely help *Maa* adapt to volatile environmental conditions. In general, the genomic differences between MAC and the *M. tuberculosis* complex seem to reflect their disparate lifestyles. *M. tuberculosis* is an obligate pathogen with a relatively stable intracellular niche and little exposure to other bacteria. In contrast, MAC are environmental organisms, which must contend with changing environmental conditions and have greater opportunity for horizontal gene transfer via interactions with disparate bacteria and phages. *Map* is capable of infecting cattle, but is also capable of survival for months in a barnyard or field *(78)*. *Maa* infects birds, thrives in drinking water and hot tubs, and is an endosymbiont of free-living protists *(79)*. Indeed, it has been suggested that the ability of *Maa* to colonize phagocytic amoebae and protozoans prefigures that hallmark of mycobacterial infections: the invasion and subversion of macrophages *(80)*.

Maa is considered to be the ancestral form of MAC, but the enormous diversity between *Maa* isolates has precluded a robust analysis of phylogeny. Isolates from diverse sources (e.g., a deer in New Zealand and a bird from the Netherlands) may be indistinguishable, yet samples from similar sources (immunocompromised people living in the same city) can differ at multiple sites and exhibit discrepancies in genome size of greater than 250 kb *(81)*.

Traditionally, ISs have been used to define species. Isolates containing IS*900* are called *Map (82)*. IS*900* negative isolates are called *Maa* when positive for IS*1245*, and *Mas* if positive for IS*901*. However, inconsistencies exist, and some nonMAC organisms may contain these mobile elements. New, genome-based approaches likely will reveal the evolutionary history of MAC *(81)*. Such methods have already contributed to the phylogeny of *Map (83)*. Long suspected on the basis of culture characteristics, it is now clear that *Map* can be subdivided into bovine (cattle) and ovine (sheep) branches *(84,85)*. However, it remains unclear if these genetic differences are related to idiosyncrasies of the host–pathogen interaction, or if they represent some geographical bias. Identification of markers restricted to individual MAC subspecies is key to the development of sensitive and specific diagnostic tests *(86)*. Reliable tests are not yet available for MAC diseases, but would be valuable for the efficient detection and treatment of conditions, such as Johne's disease *(4)*.

8. *Mycobacterium ulcerans* and *Mycobacterium marinum*

Buruli ulcer is a devastating skin disease. The ulcers are difficult to treat and can consume as much as 70% of the skin surface before causing death *(87)*. First described in the scientific literature in 1948, after an outbreak in an Australian resort town *(88)*, historical evidence suggests the disease has long been endemic in Africa. It takes its name from the Buruli region of Uganda *(89)*. The causative agent, *M. ulcerans*, is among the slowest growing mycobacteria, with an in vitro generation time of more than 30 h. It is associated with wetlands. Contaminated water and water-borne insect larvae are implicated in the infection cycle. The related organism, *M. marinum*, is a lethal pathogen of fish, amphibians, and reptiles *(90)*. In humans, *M. marinum* is responsible for skin diseases, such as swimming-pool granuloma and fish-fancier's finger. As the names suggest, this opportunistic infection also results from contact with contaminated water. However, *M. marinum* infections are rarely lethal.

The genome of *M. ulcerans* has yet to be completely sequenced, but its associated plasmid, pMUM1, has been deciphered *(91)*. This 174-kb plasmid contains genes for the biosynthesis of mycolactone, a ketolide with immunosuppressive properties.

Mycolactone is thought to be the virulence factor responsible for the Buruli ulcer, and the key difference between *M. ulcerans* and *M. marinum*. A thorough comparison of these organisms with *M. tuberculosis* has yet to be completed. The genomes of *M. marinum* and *M. ulcerans* are in the 6 Mbp range, and likely encode numerous genes required for their aquatic lifestyles. Even so, rDNA sequencing suggests that *M. marinum* and *M. ulcerans* are closely related to the *M. tuberculosis* complex. Large regions of synteny are present, and virulence-associated genes, including the RD1 region, are known to be intact *(92)*.

M. marinum has found favor as an experimental model of mycobacterial infections because it grows more quickly and is safer to work with than *M. tuberculosis (93–96)*. More importantly, *M. marinum* can be used to study genuine host–pathogen interactions. Mice are widely used to study tuberculosis, but *M. tuberculosis* is not a mouse pathogen and the mouse model, although useful, does not accurately reflect human disease. In contrast, *M. marinum* naturally infects both fish and frogs. The combination of *M. marinum* and zebrafish may prove to be a particularly useful model. Numerous genetic tools are available for their study and, like humans, zebrafish exhibit both innate and adaptive immune responses.

9. *Mycobacterium smegmatis*

Mycobacterium smegmatis was once believed to cause syphilis. It is now recognized as a harmless saprophyte, common in soil. Fast growing, it has served as a model for the study of mycobacteria *(97–99)*. Sequencing of the *M. smegmatis* mc^{2}155 laboratory strain reveals that, like the other environmental mycobacteria, it has a large genome. It also shares many of the physiological characteristics particular to the mycobacteria. However, *M. smegmatis* is quite different from the pathogenic mycobacteria and is unable to survive in macrophages *(100)*. Its genome shows little synteny with *M. tuberculosis* and many putative virulence genes are absent.

10. Mycobacteriophages

The first mycobacteriophage was described in 1947 *(101)*. Since then, several hundred have been isolated and every mycobacterial genome sequence contains at least one prophage. Phages are likely a key mediator of diversity in the MAC complex. It is not known if prophage gene products contribute to mycobacterial virulence, but considering the importance of prophage-derived toxins in other actinomycetes (e.g., the diphtheria and tetanus toxins of *Corynebacterium* spp.) their role in disease pathogenesis would not be surprising.

Mycobacteriophages have been widely employed as diagnostic tools. Phage-based strain typing has now been superseded by other methods, but luminescent reporter phages and phage replication assays still are used for rapid detection of mycobacteria in clinical and environmental samples, and to determine antibiotic resistance *(102–106)*. Phage-based systems also are used for genetic manipulation of mycobacteria, including allelic replacement and transposon-delivery *(107–111)*. Although little is known about individual phages, sequencing projects, such as those conducted at the Pittsburgh Bacteriophage Institute have revealed great diversity in both genome size and gene content *(112–117)*.

11. Mycobacterial Gene Families

Analysis of the genome sequences of different *Mycobacterium* species indicates that genome size may vary and individual genes may lack orthologs, but most functions are conserved. These include basic metabolic activities common to all prokaryotes, such as DNA replication, transcription, cell division, and small molecule biosynthesis, along with some mycobacterial-specific functions. Perhaps most importantly, several large gene families that were either previously unknown or poorly understood have been identified through the genome projects. Analysis of *Mycobacterium* genomes indicates that there is a core set of approx 200 highly conserved genes encoding mycobacterial-specific functions *(118)*. Half of these are associated with cell wall biosynthesis. Others fall into a "conserved hypothetical" category for which roles remain to be determined. Several are PE/PPE genes, and a few are classified as regulatory genes or virulence determinants, such as the *mce* genes. Not included in this core set are antigens of the 6 kDa early secretory antigenic target (ESAT)-6 and 10 kDa culture filtrate protein (CFP)-10 family, as individual loci are not perfectly conserved, and similar proteins are found in other actinomycetes. Likewise, the resuscitation promoting factors have homologs in other organisms.

11.1. Mycobacterial Lipid Metabolism

Genome sequencing revealed that a large number of *M. tuberculosis* genes (~250) encode enzymes involved in lipid metabolism. In contrast, *E. coli*, which has a similar genome size as *M. tuberculosis*, contains approx 50 enzymes involved in lipid metabolism. As mentioned earlier, the lipid-rich cell wall is a defining characteristic of mycobacteria (Fig. 2) *(119)*. The excess of lipid metabolic enzymes in *M. tuberculosis* correlate with the unusual chemical composition of the structure. The wall consists of three covalently linked polymers: peptidoglycan, arabinogalactan, and mycolic acid (Fig. 2) together with a variety of complex lipids, including lipoglycans (e.g., lipoarabinomannan, lipomannan, and the related phosphatidylinositol mannosides), glycopeptidolipids, sulfolipids, trehalose-containing glycolipids, phthiocerol dimycocerosates, phenolic glycolipids, and triacylglycerols *(120,121)*. The proportion of these lipids varies from species to species and is also affected by changing growth conditions *(122)*.

The lipid domain of mycobacterial cell wall forms an asymmetric bilayer. The outer leaflet of this bilayer contains various surface glycolipids, whereas the inner leaflet is composed exclusively of mycolic acids. The mycolic acid layer displays exceptionally low fluidity and low permeability. It is this barrier that is responsible for the natural resistance of mycobacteria to many antimicrobial agents, including antibiotics and host immune factors *(121,123,124)*. Important roles also have been suggested for the surface glycolipids, especially the multiple methyl-branched fatty acids: sulfolipids, phthiocerol dimycocerosates, phenolic glycolipids, diacylated trehaloses, and polyacylated trehaloses *(125)*. Genes for phthiocerol and phenolphthiocerol dimycocerosate synthesis are present on a *M. tuberculosis* pathogenicity island (PAI). The latter lipid is necessary for the growth of *M. tuberculosis* in the lungs of infected mice *(126)*. Another component of interest is the lipoglycan, lipoarabinomannan. The *M. tuberculosis* version of this lipid exerts immunomodulatory effects, including the downregulation of cell-mediated immunity *(127)*. Curiously, lipoarabinomannan from *M. chelonae* has

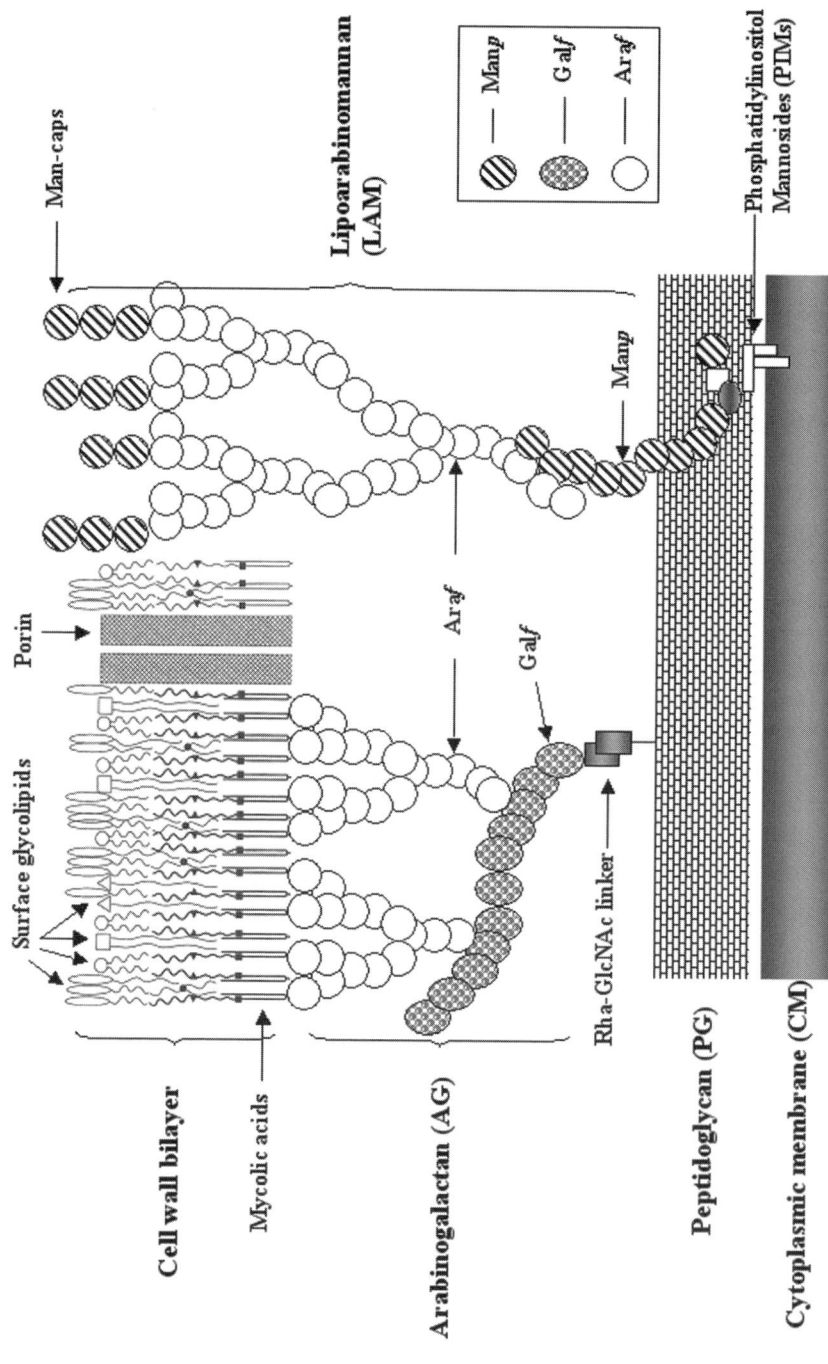

Fig. 2. Cell envelope of *Mycobacterium*. The cell wall core consists of three covalently linked polymers: peptidoglycan, arabinogalactan, and mycolic acids. The cell wall lipid domain forms an asymmetric lipid bilayer with mycolic acids constituting exclusively the inner leaflet and the extractable glycolipids occupying the outer leaflet of the bilayer. Lipoarabinomannan is thought to be associated with the cell wall through a phosphatidylinositol anchor to the cytoplasmic membrane.

no such ability. The difference appears to stem from species-specific variations in the structure of lipoarabinomannan *(127)*.

Homology searches to genes of known function in other bacteria have assisted in the identification of mycobacterial genes involved in cell wall lipid synthesis. Speculations on the role of individual genes can be tested by mutagenesis of the target gene and biochemical analysis of purified gene products. Such insight into the genetics and enzymology of cell wall biosynthesis and assembly makes it possible to identify cellular targets for the development of new drugs. Several front-line antitubercular drugs, such as isoniazid, ethambutol, and pyrazinamide, are now known to target lipid biosynthesis and cell wall assembly.

11.2. The PE and PPE Gene Families

The novel PE and PPE gene families are also highly abundant, accounting for 9% of all genes in *M. tuberculosis*. The proteins are distinguished by the eponymous PE or PPE motifs at N-terminal residues 8 and 9, or 8 to 10, respectively. Each family is further subdivided on the basis of characteristic C-terminal motifs. For example, all PE proteins have a conserved N-terminal domain (~110 amino acids). One subgroup includes short proteins with no C-terminal region. Members of the second group, PE-PGRS, have large C-terminal domains that contain multiple (sometimes hundreds) of tandem repeats of the glycine-rich motif, Gly–Gly–X (often Gly–Gly–Ala). The remaining PE proteins have substantial (100–400 amino acids) C-terminal domains, but the repeated Gly–Gly–X motif is absent. Analysis suggests that some proteins in this third group share an approx 225 amino acid motif in which the primary sequence is degenerate, but the secondary structure is conserved *(128)*.

The role of the PE/PPE genes has been the source of much speculation. One prominent idea is that they contribute to antigenic variation among strains of *M. tuberculosis*, and influence the host immune response. Comparative analysis of the PE-PGRS proteins of *M. tuberculosis* strains H37Rv and CDC1551 revealed variations resulting from frame-shift mutations, as well as in-frame insertions and deletions. Usually, the PE domains were unaffected and just the PGRS domains differed between strains. Size variations also were seen in clinical samples of *M. tuberculosis* by Western blot using PE-PGRS-specific antibody. In addition, some PE-PGRS proteins are surface exposed *(129)*. Others are antigenic and recognized by sera obtained from TB patients and those vaccinated with BCG *(130,131)*. However, their enzymatic functions, if any, are unknown and the importance of the conserved and variable regions is a mystery.

Additional evidence for the role of the PE/PPE genes in pathogenesis stems from the recent identification of three PAIs conserved among *M. tuberculosis* H37Rv, CDC1551, and *M. bovis (132)*. Prominent in PAI2 and PAI3 are PPE and PE-PGRS family genes. In vivo studies also indicate that these genes are important. For instance, when *M. marinum* resides in host granulomas or macrophages, two PE-PGRS genes are preferentially expressed *(95)*. Furthermore, disruption of PE/PPE family genes can lead to growth attenuation in the mouse model of TB *(133)*.

11.3. ESAT-6/CFP-10 Antigens

Exponentially growing *M. tuberculosis* secretes numerous proteins into the surrounding media. Two of these, ESAT-6 and CFP-10, are immunodominant antigens *(134,135)*.

However, neither protein is produced by BCG. The genes encoding ESAT-6 (*esx*A/Rv3875) and CFP-10 (*esx*B/Rv3874) belong to the RD1 region that is absent from all BCG vaccine strains. Independent deletion events also resulted in the loss of these genes from *M. microti* and the dassie bacillus *(46)*. ESAT-6 and CFP-10 are important for pathogenicity. Experimental deletion of RD1 from *M. tuberculosis* results in a diminution of virulence, whereas restoration of this region to *M. microti* enhances its virulence *(56,57, 136)*. The two proteins form a heterodimeric complex *(137)*. CFP-10 most likely acts as a chaperone. ESAT-6 has cytolytic properties and contributes to the cell-to-cell spread of mycobacteria within the host *(92,136)*. The antigens are coexpressed with genes encoding a novel type of secretion system. Work with both *M. tuberculosis* and *M. marinum* indicates that, even if the *esx*A and *esx*B genes are intact, disruption of this secretory apparatus blocks export of ESAT-6 and CFP-10, such that cytolysis does not occur *(92)*. Interaction of CFP-10, but not ESAT-6, with components of the secretion system has been demonstrated *(138)*. This entire gene cluster is conserved in *M. marinum*, *M. smegmatis*, and probably *M. kansasii* and *M. szulgai*. A homologous region is present in *M. leprae*, but contains several pseudogenes In contrast, the cluster is completely absent from *M. avium (137,139)*.

Ten paralogous gene pairs (*esx*C/*esx*D to *esx*V/*esx*W) are annotated in the *M. tuberculosis* H37Rv genome. Although expected to form heterodimeric complexes similar to ESAT-6/CFP-10, their roles have yet to be established. Several of these pairs are associated with their own versions of the novel secretion system. These clusters are variably present across mycobacterial genomes, but at least one, *esx*G/*esx*H and its associated secretion apparatus, is conserved between *M. tuberculosis*, *M. avium*, *M. smegmatis*, and *M. leprae (139)*. Related systems are present in other actinomycetes, including *Corynebacterium* and *Streptomyces*, but their importance is currently unknown *(139)*.

ESAT-6 and CFP-10 are being used in the development of new diagnostics and new vaccines. Skin testing with PPD (the purified protein derivative of *M. tuberculosis*) is a sensitive method for the diagnosis of TB. Unfortunately, it lacks specificity because false-positive reactions commonly occur among BCG vaccinated individuals. In contrast, ESAT-6 and CFP-10 are not produced by BCG. The two immunodominant antigens are recognized by sera from the majority of TB patients, but not sera from people vaccinated with BCG. As such, ESAT-6 and CFP-10 can effectively discriminate between vaccinated and infected individuals *(140)*.

Vaccine development has employed two approaches toward the antigens. The first involves deletion of *esx*A and *esx*B from *M. bovis* and *M. tuberculosis* to produce a live attenuated vaccine *(141,142)*. The opposite strategy has been to use ESAT-6 and CFP-10 for vaccination. They have been used individually, in protein cocktails, and expressed in either recombinant BCG or alternate carrier strains, such as attenuated *Salmonella (63,143–146)*. Both approaches show promise, but a deletion strain would permit continued use of the antigens as diagnostic tools.

11.4. Mammalian Cell Entry

The mammalian cell entry (MCE) proteins are putative virulence factors present in diverse mycobacteria *(147)*. Their name stems from the finding that a cloned fragment of the *M. tuberculosis* H37Rv *mce1* region permits nonpathogenic *E. coli* cells to invade cultured HeLa cells *(148)*. The canonical *mce* region is a polycistronic operon of eight genes following the order *yrbAB mceABCDEF*. However, among the four *mce* operons in *M. tuberculosis* and nine in *Maa*, some variations exist. The *M. tuberculosis mce2* operon contains two extra genes, one upstream of *yrb2A* and a second between *mce2B* and *mce2C (149)*. Consistent with a putative role in invasion, these proteins are predicted to be cell wall-associated and/or secreted. *In silico* modeling of Mce1A suggests its structure resembles that of Colicin N, a pore-forming bacterial toxin *(150)*. This attractive, albeit hypothetical, model predicts that binding of Mce1A to its cognate receptor is followed by a conformational change of the protein and perforation of the target cell membrane. The putative receptor-binding surface of the *in silico* structure corresponds to a known MceA1 epitope, recognized by a monoclonal antibody. This epitope is not conserved in MceA2, MceA3, or MceA4, implying that the proteins bind to different receptors and possibly target different cells. In agreement with this model, a truncated form of recombinant MceA1 protein, which includes the putative receptor-binding domain, promotes the uptake of latex beads by HeLa cells, whereas recombinant Mce2A protein does not *(151)*. Although the roles of individual *mce* operons have not been determined, gene expression profiling indicates that they are differentially transcribed. In synthetic media, *mce1* is strongly expressed by exponentially growing *M. tuberculosis*, but only weakly by stationary phase cells. The inverse pattern is found with *mce4*; strong expression in stationary phase but none during exponential growth. The *mce* genes are also transcribed in vivo. *M. tuberculosis* from the spleens of experimentally infected Guinea pigs were found to express *mce4*. Bacilli from the lungs of infected rabbits express *mce1*, *mce3*, and *mce4 (149)*. Appropriate spatial and temporal expression of the *mce* operons may be important events in the infectious process. Primary mouse macrophages infected with a *mce1*-defective *M. tuberculosis* strain have an altered cytokine profile and may be unable to stimulate T-cell immunity. Mice infected with the same mutant *mce1* strain develop histologically aberrant lung granulomas and are unable to control bacterial replication *(152)*. However, the impact of *mce* defects likely varies among different mycobacterial species and in different hosts. Homologs of the *mce3* operon have been deleted from both *Map* and *M. bovis*, yet these organisms remain effective pathogens in cattle.

11.5. Resuscitation Promoting Factors

The resuscitation promoting factor (Rpf) was first described in another actinomycete, *Micrococcus luteus (153)*. Following long-term culture, these cells lose viability and are unable to form colonies when plated onto fresh solid medium. However, the senescent cells are not dead, just dormant and nonculturable. Viability and colony formation by *Micrococcus* are restored by the addition of Rpf, a small protein secreted by actively growing cells. As mentioned earlier, dormancy is a feature of the bacteria persisting in TB granulomas. Moreover, *M. tuberculosis*, BCG, and *M. smegmatis* all respond to exogenously added Rpf *(154,155)*. Genome sequencing has revealed five *rpf* genes in *M. tuberculosis* and a similar number in MAC. *M. smegmatis* also produces Rpf.

The *M. tuberculosis rpf* genes have been systematically deleted *(156)*. Individually, the genes are neither essential for growth in vitro nor for in vivo infection of mice. This suggests that they have redundant or overlapping functions, although strains with multiple *rpf* deletions have yet to be tested. Analysis indicates that all *M. tuberculosis rpf* genes are transcribed during logarithmic growth. Expression patterns vary at intermediate time points, but all transcripts are also detected in 4-mo-old cultures. During the acute phase of infection, *rpf* expression is also present in *M. tuberculosis* from mouse lungs *(156)*. Additional work on the immunological impact of Rpfs is required, but they have attracted some attention as vaccine candidates *(157)*.

12. Concluding Remarks

Mycobacterial research is experiencing a renaissance. The wealth of genome sequence data, new molecular tools, plus bioinformatic, proteomic, structural, and functional genomic approaches hold substantial promise for understanding the biology of these unusual microbes, elucidating the molecular mechanisms of pathogenesis, and developing new chemotherapeutic agents and effective vaccines. These approaches ultimately should lead to the effective control of mycobacterial disease and end the scourge of TB.

References

1. Cornet, G. (1904) *Tuberculosis and Acute General Miliary Tuberculosis.* W.B. Sauders and Co., Philadelphia.
2. World Health Organization. Stop TB Annual Report. (2001) World Health Organization, Geneva, Switzerland.
3. World Health Organization (2002) Leprosy. Global situation. *Wkly. Epidemiol. Rec.* **77,** 1–8.
4. Cocito, C., Gilot, P., Coene, M., de Kesel, M., Poupart, P., and Vannuffel, P. (1994) Paratuberculosis. *Clin. Microbiol. Rev.* **7,** 328–345.
5. Cole, S. T., Brosch, R., Parkhill, J., et al. (1998) Deciphering the biology of *Mycobacterium tuberculosis* from the complete genome sequence. *Nature* **393,** 537–544.
6. Garnier, T., Eiglmeier, K., Camus, J. C., et al. (2003) The complete genome sequence of *Mycobacterium bovis. Proc. Natl. Acad. Sci. USA* **100,** 7877–7882.
7. Cole, S. T., Eiglmeier, K., Parkhill, J., et al. (2001) Massive gene decay in the leprosy bacillus. *Nature* **409,** 1007–1011.
8. Fleischmann, R. D., Alland, D., Eisen, J. A., et al. (2002) Whole-genome comparison of *Mycobacterium tuberculosis* clinical and laboratory strains. *J. Bacteriol.* **184,** 5479–5490.
9. Harmsen, D., Dostal, S., Roth, A., et al. (2003) RIDOM: comprehensive and public sequence database for identification of *Mycobacterium* species. *BMC. Infect. Dis.* **3,** 26.
10. Primm, T. P., Lucero, C. A., and Falkinham, J. O. (2004) Health impacts of environmental mycobacteria. *Clin. Microbiol. Rev.* **17,** 98–106.
11. Wayne, L. G. and Sramek, H. A. (1992) Agents of newly recognized or infrequently encountered mycobacterial diseases. *Clin. Microbiol. Rev.* **5,** 1–25.
12. Wolinsky, E. (1992) Mycobacterial diseases other than tuberculosis. *Clin. Infect. Dis.* **15,** 1–10.
13. Tortoli, E. (2003) Impact of genotypic studies on mycobacterial taxonomy: the new mycobacteria of the 1990s. *Clin. Microbiol. Rev.* **16,** 319–354.
14. Roth, A., Fischer, M., Hamid, M. E., Michalke, S., Ludwig, W., and Mauch, H. (1998) Differentiation of phylogenetically related slowly growing mycobacteria based on 16S-23S rRNA gene internal transcribed spacer sequences. *J. Clin. Microbiol.* **36,** 139–147.
15. Cloud, J. L., Neal, H., Rosenberry, R., et al. (2002) Identification of *Mycobacterium* spp. by using a commercial 16S ribosomal DNA sequencing kit and additional sequencing libraries. *J. Clin. Microbiol.* **40,** 400–406.

16. Koch, R. (1882) Die Aetiologie der Tuberkulose. *Berliner Klinischen Wochenschrift.* **15,** 221–230.

17. Salo, W. L., Aufderheide, A. C., Buikstra, J., and Holcomb, T. A. (1994) Identification of *Mycobacterium tuberculosis* DNA in a pre-Columbian Peruvian mummy. *Proc. Natl. Acad. Sci. USA* **91,** 2091–2094.

18. Zink, A. R., Sola, C., Reischl, U., et al. (2003) Characterization of *Mycobacterium tuberculosis* complex DNAs from Egyptian mummies by spoligotyping. *J. Clin. Microbiol.* **41,** 359–367.

19. Dye, C., Scheele, S., Dolin, P., Pathania, V., and Raviglione, M. C. (1999) Consensus statement. Global burden of tuberculosis: estimated incidence, prevalence, and mortality by country. WHO Global Surveillance and Monitoring Project. *JAMA* **282,** 677–686.

20. World Health Organization. Global Tuberculosis Control: Surveillance, Planning, Financing. WHO Report 2004. (2004) World Health Organization, Geneva, Switzerland.

21. Gazzard, B. (2001) Tuberculosis, HIV and the developing world. *Clin. Med.* **1,** 62–68.

22. Porter, J. D. (1996) Mycobacteriosis and HIV infection: the new public health challenge. *J. Antimicrob. Chemother.* **37,** 113–120.

23. Cosma, C. L., Sherman, D. R., and Ramakrishnan, L. (2003) The secret lives of the pathogenic mycobacteria. *Annu. Rev. Microbiol.* **57,** 641–676.

24. Smith, I. (2003) *Mycobacterium tuberculosis* pathogenesis and molecular determinants of virulence. *Clin. Microbiol. Rev.* **16,** 463–496.

25. Amer, A. O. and Swanson, M. S. (2002) A phagosome of one's own: a microbial guide to life in the macrophage. *Curr. Opin. Microbiol.* **5,** 56–61.

26. Deretic, V. and Fratti, R. A. (1999) *Mycobacterium tuberculosis* phagosome. *Mol. Microbiol.* **31,** 1603–1609.

27. Stewart, G. R., Robertson, B. D., and Young, D. B. (2003) Tuberculosis: a problem with persistence. *Nat. Rev. Microbiol.* **1,** 97–105.

28. Wayne, L. G. (1994) Dormancy of *Mycobacterium tuberculosis* and latency of disease. *Eur. J. Clin. Microbiol. Infect. Dis.* **13,** 908–914.

29. Lawn, S. D., Butera, S. T., and Shinnick, T. M. (2002) Tuberculosis unleashed: the impact of human immunodeficiency virus infection on the host granulomatous response to *Mycobacterium tuberculosis*. *Microbes. Infect.* **4,** 635–646.

30. Espinal, M. A. (2003) The global situation of MDR-TB. *Tuberculosis* **83,** 44–51.

31. Mukherjee, J. S., Rich, M. L., Socci, A. R., et al. (2004) Programmes and principles in treatment of multidrug-resistant tuberculosis. *Lancet* **363,** 474–481.

32. Nachega, J. B. and Chaisson, R. E. (2003) Tuberculosis drug resistance: a global threat. *Clin. Infect. Dis.* **36,** S24–S30.

33. Valway, S. E., Sanchez, M. P., Shinnick, T. F., et al. (1998) An outbreak involving extensive transmission of a virulent strain of *Mycobacterium tuberculosis*. *N. Engl. J. Med.* **338,** 633–639.

34. Glynn, J. R., Whiteley, J., Bifani, P. J., Kremer, K., and van Soolingen, D. (2002) Worldwide occurrence of Beijing/W strains of *Mycobacterium tuberculosis*: a systematic review. *Emerg. Infect. Dis.* **8,** 843–849.

35. van Soolingen, D., Qian, L., de Haas, P. E., et al. (1995) Predominance of a single genotype of *Mycobacterium tuberculosis* in countries of east Asia. *J. Clin. Microbiol.* **33,** 3234–3238.

36. Tsolaki, A. G., Hirsh, A. E., DeRiemer, K., et al. (2004) Functional and evolutionary genomics of *Mycobacterium tuberculosis*: insights from genomic deletions in 100 strains. *Proc. Natl. Acad. Sci. USA* **101,** 4865–4870.

37. Perna, N. T., Plunkett, G., Burland, V., et al. (2001) Genome sequence of enterohaemorrhagic *Escherichia coli* O157:H7. *Nature* **409,** 529–533.

38. Sreevatsan, S., Pan, X., Stockbauer, K. E., et al. (1997) Restricted structural gene polymorphism in the *Mycobacterium tuberculosis* complex indicates evolutionarily recent global dissemination. *Proc. Natl. Acad. Sci. USA* **94,** 9869–9874.

39. Barnes, P. F. and Cave, M. D. (2003) Molecular epidemiology of tuberculosis. *N. Engl. J. Med.* **349,** 1149–1156.

40. van Soolingen, D. (2001) Molecular epidemiology of tuberculosis and other mycobacterial infections: main methodologies and achievements. *J. Intern. Med.* **249,** 1–26.

41. Mostrom, P., Gordon, M., Sola, C., Ridell, M., and Rastogi, N. (2002) Methods used in the molecular epidemiology of tuberculosis. *Clin. Microbiol. Infect.* **8,** 694–704.

42. Brosch, R., Gordon, S. V., Marmiesse, M., et al. (2002) A new evolutionary scenario for the *Mycobacterium tuberculosis* complex. *Proc. Natl. Acad. Sci. USA* **99,** 3684–3689.

43. Gutacker, M. M., Smoot, J. C., Migliaccio, C. A., et al. (2002) Genome-wide analysis of synonymous single nucleotide polymorphisms in *Mycobacterium tuberculosis* complex organisms: resolution of genetic relationships among closely related microbial strains. *Genetics* **162,** 1533–1543.

44. Mostowy, S., Cousins, D., Brinkman, J., Aranaz, A., and Behr, M. A. (2002) Genomic deletions suggest a phylogeny for the *Mycobacterium tuberculosis* complex. *J. Infect. Dis.* **186,** 74–80.

45. Mostowy, S., Onipede, A., Gagneux, S., et al. (2004) Genomic analysis distinguishes *Mycobacterium africanum*. *J. Clin. Microbiol.* **42,** 3594–3599.

46. Mostowy, S., Cousins, D., and Behr, M. A. (2004) Genomic interrogation of the dassie bacillus reveals it as a unique RD1 mutant within the *Mycobacterium tuberculosis* complex. *J. Bacteriol.* **186,** 104–109.

47. Chen, J. M., Alexander, D. C., Behr, M. A., and Liu, J. (2003) *Mycobacterium bovis* BCG vaccines exhibit defects in alanine and serine catabolism. *Infect. Immun.* **71,** 708–716.

48. Behr, M. A. and Small, P. M. (1999) A historical and molecular phylogeny of BCG strains. *Vaccine* **17,** 915–922.

49. Crispen, R. (1989) History of BCG and its substrains. *Prog. Clin. Biol. Res.* **310,** 35–50.

50. Behr, M. A. and Small, P. M. (1997) Has BCG attenuated to impotence? *Nature* **389,** 133–134.

51. Brewer, T. F. and Colditz, G. A. (1995) Relationship between bacille Calmette-Guerin (BCG) strains and the efficacy of BCG vaccine in the prevention of tuberculosis. *Clin. Infect. Dis.* **20,** 126–135.

52. Brandt, L., Feino, C. J., Weinreich, O. A., et al. (2002) Failure of the *Mycobacterium bovis* BCG vaccine: some species of environmental mycobacteria block multiplication of BCG and induction of protective immunity to tuberculosis. *Infect. Immun.* **70,** 672–678.

53. Buddle, B. M., Wards, B. J., Aldwell, F. E., Collins, D. M., and de Lisle, G. W. (2002) Influence of sensitisation to environmental mycobacteria on subsequent vaccination against bovine tuberculosis. *Vaccine* **20,** 1126–1133.

54. Behr, M. A., Wilson, M. A., Gill, W. P., et al. (1999) Comparative genomics of BCG vaccines by whole-genome DNA microarray. *Science* **284,** 1520–1523.

55. Mahairas, G. G., Sabo, P. J., Hickey, M. J., Singh, D. C., and Stover, C. K. (1996) Molecular analysis of genetic differences between *Mycobacterium bovis* BCG and virulent *M. bovis*. *J. Bacteriol.* **178,** 1274–1282.

56. Lewis, K. N., Liao, R., Guinn, K. M., et al. (2003) Deletion of RD1 from *Mycobacterium tuberculosis* mimics bacille Calmette-Guerin attenuation. *J. Infect. Dis.* **187,** 117–123.

57. Pym, A. S., Brodin, P., Brosch, R., Huerre, M., and Cole, S. T. (2002) Loss of RD1 contributed to the attenuation of the live tuberculosis vaccines *Mycobacterium bovis* BCG and *Mycobacterium microti*. *Mol. Microbiol.* **46,** 709–717.

58. Mostowy, S., Tsolaki, A. G., Small, P. M., and Behr, M. A. (2003) The in vitro evolution of BCG vaccines. *Vaccine* **21,** 4270–4274.

59. Doherty, T. M. and Andersen, P. (2002) Tuberculosis vaccine development. *Curr. Opin. Pulm. Med.* **8,** 183–187.

60. Kumar, H., Malhotra, D., Goswami, S., and Bamezai, R. N. (2003) How far have we reached in tuberculosis vaccine development? *Crit. Rev. Microbiol.* **29,** 297–312.

61. Young, D. B. and Stewart, G. R. (2002) Tuberculosis vaccines. *Br. Med. Bull.* **62,** 73–86.

62. Horwitz, M. A. and Harth, G. (2003) A new vaccine against tuberculosis affords greater survival after challenge than the current vaccine in the guinea pig model of pulmonary tuberculosis. *Infect. Immun.* **71,** 1672–1679.

63. Pym, A. S., Brodin, P., Majlessi, L., et al. (2003) Recombinant BCG exporting ESAT-6 confers enhanced protection against tuberculosis. *Nat. Med.* **9,** 533–539.

64. Stover, C. K., Bansal, G. P., Hanson, M. S., et al. (1993) Protective immunity elicited by recombinant bacille Calmette-Guerin (BCG) expressing outer surface protein A (OspA) lipoprotein: a candidate Lyme disease vaccine. *J. Exp. Med.* **178,** 197–209.

65. Stover, C. K., de la Cruz, V. F., Fuerst, T. R., et al. (1991) New use of BCG for recombinant vaccines. *Nature* **351,** 456–460.

66. Varaldo, P. B., Leite, L. C., Dias, W. O., et al. (2004) Recombinant *Mycobacterium bovis* BCG expressing the Sm14 antigen of *Schistosoma mansoni* protects mice from cercarial challenge. *Infect. Immun.* **72,** 3336–3343.

67. Shelley, M. D., Court, J. B., Kynaston, H., et al. (2004) Intravesical Bacillus Calmette-Guerin in Ta and T1 Bladder Cancer. In: *The Cochrane Library* **4,** John Wiley and Sons, Ltd, Chichester, UK.

68. Hansen, G. H. A. (1875) On the etiology of leprosy. *Br. J. Foreign Med. Chir. Rev.* **55,** 459–489.

69. Brennan, P. J. and Vissa, V. D. (2001) Genomic evidence for the retention of the essential mycobacterial cell wall in the otherwise defective *Mycobacterium leprae. Lepr. Rev.* **72,** 415–428.

70. Behr, M. A., Schroeder, B. G., Brinkman, J. N., Slayden, R. A., and Barry, C. E. (2000) A point mutation in the *mma*3 gene is responsible for impaired methoxymycolic acid production in *Mycobacterium bovis* BCG strains obtained after 1927. *J. Bacteriol.* **182,** 3394–3399.

71. Groathouse, N. A., Rivoire, B., Kim, H., et al. (2004) Multiple polymorphic loci for molecular typing of strains of *Mycobacterium leprae. J. Clin. Microbiol.* **42,** 1666–1672.

72. Shin, Y. C., Lee, H., Walsh, G. P., Kim, J. D., and Cho, S. N. (2000) Variable numbers of TTC repeats in *Mycobacterium leprae* DNA from leprosy patients and use in strain differentiation. *J. Clin. Microbiol.* **38,** 4535–4538.

73. Chacon, O., Bermudez, L. E., and Barletta, R. G. (2004) Johne's disease, inflammatory bowel disease, and *Mycobacterium paratuberculosis. Annu. Rev. Microbiol.* **58,** 329–363.

74. Greenstein, R. J. (2003) Is Crohn's disease caused by a mycobacterium? Comparisons with leprosy, tuberculosis, and Johne's disease. *Lancet Infect. Dis.* **3,** 507–514.

75. Hermon-Taylor, J. and Bull, T. (2002) Crohn's disease caused by *Mycobacterium avium* subspecies *paratuberculosis*: a public health tragedy whose resolution is long overdue. *J. Med. Microbiol.* **51,** 3–6.

76. Novi, C., Rindi, L., Lari, N., and Garzelli, C. (2000) Molecular typing of *Mycobacterium avium* isolates by sequencing of the 16S-23S rDNA internal transcribed spacer and comparison with *IS1245*-based fingerprinting. *J. Med. Microbiol.* **49,** 1091–1095.

77. Krzywinska, E., Krzywinski, J., and Schorey, J. S. (2004) Naturally occurring horizontal gene transfer and homologous recombination in *Mycobacterium. Microbiology* **150,** 1707–1712.

78. Whittington, R. J., Marshall, D. J., Nicholls, P. J., Marsh, I. B., and Reddacliff, L. A. (2004) Survival and dormancy of *Mycobacterium avium* subsp. *paratuberculosis* in the environment. *Appl. Environ. Microbiol.* **70,** 2989–3004.

79. Falkinham, J. O., Norton, C. D., and LeChevallier, M. W. (2001) Factors influencing numbers of *Mycobacterium avium*, *Mycobacterium intracellulare*, and other Mycobacteria in drinking water distribution systems. *Appl. Environ. Microbiol.* **67,** 1225–1231.

80. Skriwan, C., Fajardo, M., Hagele, S., et al. (2002) Various bacterial pathogens and symbionts infect the amoeba *Dictyostelium discoideum. Int. J. Med. Microbiol.* **291,** 615–624.

81. Semret, M., Zhai, G., Mostowy, S., et al. (2004) Extensive genomic polymorphism within *Mycobacterium avium. J. Bacteriol.* **186,** 6332–6334.

82. Bull, T. J., Hermon-Taylor, J., Pavlik, I., El-Zaatari, F., and Tizard, M. (2000) Characterization of IS*900* loci in *Mycobacterium avium* subsp. *paratuberculosis* and development of multiplex PCR typing. *Microbiology* **146,** 2185–2197.

83. Amonsin, A., Li, L. L., Zhang, Q., et al. (2004) Multilocus short sequence repeat sequencing approach for differentiating among *Mycobacterium avium* subsp. *paratuberculosis* strains. *J. Clin. Microbiol.* **42,** 1694–1702.

84. Dohmann, K., Strommenger, B., Stevenson, K., et al. (2003) Characterization of genetic differences between *Mycobacterium avium* subsp. *paratuberculosis* type I and type II isolates. *J. Clin. Microbiol.* **41,** 5215–5223.

85. Motiwala, A. S., Strother, M., Amonsin, A., et al. (2003) Molecular epidemiology of *Mycobacterium avium* subsp. *paratuberculosis*: evidence for limited strain diversity, strain sharing, and identification of unique targets for diagnosis. *J. Clin. Microbiol.* **41,** 2015–2026.

86. Bannantine, J. P., Hansen, J. K., Paustian, M. L., et al. (2004) Expression and immunogenicity of proteins encoded by sequences specific to *Mycobacterium avium* subsp. *paratuberculosis. J. Clin. Microbiol.* **42,** 106–114.

87. van der Werf, T. S., Stinear, T., Stienstra, Y., van der Graaf, W. T., and Small, P. L. (2003) Mycolactones and *Mycobacterium ulcerans* disease. *Lancet* **362,** 1062–1064.

88. MacCallum, P., Tolhurst, J. C., Buckle, G., and Sissons, H. A. (1948) A new mycobacterial infection in man. *J. Pathol. Bacteriol.* **60,** 93–122.

89. Clancey, J. K. (1964) Mycobacterial skin ulcers in Uganda: description of a new mycobacterium (Mycobacterium Buruli). *J. Pathol. Bacteriol.* **88,** 175–187.

90. Decostere, A., Hermans, K., and Haesebrouck, F. (2004) Piscine mycobacteriosis: a literature review covering the agent and the disease it causes in fish and humans. *Vet. Microbiol.* **99,** 159–166.

91. Stinear, T. P., Mve-Obiang, A., Small, P. L., et al. (2004) Giant plasmid-encoded polyketide synthases produce the macrolide toxin of *Mycobacterium ulcerans. Proc. Natl. Acad. Sci. USA* **101,** 1345–1349.

92. Gao, L. Y., Guo, S., McLaughlin, B., Morisaki, H., Engel, J. N., and Brown, E. J. (2004) A mycobacterial virulence gene cluster extending RD1 is required for cytolysis, bacterial spreading and ESAT-6 secretion. *Mol. Microbiol.* **53,** 1677–1693.

93. Chan, K., Knaak, T., Satkamp, L., Humbert, O., Falkow, S., and Ramakrishnan, L. (2002) Complex pattern of *Mycobacterium marinum* gene expression during long-term granulomatous infection. *Proc. Natl. Acad. Sci. USA* **99,** 3920–3925.

94. Ramakrishnan, L. and Falkow, S. (1994) *Mycobacterium marinum* persists in cultured mammalian cells in a temperature-restricted fashion. *Infect. Immun.* **62,** 3222–3229.

95. Ramakrishnan, L., Federspiel, N. A., and Falkow, S. (2000) Granuloma-specific expression of *Mycobacterium* virulence proteins from the glycine-rich PE-PGRS family. *Science* **288,** 1436–1439.

96. Ramakrishnan, L., Valdivia, R. H., McKerrow, J. H., and Falkow, S. (1997) *Mycobacterium marinum* causes both long-term subclinical infection and acute disease in the leopard frog (*Rana pipiens*). *Infect. Immun.* **65,** 767–773.

97. Andrew, P. W. and Roberts, I. S. (1993) Construction of a bioluminescent mycobacterium and its use for assay of antimycobacterial agents. *J. Clin. Microbiol.* **31,** 2251–2254.

98. Mayuri, Bagchi, G., Das, T. K., and Tyagi, J. S. (2002) Molecular analysis of the dormancy response in *Mycobacterium smegmatis*: expression analysis of genes encoding the DevR-DevS two-component system, Rv3134c and chaperone alpha-crystallin homologues. *FEMS Microbiol. Lett.* **211,** 231–237.

99. Triccas, J. A., Parish, T., Britton, W. J., and Gicquel, B. (1998) An inducible expression system permitting the efficient purification of a recombinant antigen from *Mycobacterium smegmatis*. *FEMS Microbiol. Lett.* **167,** 151–156.

100. Wei, J., Dahl, J. L., Moulder, J. W., et al. (2000) Identification of a *Mycobacterium tuberculosis* gene that enhances mycobacterial survival in macrophages. *J. Bacteriol.* **182,** 377–384.

101. Gardner, G. M. and Weiser, R. S. (1947) A bacteriophage for *Mycobacterium smegmatis*. *Proc. Soc. Exp. Biol. Med.* **66,** 205–206.

102. Bardarov, S. J., Dou, H., Eisenach, K., et al. (2003) Detection and drug-susceptibility testing of *M. tuberculosis* from sputum samples using luciferase reporter phage: comparison with the Mycobacteria Growth Indicator Tube (MGIT) system. *Diagn. Microbiol. Infect. Dis.* **45,** 53–61.

103. Carriere, C., Riska, P. F., Zimhony, O., et al. (1997) Conditionally replicating luciferase reporter phages: improved sensitivity for rapid detection and assessment of drug susceptibility of *Mycobacterium tuberculosis*. *J. Clin. Microbiol.* **35,** 3232–3239.

104. Hazbon, M. H., Guarin, N., Ferro, B. E., et al. (2003) Photographic and luminometric detection of luciferase reporter phages for drug susceptibility testing of clinical *Mycobacterium tuberculosis* isolates. *J. Clin. Microbiol.* **41,** 4865–4869.

105. Riska, P. F. and Jacobs, W. R. Jr. (1998) The use of luciferase-reporter phage for antibiotic-susceptibility testing of mycobacteria. *Methods Mol. Biol.* **101,** 431–455.

106. Riska, P. F., Su, Y., Bardarov, S., et al. (1999) Rapid film-based determination of antibiotic susceptibilities of *Mycobacterium tuberculosis* strains by using a luciferase reporter phage and the Bronx Box. *J. Clin. Microbiol.* **37,** 1144–1149.

107. Bardarov, S., Bardarov, J. S. J., Pavelka, J. M. J., et al. (2002) Specialized transduction: an efficient method for generating marked and unmarked targeted gene disruptions in *Mycobacterium tuberculosis*, *M. bovis* BCG and *M. smegmatis*. *Microbiology* **148,** 3007–3017.

108. Bardarov, S., Kriakov, J., Carriere, C., et al. (1997) Conditionally replicating mycobacteriophages: a system for transposon delivery to *Mycobacterium tuberculosis*. *Proc. Natl. Acad. Sci. USA* **94,** 10,961–10,966.

109. Jacobs, W. R. Jr., Snapper, S. B., Tuckman, M., and Bloom, B. R. (1989) Mycobacteriophage vector systems. *Rev. Infect. Dis.* **11,** S404–S410.

110. Pearson, R. E., Jurgensen, S., Sarkis, G. J., Hatfull, G. F., and Jacobs, W. R. J. (1996) Construction of D29 shuttle phasmids and luciferase reporter phages for detection of mycobacteria. *Gene* **183,** 129–136.

111. Jacobs, W. R. J., Tuckman, M., and Bloom, B. R. (1987) Introduction of foreign DNA into mycobacteria using a shuttle phasmid. *Nature* **327,** 532–535.

112. Pedulla, M. L., Ford, M. E., Houtz, J. M., et al. (2003) Origins of highly mosaic mycobacteriophage genomes. *Cell* **113,** 171–182.

113. Hendrix, R. W., Smith, M. C., Burns, R. N., Ford, M. E., and Hatfull, G. F. (1999) Evolutionary relationships among diverse bacteriophages and prophages: all the world's a phage. *Proc. Natl. Acad. Sci. USA* **96,** 2192–2197.

114. Hatfull, G. F. and Sarkis, G. J. (1993) DNA sequence, structure and gene expression of mycobacteriophage L5: a phage system for mycobacterial genetics. *Mol. Microbiol.* **7,** 395–405.

115. Ford, M. E., Sarkis, G. J., Belanger, A. E., Hendrix, R. W., and Hatfull, G. F. (1998) Genome structure of mycobacteriophage D29: implications for phage evolution. *J. Mol. Biol.* **279,** 143–164.

116. Ford, M. E., Stenstrom, C., Hendrix, R. W., and Hatfull, G. F. (1998) Mycobacteriophage TM4: genome structure and gene expression. *Tuber. Lung Dis.* **79,** 63–73.

117. Mediavilla, J., Jain, S., Kriakov, J., et al. (2000) Genome organization and characterization of mycobacteriophage Bxb1. *Mol. Microbiol.* **38,** 955–970.

118. Marmiesse, M., Brodin, P., Buchrieser, C., et al. (2004) Macro-array and bioinformatic analyses reveal mycobacterial 'core' genes, variation in the ESAT-6 gene family and new phylogenetic markers for the *Mycobacterium tuberculosis* complex. *Microbiology* **150,** 483–496.

119. Brennan, P. J. (2003) Structure, function, and biogenesis of the cell wall of Mycobacterium tuberculosis. *Tuberculosis* **83,** 91–97.

120. Brennan, P. J. and Nikaido, H. (1995) The envelope of mycobacteria. *Ann. Rev. Biochem.* **64,** 29–63.

121. Liu, J. and Nikaido, H. (1999) A mutant in *Mycobacterium smegmatis* defective in the biosynthesis of mycolic acids accumulates meromycolates. *Proc. Natl. Acad. Sci. USA* **96,** 4011–4016.

122. Kolattukudy, P. E., Fernandes, N. D., Azad, A. K., Fitzmaurice, A. M., and Sirakova, T. D. (1997) Biochemistry and molecular genetics of cell-wall lipid biosynthesis in mycobacteria. *Mol. Microbiol.* **24,** 263–270.

123. Liu, J., Barry, C. E., Besra, G. S., and Nikaido, H. (1996) Mycolic acid structure determines the fluidity of the mycobacterial cell wall. *J. Biol. Chem.* **271,** 29,545–29,551.

124. Liu, J., Rosenberg, E. Y., and Nikaido, H. (1995) Fluidity of the lipid domain of cell wall from *Mycobacterium chelonae. Proc. Natl. Acad. Sci. USA* **92,** 11,254–11,258.

125. Minnikin, D. E., Kremer, L., Dover, L. G., and Besra, G. S. (2002) The methyl-branched fortifications of *Mycobacterium tuberculosis. Chem. Biol.* **9,** 545–553.

126. Cox, J. S., Chen, B., McNeil, M., and Jacobs, W. R. J. (1999) Complex lipid determines tissue-specific replication of *Mycobacterium tuberculosis* in mice. *Nature* **402,** 79–83.

127. Nigou, J., Gilleron, M., and Puzo, G. (2003) Lipoarabinomannans: from structure to biosynthesis. *Biochimie* **85,** 153–166.

128. Adindla, S. and Guruprasad, L. (2003) Sequence analysis corresponding to the PPE and PE proteins in *Mycobacterium tuberculosis* and other genomes. *J. Biosci.* **28,** 169–179.

129. Sampson, S. L., Lukey, P., Warren, R. M., van Helden, P. D., Richardson, M., and Everett, M. J. (2001) Expression, characterization and subcellular localization of the *Mycobacterium tuberculosis* PPE gene Rv1917c. *Tuberculosis* **81,** 305–317.

130. Choudhary, R. K., Mukhopadhyay, S., Chakhaiyar, P., et al. (2003) PPE antigen Rv2430c of *Mycobacterium tuberculosis* induces a strong B-cell response. *Infect. Immun.* **71,** 6338–6343.

131. Okkels, L. M., Brock, I., Follmann, F., et al. (2003) PPE protein (Rv3873) from DNA segment RD1 of *Mycobacterium tuberculosis*: strong recognition of both specific T-cell epitopes and epitopes conserved within the PPE family. *Infect. Immun.* **71,** 6116–6123.

132. Zubrzycki, I. Z. (2004) Analysis of the products of genes encompassed by the theoretically predicted pathogenicity islands of *Mycobacterium tuberculosis* and *Mycobacterium bovis. Proteins* **54,** 563–568.

133. Sassetti, C. M. and Rubin, E. J. (2003) Genetic requirements for mycobacterial survival during infection. *Proc. Natl. Acad. Sci. USA* **100,** 12,989–12,994.

134. Andersen, P., Andersen, A. B., Sorensen, A. L., and Nagai, S. (1995) Recall of long-lived immunity to *Mycobacterium tuberculosis* infection in mice. *J. Immunol.* **154,** 3359–3372.

135. Berthet, F. X., Rasmussen, P. B., Rosenkrands, I., Andersen, P., and Gicquel, B. (1998) A *Mycobacterium tuberculosis* operon encoding ESAT-6 and a novel low-molecular-mass culture filtrate protein (CFP-10). *Microbiology* **144,** 3195–3203.

136. Hsu, T., Hingley-Wilson, S. M., Chen, B., et al. (2003) The primary mechanism of attenuation of bacillus Calmette-Guerin is a loss of secreted lytic function required for invasion of lung interstitial tissue. *Proc. Natl. Acad. Sci. USA* **100,** 12,420–12,425.

137. Sorensen, A. L., Nagai, S., Houen, G., Andersen, P., and Andersen, A. B. (1995) Purification and characterization of a low-molecular-mass T-cell antigen secreted by *Mycobacterium tuberculosis. Infect. Immun.* **63,** 1710–1717.

138. Okkels, L. M. and Andersen, P. (2004) Protein-protein interactions of proteins from the ESAT-6 family of *Mycobacterium tuberculosis. J. Bacteriol.* **186,** 2487–2491.

139. Gey, V. P. N., Gamieldien, J., Hide, W., Brown, G. D., Siezen, R. J., and Beyers, A. D. (2001) The ESAT-6 gene cluster of *Mycobacterium tuberculosis* and other high G+C Gram-positive bacteria. *Genome Biol.* **2,** 1–18.

140. van Pinxteren, L. A., Ravn, P., Agger, E. M., Pollock, J., and Andersen, P. (2000) Diagnosis of tuberculosis based on the two specific antigens ESAT-6 and CFP10. *Clin. Diagn. Lab. Immunol.* **7,** 155–160.

141. Collins, D. M., Kawakami, R. P., Wards, B. J., Campbell, S., and de Lisle, G. W. (2003) Vaccine and skin testing properties of two avirulent *Mycobacterium bovis* mutants with and without an additional *esat-6* mutation. *Tuberculosis* **83,** 361–366.

142. Wards, B. J., de Lisle, G. W., and Collins, D. M. (2000) An *esat*6 knockout mutant of *Mycobacterium bovis* produced by homologous recombination will contribute to the development of a live tuberculosis vaccine. *Tuber. Lung Dis.* **80,** 185–189.

143. Brandt, L., Elhay, M., Rosenkrands, I., Lindblad, E. B., and Andersen, P. (2000) ESAT-6 subunit vaccination against *Mycobacterium tuberculosis. Infect. Immun.* **68,** 791–795.

144. Mollenkopf, H. J., Groine-Triebkorn, D., Andersen, P., Hess, J., and Kaufmann, S. H. (2001) Protective efficacy against tuberculosis of ESAT-6 secreted by a live *Salmonella typhimurium* vaccine carrier strain and expressed by naked DNA. *Vaccine* **19,** 4028–4035.

145. Mustafa, A. S. and Al-Attiyah, R. (2003) Tuberculosis: looking beyond BCG vaccines. *J. Postgrad. Med.* **49,** 134–140.

146. Olsen, A. W., Hansen, P. R., Holm, A., and Andersen, P. (2000) Efficient protection against *Mycobacterium tuberculosis* by vaccination with a single subdominant epitope from the ESAT-6 antigen. *Eur. J. Immunol.* **30,** 1724–1732.

147. Haile, Y., Caugant, D. A., Bjune, G., and Wiker, H. G. (2002) *Mycobacterium tuberculosis* mammalian cell entry operon (*mce*) homologs in *Mycobacterium* other than tuberculosis (MOTT). *FEMS Immunol. Med. Microbiol.* **33,** 125–132.

148. Arruda, S., Bomfim, G., Knights, R., Huima-Byron, T., and Riley, L. W. (1993) Cloning of an *M. tuberculosis* DNA fragment associated with entry and survival inside cells. *Science* **261,** 1454–1457.

149. Kumar, A., Bose, M., and Brahmachari, V. (2003) Analysis of expression profile of mammalian cell entry (*mce*) operons of *Mycobacterium tuberculosis. Infect. Immun.* **71,** 6083–6087.

150. Das, A. K., Mitra, D., Harboe, M., et al. (2003) Predicted molecular structure of the mammalian cell entry protein Mce1A of *Mycobacterium tuberculosis. Biochem. Biophys. Res. Commun.* **302,** 442–447.

151. Chitale, S., Ehrt, S., Kawamura, I., et al. (2001) Recombinant *Mycobacterium tuberculosis* protein associated with mammalian cell entry. *Cell Microbiol.* **3,** 247–254.

152. Shimono, N., Morici, L., Casali, N., et al. (2003) Hypervirulent mutant of *Mycobacterium tuberculosis* resulting from disruption of the *mce1* operon. *Proc. Natl. Acad. Sci. USA* **100,** 15,918–15,923.

153. Mukamolova, G. V., Kaprelyants, A. S., Young, D. I., Young, M., and Kell, D. B. (1998) A bacterial cytokine. *Proc. Natl. Acad. Sci. USA* **95,** 8916–8921.

154. Mukamolova, G. V., Turapov, O. A., Young, D. I., Kaprelyants, A. S., Kell, D. B., and Young, M. (2002) A family of autocrine growth factors in *Mycobacterium tuberculosis.* *Mol. Microbiol.* **46,** 623–635.

155. Shleeva, M., Mukamolova, G. V., Young, M., Williams, H. D., and Kaprelyants, A. S. (2004) Formation of 'non-culturable' cells of *Mycobacterium smegmatis* in stationary phase in response to growth under suboptimal conditions and their Rpf-mediated resuscitation. *Microbiology* **150,** 1687–1697.

156. Tufariello, J. M., Jacobs, W. R. Jr., and Chan, J. (2004) Individual *Mycobacterium tuberculosis* resuscitation-promoting factor homologues are dispensable for growth in vitro and in vivo. *Infect. Immun.* **72,** 515–526.

157. Yeremeev, V. V., Kondratieva, T. K., Rubakova, E. I., et al. (2003) Proteins of the Rpf family: immune cell reactivity and vaccination efficacy against tuberculosis in mice. *Infect. Immun.* **71,** 4789–4794.

10

Mycoplasma

Yuko Sasaki

Summary

Mycoplasmas are cell wall-less bacterium with small genome sizes, typically 0.6–1.4 Mb. All mycoplasma species are obligate parasites with specific hosts. Their small genomes are thought to be the result of reductive evolution from an ancestor on a bacterial phylogenetic branch with a low guanine and cytosine content (i.e., a member of the Firmicutes, such as *Clostridium* spp. and *Bacillus* spp.) adapting to obligate parasitic life. In this chapter, the features of mycoplasma/ureaplasma/phytoplasma genomes are discussed in terms of reductive evolution, a gene set for essential functions, and paralog formation under evolutionary pressure for gene reduction.

Key Words: Mollicutes; mycoplasma; reductive evolution; host-specificity; genome size; antigenic variation; self-replication; paralog; ortholog.

1. Introduction

1.1. Host Specificity of Mollicutes

Mollicutes are cell wall-less bacteria and form the taxonomic class *Mollicutes*, containing five families and nine genera including uncultivated phytoplasma *(1)*. Most mollicutes, except some acholeplasma, are obligate parasites for specific hosts. Four genera (*Entomoplasma*, *Mesoplasma*, *Spiroplasma*, and *Phytoplasma*) are both plant- and insect-associated, two genera (*Anaeroplasma* and *Asteroleplasma*) are anaerobes in the bovine rumen, and two genera (*Mycoplasma* and *Ureaplasma*) form the family *Mycoplasmataceae* and associate with many different hosts, including fish, birds, and mammals. The genus *Mycoplasma* contains greater than 100 species, each with host specificity in Pisces, Reptilia, Aves, Rodentia, Carnivora, Perissodactyla, Artiodactyla, and Primates. For example, *M. mycoides* subsp. *mycoides* SC, the first mycoplasma isolated in 1898 by Nocard and Roux and originally named a pleuropneumoniae-like organism, is the etiological agent of contagious bovine pleuropneumoniae and causes a highly contagious respiratory disease in cattle and water buffaloes with high mortality, but is harmless for other animals and birds. The pathogenicity of mycoplasma species is linked to their natural host; e.g., *M. gallisepticum* for birds, *M. pulmonis* for rodents, *M. bovis* for cattle, *M. capricolum* for sheep and goats, *M. hyopheumoniae* for pigs, phytoplasma for plants, and so on. These relationships between mollicutes and their hosts are probably the result of coevolution during geological epochs. For example, analysis of 16S–23S ribosomal RNA (rRNA) intergenic spacer regions of ureaplasma

From: *Bacterial Genomes and Infectious Diseases*
Edited by: V. L. Chan, P. M. Sherman, and B. Bourke © Humana Press Inc., Totowa, NJ

suggests the possibility of coevoluation between ureaplasmal species and host animals *(2)*. Mollicutes infection in nature also reflect such evolutionary relationships *(3–10)*.

1.2. Mycoplasma *Infection-Related Diseases in Humans*

Sixteen mollicutes species, including 13 mycoplasma species, have been isolated from either the respiratory or genitourinary tract of humans *(11)*. A pathogenic role has been described for at least six species: *M. pneumoniae*, *M. genitalium*, *M. hominis*, *M. fermentans*, *M. penetrans*, and *U. urealyticum*. In certain hosts, both mycoplasma and ureaplasma cause respiratory disease, urogenital disease, breeding disorders, and arthritis, possibly including rheumatoid arthritis. *M. pneumoniae* is the best known species and causes atypical pneumonia in humans. In rare cases, *M. pneumoniae* infection is fatal with diffuse pneumonia, respiratory distress, disseminated intravascular coagulation, and multiple organ failure *(12,13)*. *M. genitalium* is phylogenetically close to *M. pneumoniae* and is a major human pathogen causing nongonococcal urethritis. Even in recent years, human-related mycoplasmal species such as *M. amphoriforme* (Pitcher 2005) and *M. penetrans* have been newly discovered. *M. penetrans* was originally isolated from the urine of a human immunodeficiency virus (HIV)-positive patient *(14,15)*. *M. penetrans* produces an intracellular infection in urothelium *(16)* and persists long-term *(17)*. Seroepidemiological data in Europe and the United States indicated that about 20–40% of HIV-positive men infected via a homosexual route were also *M. penetrans*-positive *(18,19)*. Rapid decline in CD4-positive lymphocyte counts in *M. penetrans*-seropositive HIV-infected individuals has been observed *(20)*. The mitogenic effects of *M. penetrans* on lymphocytes of HIV-positive people imply its possible contribution for both the deterioration of the immune system and virus replication *(21,22)*. More recently, a systemic disease was reported by *M. penetrans* infection in a non-HIV-infected previously healthy young woman *(23)*. The patient, from whom *M. penetrans* HF strains were isolated, developed severe respiratory distress and abnormality in blood coagulation accompanied with increasing level of anti-cardioripin antibody, diagnosed as primary anti-phospholiplid syndrome (APS), required intensive care management, and recovered after treatment with suitable anti-*M. penetrans* antibiotics. The genome of the *M. penetrans* HF-2 strain has been completely sequenced *(24)*. Comparative genome analysis of the *Mollicutes* is discussed in **Subheading 3.1.**

2. Reductive Evolution

2.1. Reductive Evolution in Both Mutualistic and Commensal Symbiosis

Symbiotic relationships between organisms tend to coevolve. In mutualistic relationships, both organisms develop functions in a complementary manner, therefore, some genes for complemental functions are reduced in one of the two organisms. A well-known example of mutualism is *Buchnera aphidicola*, the intracellular bacteria of aphids. This bacteria is an obligate mutualistic symbiont of aphids and has a genome size about 0.6 Mb. The Buchnera's genome has genes for the biosynthesis of several amino acids essential for its host (e.g., arginine, valine, isoleucine), but lacks those amino acids that are nonessential for the host. Hence, *B. aphidicola* provides its aphid host with metabolites it cannot synthesize, and the host provides the bacteria with metabolites *B. aphidicola* cannot synthesize *(25)*. In this obligate mutualistic association, *B. aphidicola* has a reduced genome size. Another example of a reduced genome in symbiotic relationships is provided by the nucleomorph in both Cryptophyta and Chlorarachniophyta. The nucleomorph is a relic of a eukaryotic endosymbiont's nucleus encoding approx 464 putative coding sequences (CDSs; i.e., open reading frames) *(26,27)*. Such obligate endosymbionts lose gene sets for metabolism, reduce their genome size, and become organella of unicellular organisms. Some obligate parasites develop commensalistic rather than mutualistic relationships with their hosts, and evolve like symbionts in terms of genome reduction. An obligate parasite risks losing its nutrient source either by death of the host or elimination by the host immune system, subsequently it is unable to find a sensitive population to infect. For these reasons, unlike naturally occurring bacteria that occasionally infect a host, obligate parasites probably need to evolve to produce a chronic infection in the host as commensal symbionts. Mollicutes that are obligate parasites for specific hosts have lost their cell wall and the ability to synthesize cholesterol, and rely on host-dependent nutrition. Mycoplasmas have reduced their genome size to approx 1.4–0.6 Mb encoding 1000–480 CDSs. Reductive evolution in mycoplasma is similar in some ways to that of the mutualistic symbiont, *B. aphidicola*, which has a genome size of approx 0.64 Mb encoding about 550 CDSs. Both bacteria have incomplete gene sets for biosynthesis of amino acids, for the TCA cycle, and for cell envelope components. Such a small genome size is probably the result of host dependency, with the parasite being able to reduce its genome size by relying on complemental genes in the host. The rate of reductive evolution is probably rapid. Similar to the nucleomorph that has a different rate of evolution to that of the host nucleus *(28)*, the rate of evolution of mollicutes is more rapid than that of other eubacteria. On a few rapidly evolving mollicutes phylogenetic branches, such as the pneumoniae group, small-genome mycoplasmas existed *(29)*.

2.2. Reductive Evolution in Mollicutes

Based on 16S rRNA gene analysis, it has been calculated that mollicutes diverged from Gram-positive bacteria with low guanine and cytosine (G+C) contents in the late Proterozoic era of the Precambrian, about 605 million yr ago. At that time, the ancestral mollicutes probably lost cell wall synthesis, some biosynthesis and rRNA genes, and reduced their genome size. The ancestral mollicutes were probably not obligate parasites and did not require cholesterol as a nutrient for replication, similar to acholeplasma. The Genera *Acholeplasma, Anaeroplasma, Asteroplasma, Phytoplasma* (AAP) phylogenetic branch, and the *Spiroplasma, Mesoplasma, Entomoplasma, Mycoplasma, Ureaplasma* (SEM) phylogenetic branch diverged in the middle Ordovician, about 470 million years before the present. In the SEM branch, the *Spiroplasma-Mesoplasma-Entomoplasma* branch (adapting to plants and insects) split to the *Mycoplasma-Ureaplasma* branch (adapting to fish, reptiles, birds, and mammals) at the Silurian-Devonian boundary at about 410 million years ago, the time of major expansion of marine life and the appearance of the first land animals *(30)*. The genome reductions during evolution produced different genome sizes; 1.7–1.5 Mb for most of the AAP branch, 1.4–1.1 Mb for most species in the SEM branch and, the smallest, 0.8–0.6 Mb for both phytoplasma and some species in the mycoplasma-ureaplasma group *(30)*.

All mycoplasma species are on the *Mycoplasma-Ureaplasma* phylogenetic branch except the *M. mycoides* subgroup, which are within the *Spiroplasma-Mesoplasma-Entomoplasma* branch. The *Mycoplasma-Ureaplasma* phylogenetic branch consists of four groups, designated α, β, γ, and δ branches *(30)*. The most recently evolved δ branch (also designated the pneumoniae group) contains species with small genome sizes generally less than 1 Mb *(29)*, such as *M. genitalium* (0.58 Mb, 484 CDSs), *M. pneumoniae* (0.81 Mb, 689 CDSs), and *U. parvum* (0.75 Mb, 614 CDSs) (*U. urealyticum* biovar parvum was recently reclassified as a separate species). However, although species not only on other branch in mycoplasma (e.g., *M. pulmonis*, 0.96 Mb) but also on the AAP branch (e.g., onion yellow phytoplasma, a plant pathogen and symbiont for insects, *Candidatus* Pytoplasma asteris, 0.86 Mb) have smaller genomes than some members of the δ branch (e.g., *M. penetrans*, 1.3 Mb), reductive evolution probably occurred independently on several different phylogenetic branches. In each case, 0.6-0.8 Mbs is the lower genome size limit *(31)*.

2.3. The Essential Genes for Mollicutes

Study on small mollicute genomes provides on opportunity to identify the minimal essential set of genes for a functioning cell. *M. genitalium* has the smallest genome size of any cultivable organism. Based on a comparative analysis of the genomes of *M. genitalium* and *Haemophilus influenzae*, and the study of gene knockouts in *M. genitalium*, it has been proposed that 256 genes constitute the minimal gene set necessary and sufficient for sustaining a functional cell *(32–34)*, even though recent experiments re-examined by using global transposon mutagenesis for pure clonal populations indicated that 387 of 482 protein-encoding genes are essential in *M. genitalium* (Glass 2006). Subsequently, some genes in the 256 gene candidates have been found to be missing in the genome of several mycoplasma-ureaplasma species. For example, genes for the cell division protein (*ftsZ*) and chaperonins for protein folding (*groEL* and *groES*) are missing in *U. parvum*. Eight genes including *groEL*, 1-phosphofructokinase (*fluK*), and uridine kinase (*udk*) are missing in *M. pulmonis*. *udk* is also missing in *M. penetrans*, and 11 genes including *groEL* and *groES*, genes for the ribosomal protein S6 modification protein, cytidine deaminase, riboflavin kinase, thymidylate synthase, dihydrofolate reductase, and a histone-like protein are absent in *M. mycoides (24,35–39)*. In addition, nucleoside-diphosphate kinase (*ndk*) for nucleoside metabolism is absent in the genomes of all nine species that have been sequenced in mollicutes listed in Table 1 *(24,40)*. Using gene inactivation, 271 of the approx 4100 *Bacillus subtilis* genes have been found to be essential genes *(41)*. Bacillus is a low G+C Gram-positive bacterial group that probably shared a common ancestor with mollicutes *(42)*. When essential genes are compared in mollicutes and *B. subtilis*, genes for cell wall synthesis is absent in mollicutes. tRNA synthetases, the gene for threonyl-tRNA synthetase (*thrS*) is essential in mollicutes, but missing in *B. subtilis*. Two genes for a two-component system, (*yycF* and *yyc*) are essential in *B. subtilis*. However, of all the nine species mollicutes, except *M. penetrans*, this common bacterial regulation system is absent. Some pyrimidine metabolism genes (e.g., *ndk* and *udk*) are also nonessential in mollicutes and *B. subtilis*. However, genes missing in some *Mollicutes* species (e.g., *ftsZ*, *groEL*, and *groES*) are essential in *B. subtilis*. Both *Chlamidia* sp. and the archeaon *Aeropyrum pernix* also lack *ftsZ*. The functions of some missing genes are probably replaced by nonorthologous displacement *(43)*. The genome of an intracellular uncultivable species in mollicutes, onion yellow phytoplasma *Candidatus* Phytoplasma asteris, lacks genes that are essential in other organisms, such as the pentose phosphate pathway, F0F1-type ATP synthase subunits,

Table 1
Reciprocal Best Hit Genes With Either *Mycoplasma penetrans* or *Baccilus subtilis*

Species	Total number of CDSs	Number of CDS hits with *M. penetrans*	Percent of CDS hits with *M. penetrans* in each genome	Number of CDS hits with *B. subtilis*	Percent of CDS hits with *B. subtilis*/ all *B. subtilis* CDSs
M. penetrans[a]	1038	1038	100.0%	587	14.3%
M. genitalium[a]	484	383	79.1%	384	9.3%
M. mobile	633	408	64.4%	461	11.2%
U. parvum[a]	614	390	63.5%	458	11.1%
M. pneumoniae[a]	689	433	62.8%	467	11.4%
M. gallisepticum[a]	742	457	61.6%	472	11.5%
M. pulmonis	782	400	51.2%	472	11.5%
M. mycoides	985	419	42.5%	544	13.2%
Candidatus Phytoplasma asteris	754	297	39.4%	381	9.3%
Cl. perfringens	2660	503	18.9%	ND	ND
B. subtilis	4112	587	14.3%	4112	100%
Cl. acetobutilicum	4927	520	10.6%	ND	ND

[a]Species in the delta/*Pneumoniae* group.

and ATP/ADP translocation *(37)*. The missing essential genes for other bacteria in mollicutes reflect events during reductive evolution in host–parasite relationships.

3. Comparative Analysis of Closely Related Species

3.1. The M. penetrans Genome is Larger Than Other Species in the Delta/Pnemoniae Phylogenetic Branch

The sequenced genomes of mycoplasma and ureaplasma species are highly divergent and very little gene order/synteny is conserved, except between *M. genitalium* and *M. pneumoniae*. This diversity is probably because of the rapid rate of *Mycoplasma* evolution and dynamic gene rearrangement during adaptation to hosts and tissues. Of the sequenced species on the δ branch (*M. genitalium*, *M. pneumoniae*, *M. penetrans*, *M. gallisepticum*, and *U. parvum*), the genome size of *M. penetrans*, approx 1.3 Mb, is the largest. Humans are the hosts of these species, except *M. gallisepticum*, which infects birds. To clarify evolutionary events in the δ branch, we analyzed three aspects of the *M. penetrans* genome: conserved genes from an ancestral bacteria, events during gene reduction, and species-specific paralog genes developed during host adaptation *(24)*.

3.1.1. Conservation of Genes From an Ancestral Bacteria

For the ortholog search, reciprocal best-hit pairs of genes were identified by pairwise BLASTP comparisons of predicted *M. penetrans* proteins with those in other species in mycoplasma, ureaplasma and low G+C content Gram-positive bacteria. As shown in Table 1, the number of CDS hits with *B. subtilis* indicates that approx 200 orthologs of *B. subtilis* genes that are in the *M. penetrans* genome are missing in the *M. genitalium* genome. Conversely, the core proteome, defined as functionally distinct protein families in the theoretical proteome, was 847 in *M. penetrans (24)*. This core proteome is greater than that in the other species in the δ branch, indicating richer functions in *M. penetrans*. *M. penetrans* is likely to have genes concerning with a two-component system, commonly present in other bacteria for regulating gene expression under the control of sensor/regulator sets. Interestingly, there is an ortholog of the putative response regulator of *M. penetrans* in the fish-associated *M. mobile* genome *(44)*. The presence of these genes suggests that *M. penetrans* has lost fewer genes from a common ancestor with mycoplasma and *B. subtilis* than other species in the δ branch.

3.1.2. Events During Gene Reduction

Like other mycoplasma and ureaplasma species with genomes smaller than 1 Mb, *M. penetrans* also lacks some important genes for pyrimidine metabolism enzymes; e.g., *udk* that converts uridine/cytidine to uridine/cytidine monophosphate (UMP/CMP), 5'-nucleotidase for the reverse reaction to that of *udk*, and *ndk* that converts uridine/cytidine diphosphate (UDP/CDP) to uridine/cytidine triphosphate (UTP/CTP). Since phytoplasma also lacks *ndk*, *ndk* seems likely to have been lost during an early stage of evolution in mollicutes. The arginine dihydrolase operon is an example of gene rearrangement in the δ branch. In general, mollicutes have no urea cycle, although some species in the δ branch synthesize ATP via the arginine dihydrolase pathway, probably with a partial urea cycle plus arginine deiminase to convert arginine to citrulline directly. *M. penetrans* has all the genes for this pathway (i.e., arginyl-tRNA synthetase, *argS*; arginine deiminase, *arcA*; ornithine carbamoyltransferase, *arcB*; and carbamate kinase, *arcC*). These genes are located adjacently in the order *arcA-arcB-arcC*-amino acid permeases-*argS*, similar to the arrangement in *Clostridium* spp. The carbamoyl-phosphate of the end product in this pathway is converted to UMP via an orotate-related pathway that has been found only in *M. penetrans*, of the eight sequenced species in mollicutes. Arginine dihydrolase pathway-related genes have also been found in *M. pneumoniae*, although the *arcB* genes are fragmented and the related genes locate separately. Hence, this pathway in *M. pneumoniae* looks inactive. The gene set for this pathway in *M. gallisepticum* is incomplete, but it has two *arcA* genes and one *argS*. Dislocation and absence of genes in the *arc* operon may be a reflection of dynamic genome rearrangements accompanying gene reduction in both *M. pneumoniae* and *M. gallisepticum*.

3.1.3. M. penetrans-Specific Genes Developed for Adaptation

The most remarkable feature of the *M. penetrans* genome is the large number of paralogous gene families. Of 1038, 264 (25.4%) genes form 63 gene families, ranging from 2 to 44 proteins per family *(24)*. As shown in Fig. 1, the number of paralogs in bacterial genomes is correlated with genome size, with a few exceptions. These exceptions are of two types: species with very few paralogs (e.g., *Richettsia prowasekii*), and species with more paralogs than average (e.g., *Candidatus* P. asteris. and *M. penetrans*). Most *M. penetrans* paralog families appear to have evolved without horizontal transfer of genes by transposase or phage integration. A comparison of paralogs in mollicutes is shown in Table 2.

It can be concluded from these data that the larger genome size of *M. penetrans* is the result of loss of some genes in the early stage of mollicutes evolution, functional conservation after divergence from ancestral bacteria, and development of unique paralogs.

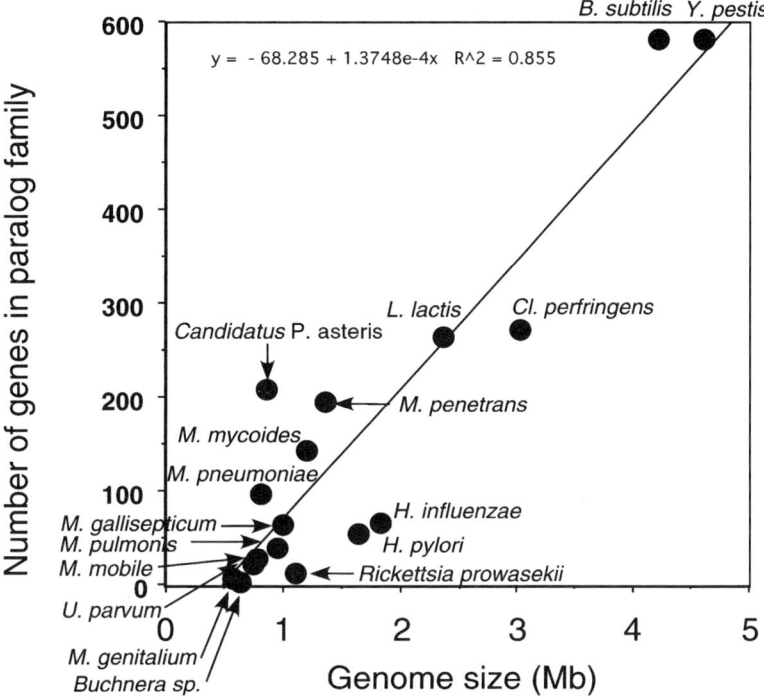

Fig. 1. Correlation between the number of paralogs and genome size in bacteria. In a BLAST score-based single-linkage clustering search using BLASTCLUST, proteins with more than 30% amino acid identity over 70% of their length were defined as paralogs. The number of genes in paralog families, including transposases, with more than three copies including transposase are counted.

Table 2
Paralogs More Than Three Copies With or Without Transposase

Species	Genome size (bp)	Total number of CDSs	Excluding transposase		Including transposase	
			Number of paralogs	Percent of paralogs in the genome	Number of paralogs	Percent of paralogs in the genome
Candidatus Phytoplasma asteris	860,631	754	203	26.9%	210	27.9%
M. penetrans	1,358,633	1038	173	16.7%	196	18.9%
M. pneumoniae	816,394	689	76	11.0%	98	14.2%
M. mycoides	1,211,703	985	72	7.3%	144	14.9%
M. gallisepticum	996,422	742	61	8.2%	66	8.9%
M. pulmonis	963,879	782	37	4.7%	40	5.1%
M. mobile	777,079	633	28	4.4%	28	4.4%
U. parvum	751,719	614	24	3.9%	24	3.9%
M. genitalium	580,074	484	7	1.4%	7	1.4%

In a BLAST score-based single-linkage clustering search using BLASTCLUST, proteins with more than 30% amino acid identity over 70% of their length were defined as paralogs.

3.2. Paralog Formation Under the Evolutionaty Pressure of Gene Reduction

3.2.1. Paralogs of the p35 Gene Family and Antigenic Variation in M. penetrans

The results of cluster analysis of paralogs in the nine sequenced mollicutes genomes show that the largest gene family is the *p35* gene family encoding the surface-exposed lipoproteins responsible for antigenic variation. This family contains 44 genes, including the unique *p35* gene. Of these 44 genes, 38 probably encode lipoproteins, because these 38 genes have a highly conserved N-terminal amino acid sequence of about 34 amino acids containing a cysteine residue that binds the fatty acid that anchors the lipoprotein to the cell membrane lipid bilayer. The remaining six genes lack conserved regions and are signal peptide-less *p35* gene homologs. The identity of each gene at the amino acids level to the actual *p35* gene ranges from 29 to 70%. These 44 genes are clustered at four different loci: (1) one with 4 genes, (2) the largest with 30 genes, (3) one with the remaining 4 genes of the 38 lipoprotein genes, and (4) one with 6 genes for the signal peptide-less *p35* homologs *(24)*. As shown in Fig. 2A, the genetic tree of the *p35* family indicates that the 30 genes in the largest locus are closely related to each other, three other genes in different loci (MYPE2690, MYPE2700, and MYPE7400) formed the same branch with the 30 genes and these are probably the results of a dynamic gene rearrangement during paralog formation, and six signal peptide-less *p35* homologs are close to the root of the tree.

3.2.2. Different Antigenic Epitopes are Regulated
by On/Off Switching of Multiple Promoters in the p35 Multi-Gene Family

Mycoplasmal lipoproteins, including the P35 lipoprotein families of *M. penetrans*, are immunodominant. The lipoproteins encoded by the *p35* gene family are phase variable and have "on and off" expression phases. The frequency of phase variation for P35 lipoprotein ranges from 1.5×10^{-2} to 4×10^{-3} per cell per generation *(45,46)*. Horino et al. found transcriptional regulation of the *p35* gene family *(47)*. Based on sodium dodecyl sulphate-polyacrylamide gel electrophoresis analysis of lipoproteins, two different sized lipoproteins (P35 and P42) of the 38 genes were expressed in *M. penetrans* strain HF-2, but P42 was not expressed in *M. penetrans* strain GTU. A 135-bp DNA region flanked by 12-bp inverted repeats upstream of the P42 gene was found to be oriented in opposite direction in strains HF-2 and GTU. The promoter-like sequence within the 135-bp sequence was confirmed by primer extension analysis. This showed the promoter is in "on" orientation in strain HF-2, and in "off" orientation in strain GTU. Moreover, in "off" orientation in strain GTU, one of the 12-bp inverted repeats upstream of the *p42* gene was partially overlapped by 16-bp inverted repeats forming a hairpin structure. This hairpin structure appears only when the 135-bp sequence is in the "off" orientation and seems to act as a terminator to block both transcription to the reverse direction and transcription of an adjacent gene upstream. These invertible promoter-like structures are found in association with 31 of the 38 genes in the *p35* gene family. Terminator-like hairpin structures are combined with promoter-like structures in 37 of the 38 genes *(47)*. One of the major antigenic epitopes in the P35 lipoprotein, FTGEAYSVWSAK, that is found in approx 66% of *M. penetrans*-seropositive individuals and one of two infected monkeys *(48)* existed only in the actual *p35* gene in the gene family. Such a gene structure suggests a large antigenic repertoire in *M. penetrans*.

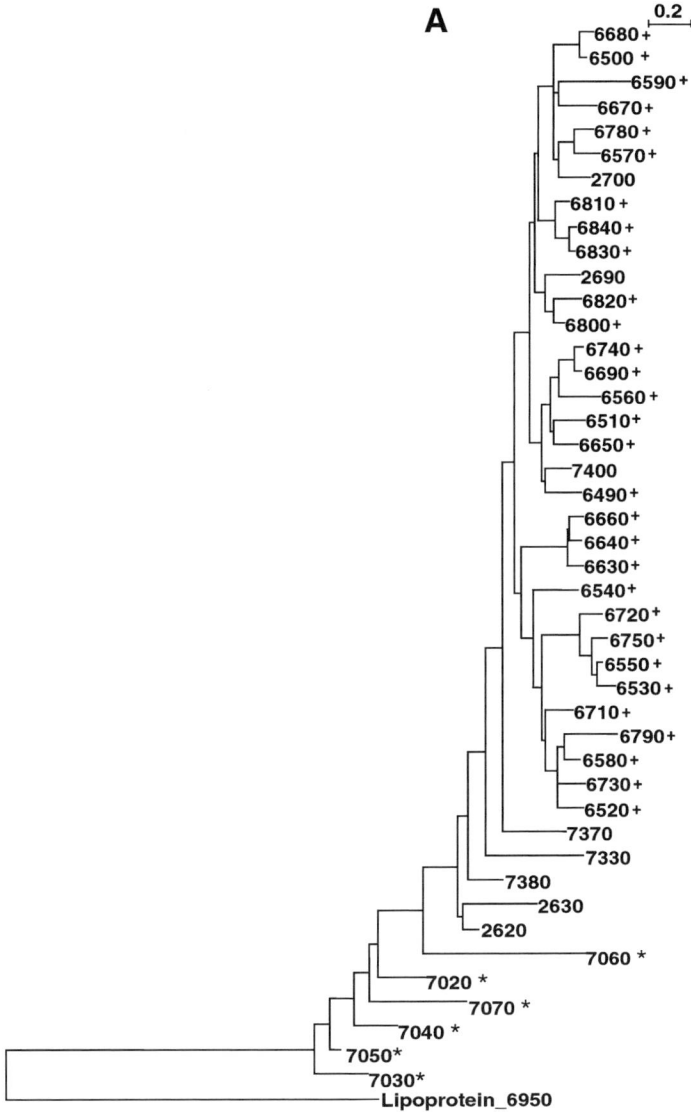

Fig. 2. (A) Gene tree for the 44 genes in the *p35* paralog family for antigenic variation-related *Mycoplasma penetrans* lipoproteins. The ClustalW program was used for sequence alignments and tree reconstructions. The number indicated is the *M. penetrans* gene ID, the MYPE number, which is related to gene location. The six signal peptide-less *p35* gene homologs are marked (*), and the 30 coding sequences at the largest locus are marked (+). The lipoprotein 6950 (*MYPE6950*) gene is an outgroup of the *p35* gene family.

3.2.3. Polymerase Slippage in Expression of Single Gene in Multi-Gene Family

Unless the structure of *p35* family in *M. penetrans* is expressible multiple antigens at a time, many other mycoplasmas have developed multi-gene family as resources of antigenic diversity for an expressing single gene in the family at a time. It means that only single gene should be translated at a time, and modification of RNA polymerase recognition site in promoters by variation in the number of di- or trinucleotides repeats,

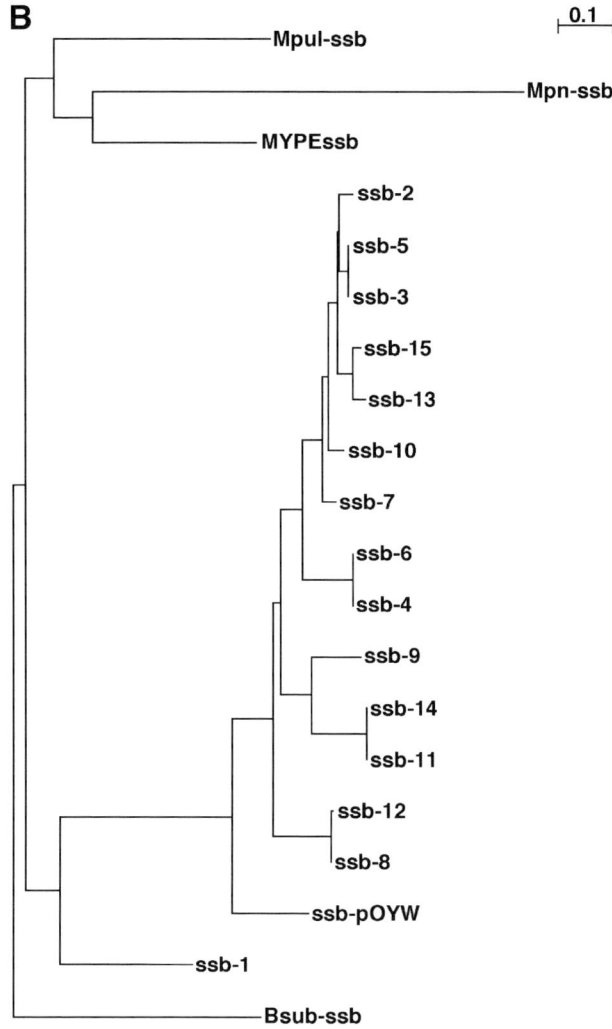

Fig. 2. (B) Gene tree for the *ssb* paralog family and pOYW-bearing *ssb* of *Candidatus* Phytoplasma asteris. Fourteen *ssb* genes in *Candidatus P. asteris* OY-M are numbered 1–14 based on gene order, with ssb-1 located in a ribosomal protein operon. The *ssb* genes from plasmid pOYW, *B. subtilis*, *M. pulmonis*, *M. pneumoniae*, and *M. penetrans* in the database are labeled ssb-pOYW, Bsub-ssb, Mpul-ssb, Mpn-ssb, and MYPE-ssb, respectively.

so called polymerase slippage, is one of the mechanisms in the reguration. For example, in *M. gallisepticum*, which is a member of the Pneumoniae group of mycoplasma and is an avian pathogen, the *vlhA* encoding haemagglutinin consists of 43 genes in five loci *(38)*, with only one of the 43 genes being expressed at a time *(49)*. Gene expression in the *vlhA* family is regulated by the number of GAA trinucleotide repeats upstream of the start codon for *vlhA*. A transcription regulatory protein may bind both upstream and downstream of GAA repeats only if the number of GAA repeats is 12 *(50)*. In the *M. gallisepticum* genome, 38 of 43 *vlhA* genes have different numbers of GAA repeats, and the unique *vlhA*3.03 gene has 12 GAA repeats. Another mechanism of antigenic variation not involving paralog genes also has evolved in mycoplasma. *M. gallisepticum* cytoadhesin GapA is an ortholog of the *M. pneumoniae* P1 adhesin and has also "on-off" switching regulated by a reversible point mutation *(51)*. Other *M. gallisepticum* adhesin-related protein PvpA has both size variation and sequence divergence in the C-terminal coding region *(52)*.

In *M. mycoides*, paralog genes, excluding transposase, consists of genes for putative variable surface proteins (11 genes), prolipoproteins (22 genes), and glycosyltransfer-ase (5 genes for *epsG* and 4 genes for *cps*) involved in capsule synthesis, a potential virulence factor (Table 2). An only known gene, *vmm*, for a phase variable lipoprotein precursor in *M. myvoides* is regulated by alteration of the number of TA repeats in the promoter region of the gene causing polymerase slippage during protein expression. Additional five gene encoding prolipoproteins with TA repeats in their promoter were also found in the *M. mycoides* genome *(39)*.

3.2.4. Small Repetitive DNA Elements and Antigenic Variation of Adhesin in M. pneumoniae and M. genitalium

Mollicutes with small genomes less than 1 Mb have very few paralogs, except *Candidatus* P. asteris. Instead of paralog formation, they have a thrifty mechanism of antigenic variation using repetitive sequences. Both the major P1 adhesin of *M. pneumoniae* and MgPa adhesin of *M. genitalium* are orthologous. The *M. pneumoniae* genome contains repetitive DNA elements, designated RepMP. RepMP sequences are located in the *p1* gene and other genes in the genome. In total, about 8% of the *M. pneumoniae* genome is composed of the RepMP sequences *(40)*. There are four different types of sequences classified into RepMP: 1, 2/3, 4, and 5. Of the 14 regions for RepMPs, only 1 region for both RepMP2/3 and RepMP4 is located in the *p1* gene, and the remaining RepMP regions exist outside of the *p1* gene. Antigenic variation of the *p1* gene via homologous recombination between RepMP regions has been hypothesized because of this genomic structure. Based on the restriction fragment length polymorphism pattern corresponding to the RepMP2/3 and RepMP4 sequences in the *p1* gene, *M. pneumoniae* clinical isolates are classified into two types, groups I and II. The isolation rate of these *M. pneumoniae* types from patients has been changed with 8- to 10-yr intervals; for example, in Japan the dominant type was group II in 1980, thereafter, it changed to group I in 1988–1990 *(53)*. During surveillance of clinical isolates, a *M. pneumoniae* strain with a new restriction fragment length polymorphism pattern of the *p1* gene has been isolated and designated as a group II variant. By sequence analysis, gene conversion between the RepMP2/3 region in the *p1* gene and in another region of the genome has been suggested *(54)*. Hybrid sequences in the RepMP2/3 region between groups I and II also have been reported *(55)*. In, *M. genitalium*, repetitive sequences are designated MgPa repeats, and show sequence similarities to the *M. pneumoniae* RepMPs except for RepMP1 *(56)*. The MgPa repeats, containing five regions (B, EF, KL, G, and LM), are probably involved in polymorphisms within sequences of MgPa adhesin in clinical isolates *(57)*. In total, 4.7% of *M. genitalium* genome is composed of the repetitive sequences contained in the *mgpa* operon and nine other elements *(58)*. This repetitive sequence-based recombination system shows the dynamics of mycoplasma evolution *(59)*.

3.2.5. Paralog Formation in Phytoplasma

As previously described in **Subheading 2.2.**, phytoplasma in the AAP phylogenetic branch diverged a long time ago from the SEM branch that contains mycoplasma have adapted to both insects and plants. Phytoplasma infects to plants through insects as vectors, and its life-style including intracellular proliferation in two different environments has probably developed multi-gene families with putative essential functions for replication. Paralogs in onion yellows phytoplasma (*Candidatus* P. asteris) have different features from those in mycoplasmas. In the small phytoplasma genome, 26% of the genome consists of paralogs (Table 2), including single-strand DNA-binding protein (*ssb*), bacterial nucleoid DNA-binding protein (*himA*), and DNA-directed RNA polymerase specialized sigma factor (*fliA*) *(37)*. One copy of *ssb* locates in the ribosomal protein operon with genes for S6 and S18, which is also observed in many bacteria. However, with high frequency, 14 paralogous *ssb* genes are adjacent to either *fliA* or *himA* paralogs. The original *Candidatus* P. asteris wild-type strain OY-W has a pOYW plasmid encoding *ssb (60)*, and the pOYW may have an evolutionary relationship to a eukaryotic single-strand DNA virus because the *rep* gene for rolling circle replication initiation protein of the pOYW contains a virus-like helicase domain of the ssDNA virus *(61)*. As shown in Fig. 2B, a gene tree suggests the possibility that plasmid-encoded *ssb* might have been the origin of the 14 paralog *ssb* genes in the phytoplasma genome.

3.3. Adaptation and Pathogenicity of Mycoplasma

As described in **Subheading 3.2.**, most of mycoplasmas's paralog genes/repetitive sequences are related to antigenic variation for lipoproteins or adhesins. Because the mycoplasma genomes lack apolipoprotein transacylase, maturation of mycoplasmal lipoproteins does not involve a phospholipidation step for adding a third fatty acid, resulting in diacylglyceryl lipoprotein. Mycoplasma genomes code only prolipoprotein diacylglyceryl transferase (*lgt*) and prolipoprotein signal peptidase (*lsp*). Mycoplasmal diacylated lipopeptides at the N-terminal of lipoproteins are recognized by human innate immunity through both Toll-like receptors 1, 2 and 6, in a different recognition pathway from other bacterial lipopolysaccharides *(62,63,* Shimizu 2004*)*. The diacylated lipopeptide (e.g., *M. fermentans* MALP-2 and *M. salivarium* FSL-1) is a powerful immune modulator, and activates monocytes/macrophages *(64–66)*, stimulates dendritic cell maturation *(67,68)*, and has cytopathic effects on lymphocytes via both necrosis and apoptosis *(69)*. Mycoplasmal surface-exposing lipoprotein also has strong antigenicity, and mycoplasmas have sophisticated antigenic variation systems for evading host adaptive immunity, as described in **Subheading 3.2.** Furthermore, mycoplasma adhesin has antigenic mimicry regions to its host's molecules. The *M. pneumoniae* P1 major adhesin has epitopes with similarity to human molecules *(70)*. In conclusion, mycoplasmas without a cell wall attach onto or invade into host cells with a cell membrane and acquire nutrition from the host cell. In this rich nutritious environment, genes encoding proteins with functions complementary to existing host proteins are probably deleted. On the other hand, mycoplasmas have to develop sophisticated mechanisms for evading the host immune system. These features stabilize adaptation of mycoplasma to their hosts.

Host-specific pathogenesis is one of the results of evolutionary relationships of obligate parasites and their host. Unlike many viruses, receptor specificity in mycoplasmal infection is not restricted; for example, the *M. pneumoniae* P1 adhesin also attaches to sheep erythrocytes. Mycoplasma contamination in cultured cell lines shows no host-specificity. The acquired immunity against *M. pneumoniae* in repeat infections may be involved in development of interstitial pneumonia in *M. pneumoniae* infection. Host-specific pathogenesis is probably regulated by either immune recognition or immune response.

Acknowledgments

The *M. penetrans* genome project was undertaken in collaboration with Jun Ishikawa in the Department of Bioactive Molecules, National Institute of Infectious Diseases, for assembly of the genome and bioinformatics; Atsushi Yamashita in the Kitasato Institute for Life Science, Kitasato University; Kenshiro Oshima, Keiko Furuya, Chie Yoshino, and Tadayoshi Shiba in the School of Science, Kitasato University, Masahira Hattori in the Kitasato Institute for Life Science, Kitasato University; the Human Genome Research Group, RIKEN Genomic Sciences Center, for whole genome sequencing; and Atsuko Horino, Tsuyoshi Kenri, and Tsuguo Sasaki in the Department of Bacterial Pathogenesis and Infection Control, National Institute of Infectious Disease, for the analysis of regulation of the *p35* gene family. The author would like to sincerely thank the collaborators for their excellent work.

References

1. Johansson, K. E. and Petersson, B. (2002) Taxonomy of *Mollicutes*, in *Molecular Biology and Pathogenicity of Mycoplasma* (Razin, S. and Herrmann, R., eds.). Kluwer Academic/Plenum Publishers, New York, NY, pp. 1–29.
2. Harasawa, R., Lefkowitz, E., Glass, J., and Cassell, G. (1996) Phylogenetic analysis of the 16S-23S rRNA intergenic spacer regions of the genus *Ureaplasma*. *J. Vet. Med. Sci.* **58,** 191–195.
3. Caudwell, A. (1984) Mycoplasma-like organisms (MLO), pathogens of the plant yellow diseases, as a model of coevolution between prokaryotes, insects and plants. *Isr. J. Med. Sci.* **20,** 1025–1027.
4. Kirchhoff, H., Schmidt, R., Lehmann, H., Clark, H. W., and Hill, A. C. (1996) *Mycoplasma elephantis sp.* nov., a new species from elephants. *J. Syst. Bacteriol.* **46,** 437–441.
5. Kirchhoff, H., Mohan, K., Schmidt, R., et al. (1997) *Mycoplasma crocodyli sp.* nov., a new species from crocodiles. *Int. J. Syst. Bacteriol.* **47,** 742–746.
6. Kobayashi, H., Runge, M., Schmidt, R., Kubo, M., Yamamoto, K., and Kirchhoff, H. (1997) *Mycoplasma lagogenitalium* sp. nov., from the preputial smegma of Afghan pikas (Ochotona rufescens rufescens). *Int. J. Syst. Bacteriol.* **47,** 1208–1211.
7. Hammond, E., Miller, C., Sneed, L., and Radcliffe, R. (2003) Mycoplasma-associated polyarthritis in a reticulated giraffe. *J. Wildl. Dis.* **39,** 233–237.
8. Messick, J. B., Walker, P. G., Raphael, W., Berent, L., and Shi, X. (2002) 'Candidatus Mycoplasma haemodidelphidis' sp. nov., 'Candidatus Mycoplasma haemolamae' sp. nov. and *Mycoplasma haemocanis* comb. nov., haemotrophic parasites from a naturally infected opossum (*Didelphis virginiana*), alpaca (*Lama pacos*) and dog (*Canis familiaris*): phylogenetic and secondary structural relatedness of their 16S rRNA genes to other mycoplasmas. *Int. J. Syst. Evol. Microbiol.* **52,** 693–698.
9. Modoff, S. (1980) Abstract of the 3rd International Conference of International Organization for Mycoplasmology, Custer, USA, p. 20.
10. Panangala, V., Stringfellow, J., Dybvig, K., et al. (1993) *Mycoplasma corogypsi* sp. nov., a new species from the footpad abscess of a black vulture, Coragyps atratus. *Int. J. Syst. Bacteriol.* **43,** 585–590.
11. Blanchard, A. and Bébéar, C. (2002) Mycoplasma of humans, in *Molecular Biology and Pathogenicity of Mycoplasmas* (Rasin, S. and Herrmann, R., eds.). Kluwer Academic/Plenum Publishers, New York.
12. Koletsky, R. J. and Weinstein, A. J. (1980) Fulminant *Mycoplasma pneumoniae* infection. Report of fatal case, and a review of the literature. *Am. Rev. Respir. Dis.* **122,** 491–497.
13. Takiguchi, Y., Shika, T., and Hirai, A. (2001) Fulminant *Mycoplasma pneumoniae* pneumonia. *Intern. Med.* **40,** 345–349.

14. Lo, S. C., Hayes, M. M., Wang, R. Y., Pierce, P. F., Kotani, H., and Shih, J. W. (1991) Newly discovered mycoplasma isolated from patients infected with HIV. *Lancet* **338,** 1415–1418.

15. Lo, S. C., Hayes, M. M., Tully, J. G., et al. (1992) *Mycoplasma penetrans* sp. nov., from the urogenital tract of patients with AIDS. *Int. J. Syst. Bacteriol.* **42,** 357–364.

16. Lo, S. C. (1992) Mycoplasmas and AIDS, in *Mycoplasmas, Molecular Biology and Pathogenesis* (Maniloff, J., McElhaney, R. N., Finch, L. R., and Baseman, J. B., eds.). American Society for Microbiology, Washington, DC, pp. 523–545.

17. Hussain, A. I., Robson, W. L., Kelley, R., Reid, T., and Gangemi, J. D. (1999) *Mycoplasma penetrans* and other mycoplasmas in urine of human immunodeficiency virus-positive children. *J. Clin. Microbiol.* **37,** 1518–1523.

18. Grau, O., Slizewicz, B., Tuppin, P., et al. (1995) Association of *Mycoplasma penetrans* with HIV infection. *J. Infect.* **172,** 672–681.

19. Wang, R. Y., Shih, J. W., Grandinetti, T., et al. (1992) High frequency of antibodies to *Mycoplasma penetrans* in HIV-infected patients. *Lancet* **340,** 1312–1316.

20. Grau, O., Tuppin, P., Slizewicz, B., et al. (1998) A longitudinal study of seroreactivity against *Mycoplasma penetrans* in HIV-infected homosexual men: association with disease progression. *AIDS Res. Hum. Retroviruses* **14,** 661–667.

21. Sasaki, Y., Honda, M., Makino, M., and Sasaki, T. (1993) Mycoplasmas stimulate replication of human immunodeficiency virus type 1 through selective activation of CD4+ T lymphocytes. *AIDS Res. Hum. Retroviruses* **9,** 775–780.

22. Sasaki, Y., Blanchard, A., Watson, H. L., et al. (1995) In vitro influence of *Mycoplasma penetrans* on activation of peripheral T lymphocytes from healthy donors or human immunodeficiency virus-infected individuals. *Infect. Immun.* **63,** 4277–4283.

23. Yáñez, A., Cedillo, L., Neyrolles, O., et al. (1999) *Mycoplasma penetrans* bacteremia and primary antiphospholipid syndrome. *Emerg. Infect. Dis.* **5,** 164–167.

24. Sasaki, Y., Ishikawa, J., Yamashita, A., et al. (2002) The complete genomic sequence of *Mycoplasma penetrans*, an intracellular bacterial pathogen in humans. *Nucleic Acids Res.* **30,** 5293–5300.

25. Shigenobu, S., Watanabe, H., Hattori, M., Sakaki, Y., and Ishikawa, H. (2000) Genome sequence of the endocellular bacterial symbiont of aphids *Buchnera sp.*APS. *Nature* **407,** 81–86.

26. Cavalier-Smith, T. (2002) Nucleomorphs: enslaved algal nuclei. *Curr. Opin. Microbiol.* **5,** 612–619.

27. Cavalier-Smith, T. (2003) Genomic reduction and evolution of novel genetic membranes and protein-targeting machinery in eukaryote-eukaryote chimaeras (meta-algae). *Philos. Trans. R. Soc. Lond. B. Biol. Sci.* **358,** 109–133.

28. Hoef-Emden, K., Marin, B., and Melkonian, M. (2002) Nuclear and nucleomorph SSU rDNA phylogeny in the Cryptophyta and the evolution of cryptophyte diversity. *J. Mol. Evol.* **55,** 161–179.

29. Maniloff, J. (1992) Phylogeny of Mycoplasmas, in *Mycoplasmas, Molecular Biology and Pathogenesis* (Maniloff, J., McElhaney, R. N., Finch, L. R., and Baseman, J. B., eds.). American Society for Microbiology, Washington, DC, pp. 549–559.

30. Maniloff, J. (2002) Phylogeny and evolution, in *Molecular Biology and Pathogenicity of Mycoplasmas* (Rasin, S. and Herrmann, R. eds.). Kluwer Academic/Plenum Publishers, New York.

31. Maniloff, J. (1996) The minimal cell genome: "on being the right size." *Proc. Natl. Acad. Sci. USA* **93,** 10,004–10,006.

32. Hutchison, C. A., Peterson, S. N., Gill, S. R., et al. (1999) Global transposon mutagenesis and a minimal Mycoplasma genome. *Science* **286,** 2089–2090.

33. Koonin, E. V. (2000) How many genes can make a cell: the minimal-gene-set concept. *Annu. Rev. Genomics Hum. Genet.* **1,** 99–116.

34. Mushegian, A. R. and Koonin, E. V. (1996) A minimal gene set for a cellular life derived by comparison of complete bacterial genomes. *Proc. Natl. Acad. Sci. USA* **93,** 10,268–10,273.

35. Chambaud, I., Heilig, R., Ferris, S., et al. (2001) The complete genome sequence of the murine respiratory pathogen *Mycoplasma pulmonis. Nucleic Acids Res.* **29,** 2145–2153.

36. Glass, J. I., Lefkowitz, E. J., Glass, J. S., Heiner, C. R., Chen, E. Y., and Cassell, G. H. (2000) The complete sequence of the mucosal pathogen *Ureaplasma urealyticum. Nature* **407,** 757–762.

37. Oshima, K., Kakizawa, S., Nishigawa, H., et al. (2003) Reductive evolution suggested from the complete genome sequence of a plant-pathogenic phytoplasma. *Nat. Genet.* **36,** 27–29.

38. Papazisi, L., Gorton, T. S., Kutish, G., et al. (2003) The complete genome sequence of the avian pathogen *Mycoplasma gallisepticum* strain Rlow. *Microbiology* **149,** 2307–2316.

39. Westberg, J., Persson, A., Holmberg, A., et al. (2004) The genome sequence of *Mycoplasma mycoides* subsp. mycoides SC type strain PG1, the causative agent of contagious bovine pleuropneumoniae (CBPP). *Genome Res.* **14,** 221–227.

40. Himmelreich, R., Hilbert, H., Plagens, H., Pirkl, E., Li, B. C., and Herrmann, R. (1996) Complete sequence analysis of the genome of the bacterium *Mycoplasma pneumoniae. Nucleic Acids Res.* **24,** 4420–4449.

41. Kobayashi, K., Ehrlich, S. D., Albertini, A., et al. (2003) Essential *Bacillus subtilis* genes. *Proc. Natl. Acad. Sci. USA* **100,** 4678–4683.

42. Kunisawa, T. (2003) Gene arrangements and branching orders of gram-positive bacteria. *J. Theor. Biol.* **222,** 495–503.

43. Koonin, E. V., Mushegian, A. R., and Bork, P. (1996) Non-orthologous displacement. *Trends. Genet.* **12,** 334–336.

44. Jaffe, J. D., Stange-Thomann, N., Smith, C., et al. (2004) The complete genome and proteome of *Mycoplasma mobile. Genome Res.* **14,** 1447–1461.

45. Neyrolles, O., Chambaud, I., Ferris, S., et al. (1999) Phase variations of the *Mycoplasma penetrans* main surface lipoprotein increase antigenic diversity. *Infect. Immun.* **67,** 1569–1578.

46. Röske, K., Blanchard, A., Chambaud, I., et al. (2001) Phase variation among major surface antigens of *Mycoplasma penetrans. Infect. Immun.* **69,** 7642–7651.

47. Horino, A., Sasaki, Y., Sasaki, T., and Kenri, T. (2003) Multiple promoter inversions generate surface antigenic variation in *Mycoplasma penetrans. J. Bacteriol.* **185,** 231–242.

48. Neyrolles, O., Eliane, J. P., Ferris, S., et al. (1999) Antigenic characterization and cytolocalization of P35, the major *Mycoplasma penetrans* antigen. *Microbiology* **145,** 343–355.

49. Glew, M. D., Browning, G. F., Markham, P. F., and Walker, I. D. (2000) pMGA phenotypic variation in *Mycoplasma gallisepticum* occurs in vivo and is mediated by trinucleotide repeat length variation. *Infect. Immun.* **68,** 6027–6033.

50. Liu, L., Panangala, V. S., and Dybvig, K. (2002) Trinucleotide GAA repeats dictate pMGA gene expression in *Mycoplasma gallisepticum* by affecting spacing between flanking regions. *J. Bacteriol.* **184,** 1335–1339.

51. Winner, F., Markova, I., Much, P., et al. (2003) Phenotypic switching in *Mycoplasma gallisepticum* hemadsorption is governed by a high-frequency, reversible point mutation. *Infect. Immun.* **71,** 1265–1273.

52. Liu, T., Garcia, M., Levisohn, S., Yogev, D., and Kleven, S. H. (2001) Molecular variability of the adhesin-encoding gene *pvpA* among *Mycoplasma gallisepticum* strains and its application in diagnosis. *J. Clin. Microbiol.* **39,** 1882–1888.

53. Sasaki, T., Kenri, T., Okazaki, N., et al. (1996) Epidemiological study of *Mycoplasma pneumoniae* infections in japan based on PCR-restriction fragment length polymorphism of the P1 cytadhesin gene. *J. Clin. Microbiol.* **34,** 447–449.

54. Kenri, T., Taniguchi, R., Sasaki, Y., et al. (1999) Identification of a new variable sequence in the P1 cytadhesin gene of *Mycoplasma pneumoniae*: evidence for the generation of antigenic variation by DNA recombination between repetitive sequences. *Infect. Immun.* **67,** 4557–4562.

55. Dorigo-Zetsma, J. W., Wilbrink, B., Dankert, J., and Zaat, S. A. J. (2001) *Mycoplasma pneumoniae* P1 type 1- and type 2-specific sequences within the P1 cytadhesin gene of individual strains. *Infect. Immun.* **69,** 5612–5618.

56. Himmelreich, R., Plagens, H., Hilbert, H., Reiner, B., and Herrmann, R. (1997) Comparative analysis of the genomes of the bacteria *Mycoplasma pneumoniae* and *Mycoplasma genitalium*. *Nucleic Acids Res.* **25,** 701–712.

57. Peterson, S., Bailey, C., Jensen, J., et al. (1995) Characterization of repetitive DNA in the *Mycoplasma genitalium* genome: possible role in the generation of antigenic variation. *Proc. Natl. Acad. Sci. USA* **92,** 11,829–11,833.

58. Fraser, C. M., Gocayne, J. D., White, O., et al. (1995) The minimal gene complement of *Mycoplasma genitalium*. *Science* **270,** 397–403.

59. Rocha, E. P. C. and Blanchard, A. (2002) Genomic repeats, genome plasticity and the dynamics of Mycoplasma evolution. *Nucleic Acids Res.* **30,** 2031–2042.

60. Kuboyama, T., Huang, C. C., Lu, X., et al. (1998) A plasmid isolated from phytopathogenic onion yellows phytoplasma and its heterogeneity in the pathogenic phytoplasma mutant. *Mol. Plant Microbe Interact.* **11,** 1031–1037.

61. Oshima, K., Kakizawa, S., Nishigawa, H., et al. (2001) A plasmid of phytoplasma encodes a unique replication protein having both plasmid- and virus-like domains, clue to viral ancestry or result of virus/plasmid recombination? *Virology* **285,** 270–277.

62. Nishiguchi, M., Matsumoto, M., Takao, T., et al. (2001) *Mycoplasma fermentans* lipoprotein M161Ag-induced cell activation is mediated by Toll-like receptor 2: role of N-terminal hydrophobic portion in its multiple functions. *J. Immunol.* **166,** 2610–2616.

63. Takeuchi, O., Kaufmann, A., Grote, K., et al. (2000) Cutting edge: preferentially the R-stereoisomer of the mycoplasmal lipopeptide macrophage-activating lipopeptide-2 activates immune cells through a toll-like receptor 2- and MyD88-dependent signaling pathway. *J. Immunol.* **164,** 554–557.

64. Muhlradt, P. F., Kiess, M., Meyer, H., Sussmuth, R., and Jung, G. (1997) Isolation, structure elucidation, and synthesis of a macrophage stimulatory lipopeptide from *Mycoplasma fermentans* acting at picomolar concentration. *J. Exp. Med.* **185,** 1951–1958.

65. Okusawa, T., Fujita, M., Nakamura, J., et al. (2004) Relationship between structures and biological activities of mycoplasmal diacylated lipopeptides and their recognition by toll-like receptors 2 and 6. *Infect. Immun.* **72,** 1657–1665.

66. Shibata, K., Hasebe, A., Into, T., Yamada, M., and Watanabe, T. (2000) The N-terminal lipopeptide of a 44-kDa membrane-bound lipoprotein of *Mycoplasma salivarium* is responsible for the expression of intercellular adhesion molecule-1 on the cell surface of normal human gingival fibroblasts. *J. Immunol.* **165,** 6538–6544.

67. Link, C., Gavioli, R., Ebensen, T., Canella, A., Reinhard, E., and Guzman, C. (2004) The Toll-like receptor ligand MALP-2 stimulates dendritic cell maturation and modulates proteasome composition and activity. *Eur. J. Immunol.* **34,** 899–907.

68. Weigt, H., Muhlradt, P. F., Emmendorffer, A., Krug, N., and Braun, A. (2003) Synthetic mycoplasma-derived lipopeptide MALP-2 induces maturation and function of dendritic cells. *Immunobiology* **207,** 223–233.

69. Into, T., Kiura, K., Yasuda, M., et al. (2004) Stimulation of human Toll-like receptor (TLR) 2 and TLR6 with membrane lipoproteins of *Mycoplasma fermentans* induces apoptotic cell death after NF-kappa B activation. *Cell Microbiol.* **6,** 187–199.

70. Jacobs, E., Bartl, A., Oberle, K., and Schiltz, E. (1995) Molecular mimicry by *Mycoplasma pneumoniae* to evade the induction of adherence inhibiting antibodies. *J. Med. Microbiol.* **43,** 422–429.

References Added in Proof

Glass, J., Assad-Garcia, N., Alperovich, N., et al. (2006) Essential genes of a minimal bacterium. *Proc. Natl. Acad. Sci. USA* **13,** 425–430.

Pitcher, D. G., Windsor, D., Windsor, H., et al. (2005) *Mycoplasma amphoriforme* sp. nov., isolated from a patient with chronic bronchopneumonia. *Int. J. Syst. Evol. Microbiol.* **55,** 2589–2594.

Shimizu, T., Kida, Y., and Kuwano, K. (2004) Lipid-associated membrane proteins of *Mycoplasma fermentans* and *M. penetrans* activate human immunodeficiency virus long-terminal repeats through Toll-like receptors. *J. Immunol.* **113,** 121–129.

11

Genome Comparisons of Diverse
Staphylococcus aureus Strains

Martin J. McGavin

Summary

As with other pathogens, the genome of *Staphylococcus aureus* can be subdivided into core and accessory segments, comprising roughly 75 and 25% of the genome. Particular attention is given to the MRSA252 strain, which is phylogenetically distinct from other *S. aureus* genomes, is epidemic in the United Kingdom and North America, and is closely related to methicillin-susceptible clinical isolates that are hypervirulent in musculoskeletal infection models. This strain contains a number of unique or unusual small-scale variations compared with other genomes, and their potential significance as mediators of virulence is discussed. With respect to horizontally transferred virulence determinants of the core genome, another focus is to assess the integration sites of toxin-carrying prophage, in terms of a potential cost–benefit relationship. The Staphylococcal Cassette Chromosome is discussed in terms of its role in promoting the evolution of antibiotic resistance, and as a potential mediator of genetic diversity and gene shuffling between species. The association of pathogenicity islands (SaPI) with temperate prophage is considered, with emphasis on a segment of the genome flanking the *coa* (coagulase) allele, and the diversity of SaPI content, such that there does not appear to be a single SaPI that is associated with the majority of virulent strains. The vSaα and vSaβ genomic islands, which can be variable in composition, but common to all strains, are discussed in terms of how they may provide insight into the infection strategy of *S. aureus*, and how vSaβ may be coevolving in association with antibiotic-resistant hospital-adapted strains.

Key Words: *Staphylococcus aureus*; MRSA; virulence; antibiotic resistance; genome structure and evolution; pathogenicity island; horizontal transfer; genetic diversity; phage lysogeny.

1. Introduction

Staphylococcus aureus has achieved notoriety as the microorganism for which more complete genome sequences are now available compared with any other bacterial species. Currently, six *S. aureus* genomes have been published *(1–4)*, and a seventh strain, *S. aureus* 8325, has been completed but not annotated (www.genome.ou.edu/staph.html). The first *S. aureus* genome sequences were published in 2001 *(4)*, 6 yr after *Haemophilus influenzae*, the first microorganism to have its entire genome sequence determined *(5)*. In terms of the justification for this remarkable progress, *S. aureus* can be described as an enigmatic opportunistic pathogen, living as a commensal of the human nose in 30 to 70% of the population; about 20% of the population are always colonized, whereas 60% are transiently colonized, and 20% never carry the organism *(6,7)*. However,

From: *Bacterial Genomes and Infectious Diseases*
Edited by: V. L. Chan, P. M. Sherman, and B. Bourke © Humana Press Inc., Totowa, NJ

it also carries a plethora of genes encoding secreted toxins and tissue-degrading enzymes, such that once it contaminates a breach in the dermal layer or mucous membranes, it can go on to colonize and infect virtually every tissue and organ system of the body, causing diseases ranging from relatively mild skin and soft tissue infections to severe osteomyelitis and septic arthritis and life threatening conditions, such as infective endocarditis and hemolytic pneumonia. Although the majority of *S. aureus* infections are not of an obviously toxigenic nature, the ability of specific strains to cause toxin-mediated diseases is well documented, including toxic shock syndrome, food poisoning, and scalded skin syndrome *(8)*. Particularly alarming is the association of strains harboring the Panton Valentin Leukotoxin (PVL) with a new syndrome of high mortality community-acquired necrotizing pneumonia in children, as well as severe boils and skin infections in prisons, gymnasiums, and the homosexual community *(9,10)*.

Approximately 40–60% of all hospital-acquired *S. aureus* are now resistant to methicillin (methicillin-resistant *S. aureus*; MRSA) across several industrialized nations. These isolates are typically also resistant to multiple classes of antimicrobial agents, and have historically been treated with vancomycin, the antibiotic of last resort when physicians are confronted with MRSA. However, recent reports have documented high level vancomycin-resistant *S. aureus* (VRSA) isolates, attributed to acquisition of the *van*A gene from enterococci *(11–13)*. This raises the spectre of an approaching postantibiotic era where it will no longer be possible to resolve *S. aureus* infections with antimicrobial therapy. These factors all contribute to the focus of genome sequencing projects on *S. aureus*.

2. Description of Strains With Sequenced Genomes

The wealth of information made available through genome sequencing projects is enormous, and major findings have been summarized in recent reviews *(7,14,15)*. A brief description of the strains for which genome sequences are available is presented in Table 1. Strains N315 and Mu50 are both MRSA and closely related to one another *(4)*. The Mu50 strain also harbors an intermediate susceptibility to vancomycin (VISA) phenotype. Strain MW2 *(2)* is a community-acquired MRSA originating from a case of fatal pediatric bacteremia in North Dakota, and is closely related to a community-acquired methicillin susceptible *S. aureus* strain MSSA476 *(3)*, isolated from a patient with osteomyelitis in the United Kingdom. Strain MRSA252 *(3)* is representative of the epidemic EMRSA-16 clonal group that is responsible for 50% of the MRSA infections in the United Kingdom *(16)*. This is also one of the dominant MRSA clones found in the United States *(17)*, and by pulsed-field gel electrophoresis it is indistinguishable from the CMRSA-4 strain that is epidemic in Canadian hospitals *(18)*. The COL strain is an early isolate of MRSA originating from the UK in 1961, and NCTC8325 is a common laboratory strain dating back to 1949.

Multi-locus sequence typing (MLST) has illustrated the relatedness of strains for which genome sequences have been completed *(3,14,19)*; NCTC8325 is closely related to COL, whereas N315 and Mu50 belong to the same ST-5 clonal type. Similarly, the community-acquired MW2 and MSSA476 strains belong to the ST-1 clonal type. These pairs of closely matched strains all originate from a common branch point on the MLST-dendrogram, whereas MRSA252, representative of the epidemic EMRSA-16, is the most divergent of the sequenced strains.

Table 1
Details of the Seven Sequenced *Staphylococcus aureus* Strains

Strain[a]	Source	Year	Comments	References
N315	Pharynx; Japan	1982	Hospital-acquired MRSA	*4*
Mu50	Wound; Japan	1997	Hospital-acquired VISA; related to N315	*4*
MW2	Fatal pediatric bacteremia; North Dakota, USA	1998	Typical USA community-acquired MRSA, PV-toxin positive	*2*
MRSA252	Fatal bacteremia; Oxford, UK	1997	Typical UK hospital-acquired epidemic MRSA (EMRSA-16)	*3*
MSSA476	Osteomyelitis; Oxford, UK	1998	Community-acquired MSSA	*3*
COL	Colindale, UK	1961	Early MRSA	*1*
NCTC8325	Colindale; UK	Earlier than 1949	Laboratory strain parent of nonlysogenic 8325-4; can be genetically manipulated	http://www. genome.ou. edu/staph. html

[a]The GenBank accession numbers of the annotated genomes are: N315 chromosome, NC_002745; Mu50 chromosome, NC_002758; MW2 chromosome, NC_003923; MRSA252 chromosome, NC_002952; MSSA476 chromosome, NC_002953; COL chromosome, NC_002951.

MRSA, methicillin-resistant *S. aureus*; MSSA, methicillin-sensitive *S. aureus*; PV-toxin, Panton-Valentine leukocidin toxin; VISA, vancomycin intermediate level-resistant *S. aureus*.

3. Genome Organization

The *S. aureus* strains listed in Table 1 possess a genome ranging in size from 2,799,802 to 2,902,619 base pairs. The genomes are dispersed with regions of small and large-scale difference, the latter of which are primarily responsible for the size variation of the individual genomes. Small-scale variation has been defined as genetic changes that affect either an individual gene or fewer than 10 protein coding sequences (CDSs), and these include the formation of pseudogenes by insertion sequence integration, point mutations, and variation in polymeric nucleotide repeats *(3)*. Large-scale variation involving more than 10 CDSs is associated with horizontally acquired DNA, and these segments comprise the "accessory" genome. Many of the genes contained in these regions are associated with virulence and drug resistance. In addition to relatively small integrated plasmids and transposons, the accessory genome also includes temperate bacteriophage, genomic islands (GEIs), pathogenicity islands, and chromosomal cassettes.

Excluding the accessory elements, the core genome is highly conserved and colinear in the different strains. Up to 52% of *S. aureus* genes find their closest homologs in either *Bacillus subtilis* or *Bacillus halodurans (4)*, and the genomes of *S. aureus* and *B. subtilis* show certain degrees of homology between 3 and 9 o'clock of the circular chromosome *(15)*. Genes present in the vertically transmitted core domain, which comprises approx 75% of any *S. aureus* genome, are mostly associated with central metabolism and housekeeping functions, and the similarity of individual genes among the five genomes is typically 98–100% at the amino acid level *(14)*. With this highly conserved core genome, MLST has been used in parallel with genome sequence data to provide a greater understanding of factors that contribute to genome evolution *(14,19)*. MLST analysis of 334 *S. aureus* isolates recovered from a well-defined population over a limited time span suggested that point mutations give rise to new alleles at least 15-fold more frequently than does recombination *(20)*. Thus, accumulation of point mutations is likely to be a significant force in the evolution of different *S. aureus* clonal complexes, and may provide a means for *S. aureus* to alter its phenotypic traits in response to selective pressure. Accordingly, strains N315 and Mu50 belong to the same ST-5 clonal complex, and exhibit less than 0.08% overall nucleotide sequence difference in their core genomes, but differ in their susceptibility to vancomycin *(4)*. Similarly, the community-acquired strains MW2 and MSSA476 belong to the ST-1 MLST type, and exhibit only 285 single nucleotide polymorphisms within functional coding regions of their core genomes *(3)*.

The core genome is supplemented with genes that are species-specific, but not essential for growth and survival. Among these are several virulence genes not carried by other species of *Staphylococcus (14)*, including coagulase (*coa*), which differentiates *S. aureus* from the coagulase-negative *staphylococci*, the immunoglobulin G-binding proteins Spa (protein A) and Sbi, α and γ hemolysins (*hla* and *hlg*), *sir* and *sbn* encoding siderophore biosynthesis and transport, superoxide dismutase (*sod*M), fibrinogen-binding protein (*fib*), capsular polysaccharide biosynthesis (*cap*), an alternate sortase (*srt*B) involved in anchoring of iron-regulated cell surface proteins to the cell wall, and *ssp*, an operon encoding secreted serine and cysteine protease functions. These virulence factors encoded by the core chromosome, together with other genes encoding essential metabolic and catabolic functions necessary for in vivo growth and survival, are likely to comprise a minimal pathogenic unit that contributes to the majority of *S. aureus* infections, whereas a number of accessory functions contribute to specific toxin-related disease syndromes.

The "accessory" genome accounts for approx 25% of the *S. aureus* chromosome, consisting primarily of large-scale regions of difference interspersed randomly throughout the core genome, and which frequently carry antibiotic resistance determinants or virulence-related functions. Consequently, the emerging view of the *S. aureus* genome is that of a conserved core genome scattered with variable regions that have been acquired through horizontal transfer. These considerations reflect two contrasting forces that dictate the overall gene content and functionality in *S. aureus*; the rapid loss or gain of accessory genes by horizontal transfer, and conservation of the core genes by vertical inheritance *(20)*. This is proposed to represent a two-tier evolution, ensuring that *S. aureus* harbors a stable core genome that is shaped by long-term selection, but is also capable of rapidly accessorizing to reshape itself in a more virulent or drug-resistant guise *(14)*. The duration of this chapter shall focus on examples of variation in the core and accessory genome, and a discussion of their potential significance.

4. Variation in the Core Genome

4.1. Pseudogenes

With the propensity of *S. aureus* to generate altered alleles via point mutation, rather than genetic recombination, the analysis of pseudogene content and associated phenotypic traits deserves closer scrutiny. The first tabulation of pseudogenes in sequenced genomes identified 66 pseudogenes in MRSA252 *(3)*. The majority of these are attributed to nonsense mutations and frameshifts. The pseudogenes of MRSA252 include 11 putative membrane or lipoproteins, two cell wall anchored proteins, and eight genes encoding known or putative exported proteins, the most significant being a nonsense mutation within α hemolysin (SAR1136) which is the major lethal toxin of *S. aureus* and one of the virulence factors encoded by the core genome. The amber nonsense mutation occurs at codon 112 of the 319 amino acid protein, and is certain to result in a nonfunctional protein. It will be of interest to determine if this same mutation appears in other isolates of this epidemic clone. The two-cell wall anchored genes, SAR1809 and SAR2208, that are annotated as pseudogenes encode previously described proteins, Haptoglobin receptor A (HarA) and FmtB (factor that affects methicillin resistance and autolysis), respectively. HarA has been described in terms of its haptoglobin and hemoglobin binding activity thought to promote iron acquisition *(21)*, whereas FmtB was first described as being required for optimal expression of methicillin resistance, but this phenotype is manifested through an uncertain mechanism *(22)*.

Analysis of pseudogene content has been employed in an effort to resolve the basis of the VISA phenotype of strain Mu50, in relation to its closely related susceptible counterpart, strain N315. A common feature of VISA strains is thickened cell walls and reduced crosslinking of glycan chains, which is believed to facilitate the intermediate resistance phenotype by binding and immobilizing vancomycin. As this phenotype can be created in vitro through a simple selection procedure, it is believed that the VISA phenotype is driven by point mutations that occur, or are selected for, in response to vancomycin. A proteome comparison of N315 and Mu50 identified 17 CDSs that were disrupted in the genome of Mu50, but not in N315 *(23)*. On resequencing to verify the published genome sequences, it was found that only four were actually disrupted in Mu50, whereas the remainder encoded predicted products that were identical to those of N315 or MW2, and the four genes that were disrupted in Mu50 did not appear to be common to other VISA isolates.

These studies illustrate the dynamic process and close mutual relationship between genome annotation and genome driven research. With respect to the VISA phenotype, although it first appeared that this was closely associated with the Clonal Complex CC5 identified by MLST analyses *(19)*, recent findings suggest that VISA phenotypes can be associated with any of the five major MLST complexes that are used to differentiate MRSA lineages *(24)*. This provides testament to the relative ease with which *S. aureus* can alter its phenotypic traits via point mutation, and the phenotypic complexity that arises from subtle genotypic alterations.

4.2. Genomic Islets

These elements are categorized as regions of small-scale variation affecting an individual gene or a small number of genes, and not generally associated with point mutations or insertion sequence integration *(3,15)*. The best description of genomic islets is for the MRSA252 strain, which harbors 41 islet-encoded genes that are not present in other sequenced *S. aureus* genomes. These include 24 CDSs associated with islets that harbor 3 or less CDSs, whereas the remaining 17 are associated with two regions of 7 and 10 CDSs *(3)*. These islet-encoded CDSs are not flanked by any obvious features associated

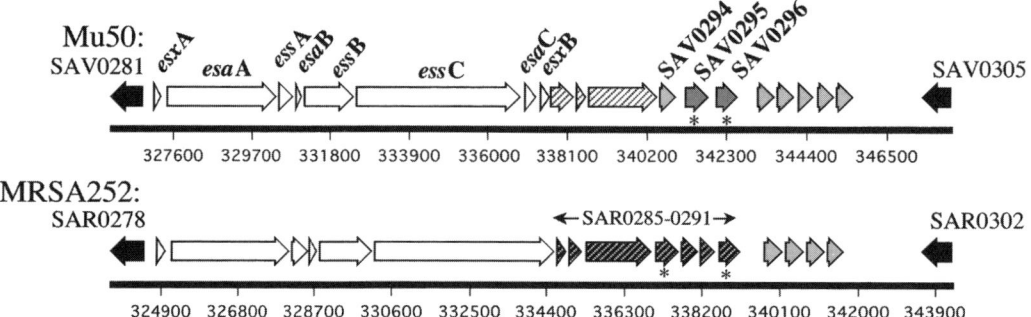

Fig. 1. Comparison of the early secreted antigen target 6 kDa (ESAT-6)-like locus and flanking regions of methicillin-resistant *Staphylococcus aureus* (MRSA) strains, Mu50 and MRSA252. The numbers on the scale bar correspond to the coordinates of the nucleotide sequences in the respective genomes. In strain Mu50, the protein coding sequences (CDSs) SAV0281 and SAV0305 (black), and their respective orthologs in MRSA252 (SAR0278 and SAR0302) are annotated as *ssa*A (homolog of secretory antigen precursor) and *nir*C (formate transporter) respectively, and flank a variable region of the genome that harbors the ESAT-6-like locus. The ESAT-6 locus (white) of strain Mu50 (SAV0282-SAV0290) and corresponding gene names are as defined by Burts et al. *(27)*. The diagonally hatched CDSs of strain Mu50 flanking the ESAT-6-like locus (SAV0291-SAV0293) are present in other sequenced genomes, but absent from MRSA252. In the ESAT-6-like locus of MRSA252, the terminal *esa*C and *esx*B genes are absent, but are replaced by similarly sized proteins encoded by the adjacent genomic islet, SAR0285-SAR0291 (gray shaded and diagonally hatched). Although this genomic islet is unique to MRSA252, two genes represented by SAR0288 and SAR0291 marked by asterisks, show approx 20% amino acid identity with SAV0295 and SAV0296 of strain Mu50, also marked with asterisks. Preceding these genes in Mu50, SAV0294 encodes a gene that is duplicated four times downstream of SAV0296 (shaded light gray). Tandem repeats of homologous genes in the MRSA252 strain are also shaded gray (SAR0293-0296).

with mobile genetic elements, and generally do not encode functions associated with virulence or antibiotic resistance. In contrast, many are predicted to encode metabolic and transport functions, which are suggested to enhance the metabolic repertoire of MRSA252. One notable exception may be the genomic islet defined by SAR0285 through SAR0291 of MRSA252, as described in **Subheading 4.2.1.**

4.2.1. The SAR0285/0291 Genomic Islet of MRSA252

This genomic islet encoding predicted proteins of unknown function is unique to MRSA252. However, it lies adjacent to a gene cluster that is conserved in Gram-positive bacteria *(25)*, which facilitates a specialized secretion pathway in *Mycobacterium tuberculosis* and *S. aureus (26,27)*. In *M. tuberculosis*, this system promotes secretion of ESAT-6 (early secreted antigen target 6 kDa) and CFP-10 (culture filtrate antigen 10 kDa) encoded by the respective *esx*A and *esx*B genes, both of which are important mediators of virulence, serving to subvert macrophage function and promote lysis of infected cells *(26,28)*. A single such ESAT-6-like cluster is a component of the core genome of *S. aureus*, but in MRSA252 it is adjacent to the SAR0285/0291 genomic islet, and displays a few differences compared with other strains (Fig. 1).

In the Mu50 genome, the CDSs SAV0282 (*esx*A) and SAV0290 (*esx*B) represent the boundaries of the ESAT-6-like locus. Although EsxA and EsxB do not harbor signal peptides, they appear in the culture supernatant of *S. aureus*, and three gene products (SAV0284, *ess*A; SAV0286, *ess*B; and SAV0287, *ess*C), each possessing one or more predicted membrane spanning domains, are essential for their secretion *(27)*. Mutation of either *esx*A, *esx*B, or *ess*C in *S. aureus* Newman resulted in a two- to three-log reduction in virulence in a kidney and liver abscess infection model *(27)*. The ESAT-6-like locus in all other sequenced genomes of *S. aureus* is identical to that of Mu50, with the exception of MRSA252, where the terminal *esa*C and *esx*B genes are absent, but appear to be replaced by genes of similar size and organization in the adjacent genomic islet that is unique to this strain (Fig. 1).

Several of the islet genes (i.e., SAR0285, SAR0286, SAR0289, SAR0290) encode small proteins (~9–17 kDa) that may be potential substrates for secretion by the adjacent ESAT-6-like cluster. Two gene products (SAR0288 and SAR0291) also share 19 to 20% amino acid identity toward similarly sized proteins encoded by SAV0295 and SAV0296 in the genome of *S. aureus* Mu50, just downstream of the ESAT-6-like locus. These genes are preceded by SAV0294, encoding a 19.8 kDa acidic protein that shares between 68 and 95% identity toward proteins encoded by sequential downstream genes SAV0298–SAV0302. In MRSA252, four tandem genes SAR0293–0296 downstream of the SAV0285–SAV0291 genomic islet also share high homology with SAV0294 from strain Mu50. The genome of MSSA476 carries nine orthologs of SAV0294 (SAS0269 and SAS0272–SAS0279), which share from 63 to 89% identity toward SAV0294. The same genes are maintained in the closely related strain MW2, whereas in strain COL there are 11 genes encoding proteins that share from 66 to 95% homology toward the product of SAV0294, all of them clustered after the conserved ESAT-6-like locus.

What forces have contributed to this gene apparently having undergone variable numbers of duplication events in the genomes of different *S. aureus* strains? Apart from its homologs in the different annotated genomes of *S. aureus*, the SAV0294 product displays greatest homology toward a 62 amino acid protein, annotated as the site of integration of transposon, Tn552, in *S. aureus* strain NCTC9789 *(29)*, and described as being the third open reading frame in a series of three tandemly repeated genes. Although none of the sequenced genomes reveal a copy of Tn552 at this site, it may represent a hot spot for transposon integration and excision, resulting in multiple duplications of the target site. Factors contributing to the duplication of SAV0294, the relationship of this and neighboring genes with the conserved ESAT-6-like locus, as well as the identification of proteins that are secreted by the ESAT-6 system, and their contributions to virulence of *S. aureus* will be an interesting area of future research.

4.3. Coagulase and Collagen Adhesin Genomic Islets

A more liberal definition of genomic islets includes coagulase *(15)*, a hallmark enzyme for species identification of *S. aureus*, defined earlier in this chapter as a component of the core genome. The basis of *coa* definition as an islet is owing to the extensive allelic variation among different strains, such that different allotypes share only 65.9–88.4% of their amino acids. Reportedly, certain *coa* alleles are distributed across several different genetic backgrounds, whereas two distinct *coa* alleles can be found in strains that are closely related. This is suggested to indicate that the *coa* gene may be transferred from cell to cell through a lateral gene transfer mechanism. Through the same rationalization, the collagen adhesin (*cna*) gene encoding a collagen-binding adhesin should also

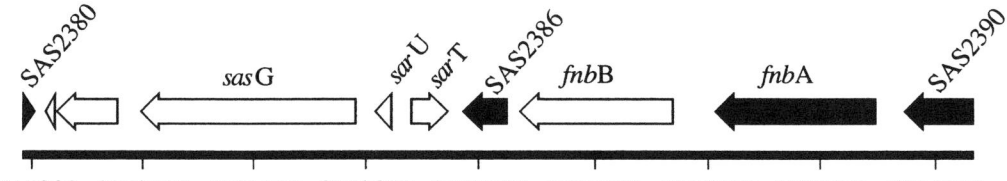

Fig. 2. The genomic adhesin islet of MSSA476 and its differentiation from MRSA252. The black-shaded protein coding sequences are contiguous in MRSA252, an arrangement that appears to be unique to this strain when compared with other sequenced genomes. In MSSA476 and other sequenced genomes, additional genes (white) are present. These include a duplication of the *fnb* adhesin, an additional adhesion protein encoded by *sas*G, and divergently transcribed transcriptional regulators, *sar*U and *sar*T. Additional details are provided in **Subheading 4.3.1.**

be classified as a genomic islet, on the basis of it having a limited strain distribution and no obvious association with a mobile genetic element *(30)*. As this same feature occurs with other genes that encode adhesion-related functions, this is described in more detail in **Subheading 4.3.1.**, which is dedicated toward islets that encode adhesion proteins.

4.3.1. Adhesion Islets

The collagen adhesin Cna of *S. aureus* was first described as being expressed by strains associated with cartilage infections, but not isolates that caused soft tissue infection, and this correlated with the presence or absence of the *cna* gene *(31)*. The *cna* gene is present in the genomes of MRSA252, as well as the closely related MW2 and MSSA476 strains, but is not present in the other sequenced genomes. As represented by MRSA252, *cna* (SAR2774) is flanked by genes that are contiguous in the genomes of strains that lack *cna*. Thus, *cna* is inserted precisely between two genes that are components of the core genome, and hence its designation as a genomic islet. In addition to its association with strains isolated from patients with cartilage infections, there is also a strong association of *cna* with strains isolated from patients with community-acquired necrotizing pneumonia *(32)*. Therefore, acquisition of this single gene may significantly enhance the disease potential of *S. aureus*.

The Cna adhesion protein is a member of the MSCRAMM family of proteins in Gram-positive bacteria, which promote adherence to the tissue extracellular matrix *(33)*. The fibronectin-binding proteins, FnBPA and FnBPB, were the first well described members of the MSCRAMM family, encoded by tandem genes *fnb*A and *fnb*B on the chromosome of *S. aureus* strain 8325-4 *(34)*. Prompted by a report that strain 879R4S possessed only one *fnb* gene *(35)*, we conducted an analysis of MRSA and MSSA clinical isolates for polymorphisms in the *fnb* locus. We found that strains harboring a single *fnb* gene were unusual, but this attribute was associated with a strain of MRSA that is epidemic in Canada, designated as CMRSA-4 *(36)*, which was later found to be closely related to the EMRSA-16 strain that is epidemic in the UK *(18)*. Among the sequenced *S. aureus* genomes, MSRA252 representative of the EMRSA-16 strain, is the only sequenced genome to possess a single *fnb* gene (SAR2580) corresponding to *fnb*A. This segment of the genome is another key feature that differentiates MRSA252 from other sequenced genomes (Fig. 2).

In MSSA476 and other strains, the tandem *fnb*B (SAS2387) and *fnb*A (SAS2388) genes are highly homologous to one another, suggesting a gene duplication event. These genes are flanked by SAS2386, encoding a putative nucleotidyl transferase function, and SAS2390, encoding a gluconate permease (*gnt*P). The same flanking genes are present in MRSA252 (SAR2579 and SAR2582), but with only one *fnb* gene corresponding to *fnb*A in the intervening sequence. Three-prime of the tandem *fnb* genes in MSSA476, the CDSs represented by the SAS2386 nucleotidyl transferase function, and SAS2380, which encodes a small transmembrane protein, define the outside boundaries of what could be described as a genomic islet that is present in all sequenced genomes, with the exception of MRSA252, in which the respective orthologs of these genes (SAS2579 and SAS2578) are contiguous.

The genes encoded by this islet include divergently transcribed transcriptional regulators SAS2385 (*sar*T) and SAS2384 (*sar*U); both belonging to the staphylococcal accessory regulator (Sar) family of proteins that modulate expression of cell surface and secreted virulence factor expression in *S. aureus (37)*. SarT represses expression of both α hemolysin *(38)* and *sar*U *(39)*, but also induces expression of *sar*S *(40)* that is encoded elsewhere on the chromosome, and is described as an inducer of staphylococcal protein A expression *(40,41)*. Additionally, *sar*T and *sar*U, together with *sar*A *(42)*, contribute to regulation of the accessory gene regulator (*agr*) locus of *S. aureus*, and *agr* in turn plays a major role in modulating expression of both cell surface and secreted virulence factors *(43)*. Therefore, it seems likely that MRSA252 and related strains that lack this genomic islet should display significant differences in the regulatory circuits that control expression of virulence factors, compared with other strains.

The divergently transcribed *sar*T–*sar*U transcriptional regulators are followed by SAS2383 (*sas*G), encoding a cell wall anchored surface protein *(44)*. The SasG protein bears significant homology toward accumulation associated protein of *S. epidermidis* that has been implicated in biofilm development *(45)*, and toward plasmin-sensitive surface protein, Pls, that is unique to certain strains of MRSA *(46)*. All three proteins promote adhesion of *S. aureus* to nasal epithelial cells *(44)*. In addition, the Pls protein expressed by certain strains of MRSA masks the function of other adhesion proteins *(46,47)*, and promotes adhesion to cellular glycolipids *(48)*. Therefore, there are several potential mechanisms through which SasG might promote adhesion and colonization functions of *S. aureus*.

Another segment of the *S. aureus* genome that is dedicated toward adhesion functions and further emphasizes the unique nature of the MRSA252 genetic background is the *sdr*CDE locus, representing another cluster of genes belonging to the MSCRAMM family. The *sdr* genes are differentiated from other MSCRAMMs on the basis of their encoding members of the serine-aspartic acid (SD)-repeat (sdr) family of proteins, characterized by long stretches of SD dipeptide repeats in their C-terminal segments *(49)*. The fibrinogen-binding proteins ClfA(SdrA) and ClfB(SdrB) were the first two such characterized proteins. In MSSA476, *sdr*A (SAS0752) and *sdr*B (SAS2516) are located in different quadrants of the genome, and the same arrangement is observed in all other sequenced genomes. In contrast, *sdr*CDE are located in tandem in the first quadrant of all published *S. aureus* genome sequences, with the exception of MRSA252, which harbors only the *sdr*CE components. In MRSA252, the *sdr*E ortholog SAR0567 is annotated as a bone sialoprotein-binding protein, Bbp. Excluding the SD repeat segment, the SAR0567 product is 98% identical to the Bbp protein first described in an osteo-myelitis isolate, strain O24 *(50)*, which confers binding of bone sialoprotein. However, as the SAR0567 gene product is 83–84% identical to *sdr*E gene products in all other sequenced *S. aureus* genomes, it appears that Bbp is an ortholog of *sdr*E. As the preced-ing gene SAR0566 is annotated as *sdr*C, it is clear that this segment of the MRSA252 genome differs from other strains because of the absence of *sdr*D.

5. Correlation of Small-Scale Genetic Variation and Phenotypic Traits

To what extent can the described small-scale genetic variations promote significant differences in phenotypical traits? It has been reported that methicillin susceptible *S. aureus* clinical isolates UAMS-1 and UAMS-601 are hypervirulent in musculoskeletal infection models (osteomyelitis and septic arthritis) compared with other strains, and comparative genomics has revealed these strains to be closely related to MRSA252 *(51)*. As with MRSA252, these isolates harbor the *cna* genomic islet but carry only a single *fnb* gene, and also lack the *sas*G genomic islet previously described. Whether or not these features directly contribute to the feral nature of these strains in musculoskeletal infection, or are part of a broader genotypic and phenotypic combination that promotes these infections, remains to be determined. It is interesting to note that α hemolysin (α toxin) could not be detected in culture supernatants of either UAMS-1 or UAMS-601 using a monoclonal antibody *(52)*, and that the *hla* gene of MRSA252 is annotated as a pseudogene *(3)*. The *sas*G genomic islet, which contains *sar*T, a repressor of α hemoly-sin expression, is absent from strains of this distinct genetic background, on which basis it would be expected that these strains should produce enhanced levels of this toxin. Could the occurrence of *hla* as a pseudogene in MRSA252 be an adaptation to the absence of *sar*T? Potentially, enhanced expression of α toxin would result in greater morbidity and less opportunity for transmission, in which case a nonsense mutation in the *hla* gene might be a compensatory mutation that allows greater transmissibility.

MRSA252 also harbors more unique genes compared with other strains that have been sequenced thus far; oddly, these are primarily associated with metabolical func-tions, rather than virulence, suggesting that MRSA252 has a greater metabolic reper-toire compared with other strains *(3)*. A notable difference is the SAR0285–SAR0291 genomic islet that is adjacent to the ESAT-6-like locus, suggesting that this strain could secrete a different subset of ESAT effecter proteins compared with other strains. Perhaps the MRSA252 background represents a more primal version of the *S. aureus* genome, which failed to acquire the SasG genomic adhesion islet, and has not undergone gene duplication events in the *fnb*A and *sdr*CE loci that are characteristic of other strains. Collectively, these considerations suggest that the success of this strain, both as an MRSA and a methicillin-sensitive *S. aureus* that is known for virulent musculoskeletal infections, may be attributed primarily to small-scale variations in the genome.

6. The Accessory Genome of *S. aureus*
6.1. Bacteriophage

As with other pathogenic bacteria, phage conversion is an important mechanism in the evolution of virulent strains of *S. aureus*. Examples of both positive and negative lyso-genic conversion are described using what has been accepted as standardized nomencla-ture for the *S. aureus* prophage *(2,14)*. The exfoliative toxin A (ETA) responsible for scalded skin syndrome is carried by the φSa1 family of *S. aureus* prophage *(53)*. Although none of the sequenced genomes harbors *eta*, strain Mu50 is lysogenized with a φSa1 variant that lacks the *eta* toxin gene. The φSa1 prophage of Mu50 is inserted after nucleotide 917,507 of the genome, and the *att*P site of the prophage occurs immediately after a gene that is highly conserved in all strains. This gene, designated as *suf*B (SACOL0918) in strain COL is homologous to a family of proteins that participates in assembly of proteins containing iron-sulfur clusters under conditions of oxidative stress and iron limitation, and is therefore a potential virulence factor. The gene encoding β-toxin (*hlb*), is expressed mainly by isolates of animal origin, but is disrupted in most human isolates, including the genomes of MRSA252, N31T, MuTO, MW2, and MSSA476, by the φSa3 family of *S. aureus* bacteriophage, all of which carry the *sak* gene encoding staphylokinase, which is a plasminogen activator. In addition to *sak*, most members of the φSa3 family also carry *sea*, encoding enterotoxin A, which is a major cause of food poisoning, and other enterotoxins (i.e., *sep*, *seg*2, *sek*2) may also be present *(3,14)*.

In contrast to the ubiquitous distribution of φSa3 amongst the sequenced genomes, the φSa2 prophage, which carries the PVL leukotoxin, appears only in the genome of the community-acquired MRSA strain MW2, although a φSa2 variant that lacks the PVL toxin is present in MRSA252. The clinical implications of this toxin, which promotes severe invasive infections, have been described in **Subheading 1.** of this chapter. The identical *att*L and *att*R sites of the φSa2 prophage genome *(54)* correspond to nucleotides 1554771–1554799 of the MSSA476 genome, which does not carry the φSa2 prophage. This site occurs near the C-terminus of SAS1429, encoding a 720 amino acid hypothetical protein of no known function. In the genome of strain MW2, the ortholog of SAS1429 is disrupted by the φSa2 prophage, and annotated as two CDSs (MW1377 and MW1433), with the intervening genes corresponding to the φSa2 prophage. Although SAS1429 bears no obvious signal peptide or domains of known function, it is flanked by genes SAS1426-28 and SAS1430, each of which harbor putative lipoprotein signal peptides, and are highly homologous to one another. This gene cluster is preceded by the previously described staphylococcal respiratory response *srr*AB operon (SAS1432 and SAS1431) *(55,56)*. The *srr*AB operon encodes a two-component signal transduction pathway that controls metabolic processes and energy transduction in response to oxygen *(57)*, and also regulates the oxygen-dependent expression of toxic shock syndrome toxin and other virulence factors *(55,56)*. It is intriguing that the φSa2 prophage that carries the PVL toxin integrates adjacent to a gene locus that appears to have an important role in regulation of toxin expression.

The function of this gene cluster, and the consequence of φSa2 prophage integration within SAS1429, is an area that deserves further attention. Although the enhanced disease severity associated with acquisition of the PVL toxin can be profound, the toxin has not yet appeared among endemic and epidemic hospital-acquired MRSA strains, suggesting that the transmissibility of strains that harbor this toxin is limited. Potentially, SAS1429 and flanking genes encoding lipoproteins may have a role in mediating host interactions or evasion of defense mechanisms, and disrupted expression by integration of the φSa2 prophage may ultimately reduce the transmissibility. The epidemic MRSA252 strain harbors a variant φSa2 prophage that lacks the PVL toxin. In this situation, the multiple resistance phenotype of MRSA252 could facilitate transmission within the hospital environment.

6.2. The Staphylococcal Cassette Chromosome Element

The Staphylococcal Cassette Chromosome (SCC) designates a family of mobile genetic elements including, but not limited to, SCC*mec*, which confers resistance to methicillin and other antimicrobial agents. Five versions of SCC*mec* are found in *S. aureus*, at the same location near the *oriC* replication origin of the chromosome *(15,58–61)*. All SCC*mec* elements carry the *mec*A gene encoding the alternate penicillin-binding protein, PBP2a, which confers resistance to the β-lactam family of antibiotics. The type I SCC*mec* carried by *S. aureus* COL is representative of the earliest arising SCC*mec* elements, dating back to the emergence of methicillin resistance in 1961, and, therefore, conveys resistance only to β-lactam antibiotics. Types II and III SCC*mec* also carry various integrated plasmids and transposons, which harbor additional resistance determinants to non-β-lactam antibiotics, including aminoglycosides, macrolides, tetracycline, and heavy metals. Types IV and V SCC*mec* are significantly smaller than types I to III, and as with type I, do not carry resistance genes other than *mec*A. So far, these elements have been detected only in community-acquired MRSA, and it is suggested that their small size renders them amenable to horizontal transfer via transducing phage. MLST analyses have established that different categories of SCC*mec* are carried by strains of the same genetic background, supporting the conclusion that *S. aureus* has acquired SCC*mec* on multiple occasions *(19)*.

Although SCC elements typically carry *mec*A, there are some interesting variants. The closely related strains MW2 and MSSA476 both harbor an SCC element *(2,3)*. That of MW2 is a typical type IV SCC*mec*, whereas MSSA476 possesses a novel element designated SCC$_{476}$ that lacks *mec*A, but harbors a putative fusidic acid resistance determinant, *far*1, encoded by SAS0043 *(3,14)* that is not present in other sequenced *S. aureus* genomes. Its assignation as a fusidic acid resistance determinant is based on it displaying 42% amino acid identity with the product of a poorly characterized fusidic acid resistance gene *far*1/*fus*B encoded by *S. aureus* plasmid pUB101 *(62)*. Fusidic acid is employed as a topical agent for treatment of superficial *S. aureus* skin infections, including impetigo and atopic dermatitis. A gene that is similar or identical to the *far*1/ *fus*B determinant of pUB101 has been identified on the chromosome of a single *S. aureus* clonal type associated with superficial skin infections in Norway and Sweden, and it was suggested that the emergence and dissemination of this strain could be associated with the increased use of topical fusidic acid *(63–65)*. Potentially, the SCC$_{476}$ element of MSSA476 is an adaptation that promotes dissemination of this strain in the community. Intriguingly, the fusidic acid resistance determinants of SCC$_{476}$ and pUB101 both possess domains that are present in a family of fibronectin-binding proteins described in *Listeria* species *(66,67)*, which have been described as promoting adherence to fibronectin-containing surfaces. Although there is no obvious signal peptide present in the *far*1 encoded product of SCC$_{476}$, the same is also true for this protein family in *Listeria*. Therefore, a dual function in promoting fusidic acid resistance and adherence to the epithelium is an interesting possibility.

Another SCC element that lacks *mec*A was found to carry genes encoding capsule biosynthesis, designated SCC*cap* *(68)*, and a gene encoding a novel staphylococcal enterotoxin was immediately adjacent to the SCC*cap* locus. Similarly, two genes with homology to staphylococcal and streptococcal enterotoxins/exotoxins are also observed outside the right boundary of the respective SCC$_{476}$ and SCC*mec* elements of the community-acquired MSSA476 (SAS0051-0052) and MRSA MW2 (MW0051-0052) strains *(2,3)*. This feature is not observed in the hospital-adapted MRSA strains MRSA252, N315, and Mu50. These associations suggest that the community-acquired MRSA strain MW2 could have first been a successful community-acquired MSSA strain that carried a non-*mec*A type SCC element, which subsequently facilitated acquisition of SCC*mec*. Indeed, it has been suggested that these novel non-*mec*A carrying SCC elements could serve as reservoirs for shuffling of genes between staphylococcal species, contributing to the diversity of various SCC elements *(69)*. This is based on the description of a SCC element of *S. epidermidis* that carried genes encoding homologs of penicillin-binding protein, *pbp*4, and a gene, *tag*F, encoding a product predicted to be involved in teichoic acid biosynthesis, a major component of the Gram-positive cell wall *(69)*. Specifically, in *S. epidermidis* and *B. subtilis*, proteins encoded by *tag*F genes promote polymerization of the polyglycerol-phosphate teichoic acid polymer *(70,71)*.

Analysis of the type I SCC*mec* element, the earliest originating member of the SCC*mec* family carried by *S. aureus* COL, appears to support this contention. This element possesses a gene, *pls* (SA0050), encoding a cell surface adhesion protein, Pls, that most closely resembles the accumulation-associated protein of *S. epidermidis*. The Pls protein is expressed as a glycoprotein in MRSA *(48,72)*, and the *pls* gene within the SCC*mec* element of COL is flanked by genes that could be involved in glycosylation of Pls (Fig. 3A). The *pls* gene is followed by SA0051, encoding a putative 64-kDa protein with 69% amino acid identity toward a protein encoded by SE2394 in the genome of *S. epidermidis* that is annotated as a polyglycerol-phosphate α-glucosyltransferase of *S. epidermidis (73)*. The basis of this assignation for the *S. epidermidis* protein is not clear, because neither it nor the protein encoded by SA0051, possess any protein family domains that would identify it as a glucosyl transferase. However, SA0051 does harbor a consensus ATP/GTP binding site motif A (P-loop), whereas neighboring SA0052 encodes a protein with a type 1 protein family glucosyl transferase domain (PF00534). This is followed by SA0053, also transcribed on the minus strand, encoding a putative muramyl ligase, a protein involved in the sequential assembly of the pentapeptide substituent of *N*-acetyl muramic acid in the biosynthesis of peptidoglycan. Although this gene is annotated as a pseudogene in *S. aureus* COL, its presence, together with other genes encoding proteins that are homologous to enzymes involved in teichoic acid biosynthesis, is suggestive of a function in glycosylation of the Pls protein, perhaps by decoration with a teichoic acid like modification.

This type of arrangement is not unique to type I SCC*mec*, being observed also downstream of the *sdr*CDE gene cluster (Fig. 3B) that was described earlier in this chapter (*see* **Subheading 4.3.1.**). The MW0519 and MW0520 CDSs downstream of *sdr*E (MW0518) in strain MW2 have the same transcriptional organization and similar annotations as the SA0052 and SA0053 genes that follow *pls* in the SCC*mec* element of *S. aureus* COL. Both genes are annotated as encoding poly glycerol-phosphate α-glucosyl-transferase proteins. A similar arrangement is observed in the genome of *S. epidermidis* strain ATCC 12228 *(73)* (Fig. 3C) immediately downstream of a gene SE2395, encoding a surface protein of the SD-repeat family that when compared with proteins of *S. aureus*, most closely resembles SdrD within the *sdr*CDE gene cluster of the different sequenced genomes. In view of these considerations, the biochemistry and significance of cell surface protein glycosylation in *S. aureus* is an area that is deserving

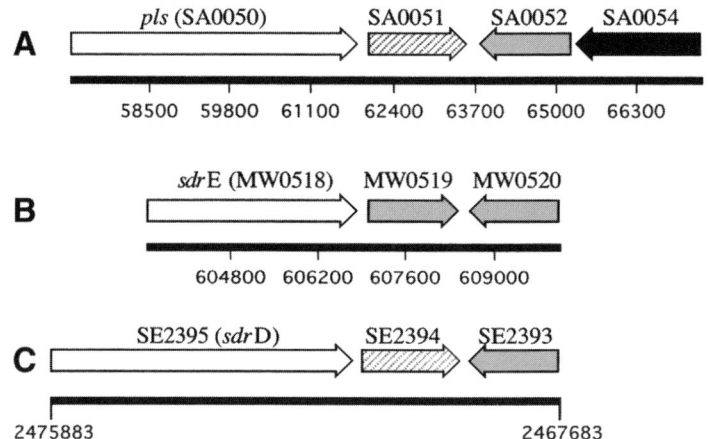

Fig. 3. Comparison of the plasmin-sensitive surface protein (Pls) locus and 3' flanking region within the type I SCC*mec* element of methicillin-resistant *Staphylococcus aureu* (MRSA) strain COL (**A**), the *sdr*CDE locus, and 3' flanking region of community-acquired MRSA strain MW2 (**B**), and a gene locus with similar organization from *Staphylococcus epidermidis* strain ATCC 12228 (**C**). Numbers below the scale correspond to the nucleotide coordinates in each genome. For *S. epidermidis*, the reverse and complement of the minus strand genomic DNA was translated, to maintain the same format as in (**A**) and (**B**). Genes encoding cell surface adhesion proteins of the serine-aspartic acid-repeat family are white. For clarity, the *sdr*CD genes are omitted from (**B**). When compared with proteins of *S. aureus*, the SE2395 coding sequence (CDS) of *S. epidermidis* most closely resembles the SdrD protein (approx 37% identity) encoded within the *sdr*CDE locus of different strains. The transcriptional organization and annotation of genes immediately downstream of each adhesin-encoding gene is similar in each of the different loci. Gray-shaded CDSs (SA0052, MW0519, MW0520, and SE2393) designate proteins that possess a C-terminal type 1 glucosyl transferase protein family domain (PF00534), and are annotated as polyglycerol-phosphate α-glucosyl transferase proteins on the basis of homology to enzymes involved in polymerization of the polyglycerol-phosphate teichoic acid polymer. The SA0052 CDS of *S. aureus* COL displays 39, 48, and 59% amino acid sequence identity, respectively, toward MW0519, MW0520, and SE2393. The cross-hatched CDS (SA0051) of *S. aureus* COL contains no annotation. It does not harbor an obvious type I glucosyl transferase domain, but possesses a consensus ATP-GTP binding motif A (P loop) and is 69% identical to SE2394 of *S. epidermidis*, which also possesses a consensus ATP-GTP binding motif. Although SE2394 is annotated as a polyglycerol-phosphate α-glucosyl transferase, this appears to be an error, as SE2394 does not possess an obvious glucosyl transferase domain, and there is no obvious homology toward the neighboring SE2393, which is correctly annotated as a glucosyl transferase. However, the ATP-GTP binding motif may indicate a role for this protein, as well as SA0051 in satisfying an energy requirement for sugar polymerization. The black-shaded SA0054 CDS of *S. aureus* COL is not present in the other loci and encodes a protein that harbors a murine ligase protein family domain (PF01225), which is present in a family of proteins that assemble the pentapeptide modification of *N*-acetyl muramic acid in the synthesis of peptidoglycan.

of attention, as is the role of SCC elements in promoting the shuffling of genes between *Staphylococcal* species, and phenotypic modification of the host strains.

6.3. Pathogenicity Islands

The description of *S. aureus* pathogenicity islands (SaPI) is a relatively recent development (74–76). They can be broadly categorized into four groups on the basis of integrase homology and insertion site, and so far no strain has been found to carry more than one copy of each *(14)*. The nomenclature is complicated by the close association of different SaPIs with one another, but standardized terminology has been proposed *(14)*, which is used here. The best-characterized SaPI belongs to the SaPI2 family, most of which carry *tst*, encoding toxic shock syndrome toxin. In addition to *tst*, the SaPI2 family can also carry genes encoding enterotoxins. Among the sequenced genomes, SaPI2 occurs only in the related N315 and Mu50 strains, in close proximity to the φSa3 prophage discussed earlier in **Subheading 6.1.** Accordingly, phage-encoded functions can promote the autonomous replication of excised SaPI2 and its packaging into phage particles *(77)*, a factor that probably promotes the dissemination of SaPI2.

Similar considerations probably apply to SaPI1, which is present in *S. aureus* COL but not in other sequenced genomes. SaPI1 carries a gene encoding a putative β-lactamase (*ear*) encoded by SACOL0908, and several enterotoxins. The β-lactamase encoded by SACOL0908 is in close proximity to the *suf*B gene encoded by SACOL0918, a component of the core genome which contains the attachment site for the φSa1 prophage discussed earlier in **Subheading 6.1.** Strain COL carries SaPI1, but is not lysogenized by φSa1, whereas Mu50 is lysogenized by φSa1 at 0.92 Mbp of the genome, but does not carry SaPI1. However, Mu50 carries a member of the SaPI3 family at approx 0.88 Mbp of the genome, in close proximity to the *coa* allele located at 0.89 Mbp. This approx 40 kb segment of the genome may be a hot spot for horizontal transfer of virulence factors, with the potential of harboring two different SaPI families, in addition to the integration site for the φSa1 prophage family, which can carry the *eta* exfoliative toxin. This may explain why closely related strains can have distinct *coa* alleles, whereas apparently nonrelated strains can share the same *coa* allele as discussed in **Subheading 4.3.** *(15)*. Potentially, the *coa* allele is carried along in the en masse horizontal transfer of genomic DNA catalyzed by the close association of pathogenicity islands and prophages.

The SaPI3 element of Mu50, first annotated as GEI SaGIm *(4)*, is closely related to SaPI2, but encodes a different integrase function, and as noted, localizes to a different segment of the genome. It does not carry any toxin-encoding genes, but encodes a homolog of the *fhu*D, ferric hydroxamate uptake transporter, which is an important factor in iron acquisition *(78,79)*. In contrast, MW2 harbors a different member of the SaPI3 family at the same relative genomic coordinates (0.84 Mbp); this element lacks *fhu*D, and instead carries genes encoding a putative β-lactamase (*ear*) and two enterotoxins *(2)*. Thus, this member of the SaPI3 family carries a similar complement of resistance determinants and enterotoxins as described for SaPI1 of strain COL. This is the only pathogenicity island carried by MW2, whereas the related MSSA476 does not carry any pathogenicity islands. MRSA252 carries a single pathogenicity island described as a member of the SaPI4 family; in this situation encoding several hypothetical proteins, but none with obvious virulence associated functions. However, it should be noted that this strain is closely related to the epidemic EMRSA-16 strain, which commonly carries *tst* *(80)*. Therefore, other isolates belonging to this clonal type probably carry SaPI2.

From these considerations, it is evident that pathogenicity island carriage in *S. aureus* is quite variable; MSSA476 does not carry any, whereas the SaPI4 element of MRSA252 has no obvious virulence factors, and other strains will carry a common SaPI that harbors different virulence determinants. The SaPI2 element that carries *tst* probably has the greatest capacity to exert a dramatic influence over the virulence phenotype of the host strain. Oddly, whereas approx 20% of *S. aureus* isolates carry *tst*, the actual clinical incidence of toxic shock syndrome is quite low *(14)*. This represents a major difference in pathogenicity island function of *S. aureus*, compared with Gram-negative pathogens such as *E. coli*, *Salmonella*, and *Yersinia* species, in which pathogenicity islands are much more likely to carry genes that are essential components of the infection strategy.

6.4. GEIs

The νSaα and νSaβ GEIs are located at approximately two o'clock (νSaα) and eight o'clock (νSaβ) on every *S. aureus* genome that has thus far been sequenced. They are differentiated from the SaPI2 pathogenicity island that carries *tst* on the basis of harboring a truncated transposase, and no obvious association with a temperate prophage. Both elements are characterized by extensive duplication of the putative virulence genes that are carried, and have therefore also been referred to as "gene nurseries." A third distinguishing feature is that both elements carry a DNA methylase (*hsd*M) and a flanking *hsd*S gene that probably dictates the specificity of methylation. It has been suggested that these genes facilitate the stable carriage of the νSaα and νSaβ, such that the genomic DNA of the host strain will undergo digestion by restriction endonucleases if these GEIs are excised from the chromosome *(4)*.

The central *hsd*M/*hsd*S genes of νSaα are flanked on the 5' side by 9–11 tandem repeats of *set* genes with high homology to one another, and also with other *S. aureus* exotoxin superantigens, whereas on the 3' side there are from 3 to 9 tandem copies of *lpl* genesencoding putative lipoproteins, but of unknown contribution to virulence. Although its role in virulence has not been proven, this GEI may provide insight into the virulence strategy of *S. aureus*, which unlike other species of pathogenic bacteria, such as *Neisseria* and *Salmonella*, has not evolved sophisticated mechanisms for evasion of the host immune response as the main strategy for survival. Rather, *S. aureus* seems to specialize in provoking a strong regional inflammatory response leading to abscess formation, and localization within an abscess could be advantageous in that the secreted exototoxins would become concentrated and efficiently destroy leukocytes *(4)*. Potentially, the multiple *lpl* genes within νSaα encoding putative cell surface lipoproteins could also promote some capacity for antigenic variation and evasion of humoral immunity. In this capacity, the virulence factors encoded by νSaα could be seen to promote key aspects of the *S. aureus* virulence strategy.

In the νSaβ GEI, the *hsd*S/*hsd*M genes are located at the left end of the element, adjacent to a truncated or frameshifted transposase, and these genes are followed by up to five tandem *spl* genes, encoding putative serine proteases, SplA-SplF. Each of these proteins possesses a histidine active site signature sequence (Prosite PS00672) that designates them as members of the V8 family of serine proteases. The V8 protease, now designated as SspA *(81)* is a glutamyl endopeptidase that is encoded by the core genome of *S. aureus*, as discussed earlier in this chapter in **Subheading 3.** Exfoliative toxins A and B, which are highly selective glutamyl endopeptidases, also possess the histidine active site signature sequence of the V8 family. Furthermore, the Spl proteins, as well as SspA and the EtA and EtB toxins, share significant homology to one another. In spite of this potential association with toxin function, the tandem *spl* genes can be deleted from the chromosome of *S. aureus* strain RN6390 with no obvious reduction in virulence in a tissue abscess infection model *(82)*, even though the Spl proteins are normally expressed and secreted by this strain in vitro.

Other elements of vSaβ are variable, and differentiate vSaβ into types I and II *(2)*. In type I vSaβ, the *spl* genes are followed closely by *luk*D and *luk*E, encoding leukotoxins, and these are in turn followed by a variable number of genes encoding putative enterotoxins, depending on the host strain. The hospital-adapted MRSA strains N315, Mu50, and MRSA252 all carry a type I vSaβ element. In contrast, the community-acquired MSSA476 and MW2 possess a type II vSaβ, in which the *luk*DE leukotoxin components are separated from the *spl* locus by insertion of a cluster of genes that most closely resemble those involved in synthesis of the peptide lantibiotic, epidermin, by *S. epidermidis*. First identified in the community-acquired MRSA strain MW2, it was suggested that the type II vSaβ is important for strain MW2 to compete with other natural flora for successful colonization *(2)*. That the same element is also present in MSSA476 appears to strengthen this contention. Intriguingly, MRSA strain COL also harbors a type II vSaβ. However, COL differs from the hospital-adapted MRSA that harbor type I vSaβ, because it was one of the earliest originating MRSA strains dating back to 1961, prior to the emergence of multiply resistant nosocomial strains. This would suggest that as MRSA evolved as nosocomial pathogens, the acquisition of multiple antibiotic resistance mechanisms displaced the need for bacteriocins as a means of competing with resident flora, and the bacteriocin locus within the vSaβ was replaced with genes of greater utility, including leukotoxins and exotoxins. Along this note, the association of the *spl* genes with leukotoxins and exotoxins in the type I vSaβ would appear to implicate an offensive role for these serine proteases in terms of disease severity. However, if type I vSaβ evolved from type II, then the original association of the Spls would be with the bacteriocin locus, and this thought should influence our considerations in suggesting functions for the Spl proteins. Potentially, the Spl proteases could be involved in proteolytic inactivation of epidermin-type lantibiotics secreted by different species of coagulase-negative staphylococci and other resident microflora.

Acknowledgments

The author would like to acknowledge the perseverance, dedication, and remarkable achievements of the *S. aureus* genome project leaders: Keiichi Hiramatsu, Matthew Holden, Steven Gill, John Iandolo, and their coworkers. Thanks to Steven Gill and colleagues at The Institute for Genomics Research for making unedited genome sequence data available for viewing and downloading, long before other sequence data became available.

References

1. Gill, S., Fouts, D. E., Archer, G. L., et al. (2005) Insights on evolution of virulence and resistance from the complete genome analysis of an early methicillin-resistant Staphylococcus aureus strain and a biofilm-producing methicillin-resistant Staphylococcus epidermidis strain. *J. Bacteriol.* **187,** 2426–2438.
2. Baba, T., Takeuchi, F., Kuroda, M., et al. (2002) Genome and virulence determinants of high virulence community-acquired MRSA. *Lancet* **359,** 1819–1827.
3. Holden, M. T., Feil, E. J., Lindsay, J. A., et al. (2004) Complete genomes of two clinical *Staphylococcus aureus* strains: evidence for the rapid evolution of virulence and drug resistance. *Proc. Natl. Acad. Sci. USA* **101,** 9786–9791.

4. Kuroda, M., Ohta, T., Uchiyama, I., et al. (2001) Whole genome sequencing of meticillin-resistant *Staphylococcus aureus*. *Lancet* **357,** 1225–1240.

5. Fleischmann, R. D., Adams, M. D., White, O., et al. (1995) Whole-genome random sequencing and assembly of *Haemophilus influenzae* Rd. *Science* **269,** 496–512.

6. Peacock, S. J., de Silva, I., and Lowy, F. D. (2001) What determines nasal carriage of *Staphylococcus aureus*? *Trends Microbiol.* **9,** 605–610.

7. Foster, T. J. (2004) The *Staphylococcus aureus* "superbug". *J. Clin. Invest.* **114,** 1693–1696.

8. Jarraud, S., Mougel, C., Thioulouse, J., et al. (2002) Relationships between *Staphylococcus aureus* genetic background, virulence factors, *agr* groups (alleles), and human disease. *Infect. Immun.* **70,** 631–641.

9. Anonymous. (2003) Emergence of PVL-producing strains of *Staphylococcus aureus*. *CDR Weekly* **13,** 5–6.

10. Gillet, Y., Issartel, B., Vanhems, P., et al. (2002) Association between *Staphylococcus aureus* strains carrying gene for Panton-Valentine leukocidin and highly lethal necrotising pneumonia in young immunocompetent patients. *Lancet* **359,** 753–759.

11. Anonymous (2002) From the Centers for Disease Control. *Staphylococcus aureus* resistant to vancomycin—United States, 2002. *JAMA* **288,** 824–825.

12. Anonymous (2002) Vancomycin-resistant *Staphylococcus aureus*—PA, October 11, 2002. *MMWR. Morb. Mortal. Wkly. Rep.* **51,** 902.

13. Weigel, L. M., Clewell, D. B., Gill, S. R., et al. (2003) Genetic analysis of a high-level vancomycin-resistant isolate of *Staphylococcus aureus*. *Science* **302,** 1569–1571.

14. Lindsay, J. A. and Holden, M. T. (2004) *Staphylococcus aureus*: superbug, super genome? *Trends Microbiol.* **12,** 378–385.

15. Hiramatsu, K., Watanabe, S., Takeuchi, F., Ito, T., and Baba, T. (2004) Genetic characterization of methicillin-resistant *Staphylococcus aureus*. *Vaccine* **22(Suppl 1),** S5–S8.

16. Johnson, A. P., Aucken, H. M., Cavendish, S., et al. (2001) Dominance of EMRSA-15 and -16 among MRSA causing nosocomial bacteraemia in the UK: analysis of isolates from the European Antimicrobial Resistance Surveillance System (EARSS). *J. Antimicrob. Chemother.* **48,** 143–144.

17. McDougal, L. K., Steward, C. D., Killgore, G. E., Chaitram, J. M., McAllister, S. K., and Tenover, F. C. (2003) Pulsed-field gel electrophoresis typing of oxacillin-resistant *Staphylococcus aureus* isolates from the United States: establishing a national database. *J. Clin. Microbiol.* **41,** 5113–5120.

18. Simor, A. E., Ofner-Agostini, M., Bryce, E., McGeer, A., Paton, S., and Mulvey, M. R. (2002) Laboratory characterization of methicillin-resistant *Staphylococcus aureus* in Canadian hospitals: results of 5 years of National Surveillance, 1995–1999. *J. Infect. Dis.* **186,** 652–660.

19. Enright, M. C., Robinson, D. A., Randle, G., Feil, E. J., Grundmann, H., and Spratt, B. G. (2002) The evolutionary history of methicillin-resistant *Staphylococcus aureus* (MRSA). *Proc. Natl. Acad. Sci. USA* **99,** 7687–7692.

20. Feil, E. J., Cooper, J. E., Grundmann, H., et al. (2003) How clonal is *Staphylococcus aureus*? *J. Bacteriol.* **185,** 3307–3316.

21. Dryla, A., Gelbmann, D., von Gabain, A., and Nagy, E. (2003) Identification of a novel iron regulated staphylococcal surface protein with haptoglobin-haemoglobin binding activity. *Mol. Microbiol.* **49,** 37–53.

22. Komatsuzawa, H., Ohta, K., Sugai, M., et al. (2000) Tn551-mediated insertional inactivation of the *fmt*B gene encoding a cell wall-associated protein abolishes methicillin resistance in *Staphylococcus aureus*. *J. Antimicrob. Chemother.* **45,** 421–431.

23. Wootton, M., Avison, M. B., Bennett, P. M., Howe, R. A., MacGowan, A. P., and Walsh, T. R. (2004) Genetic analysis of 17 genes in *Staphylococcus aureus* with reduced susceptibility to vancomycin (VISA) and heteroVISA. *J. Antimicrob. Chemother.* **53,** 406–407.

24. Howe, R. A., Monk, A., Wootton, M., Walsh, T. R., and Enright, M. C. (2004) Vancomycin susceptibility within methicillin-resistant *Staphylococcus aureus* lineages. *Emerg. Infect. Dis.* **10,** 855–857.

25. Pallen, M. J. (2002) The ESAT-6/WXG100 superfamily—and a new Gram-positive secretion system? *Trends Microbiol.* **10,** 209–212.

26. Stanley, S. A., Raghavan, S., Hwang, W. W., and Cox, J. S. (2003) Acute infection and macrophage subversion by *Mycobacterium tuberculosis* require a specialized secretion system. *Proc. Natl. Acad. Sci. USA* **100,** 13,001–13,006.

27. Burts, M. L., Williams, W. A., Debord, K., and Missiakas, D. M. (2005) EsxA and EsxB are secreted by an ESAT-6-like system that is required for the pathogenesis of *Staphylococcus aureus* infections. *Proc. Natl. Acad. Sci. USA* **102,** 1169–1174.

28. Hsu, T., Hingley-Wilson, S. M., Chen, B., et al. (2003) The primary mechanism of attenuation of bacillus Calmette-Guerin is a loss of secreted lytic function required for invasion of lung interstitial tissue. *Proc. Natl. Acad. Sci. USA* **100,** 12,420–12,425.

29. Rowlands, S. J., Dyke, K. G. H., Curnock, S. P., Boocock, M. R., and Stark, W. M. (2000) Determination of the integration site of Tn552 in *Staphylococcus aureus* NCTC9789. Unpublished; GenBank Accession CAB72943.

30. Gillaspy, A. F., Patti, J. M., Pratt, F. L. Jr., Iandolo, J. J., and Smeltzer, M. S. (1997) The *Staphylococcus aureus* collagen adhesin-encoding gene (*cna*) is within a discrete genetic element. *Gene* **196,** 239–248.

31. Switalski, L. M., Patti, J. M., Butcher, W., Gristina, A. G., Speziale, P., and Hook, M. (1993) A collagen receptor on *Staphylococcus aureus* strains isolated from patients with septic arthritis mediates adhesion to cartilage. *Mol. Microbiol.* **7,** 99–107.

32. de Bentzmann, S., Tristan, A., Etienne, J., Brousse, N., Vandenesch, F., and Lina, G. (2004) *Staphylococcus aureus* isolates associated with necrotizing pneumonia bind to basement membrane type I and IV collagens and laminin. *J. Infect. Dis.* **190,** 1506–1515.

33. Patti, J. M. and Hook, M. (1994) Microbial adhesins recognizing extracellular matrix macromolecules. *Curr. Opin. Cell. Biol.* **6,** 752–758.

34. Jonsson, K., Signas, C., Muller, H. P., and Lindberg, M. (1991) Two different genes encode fibronectin binding proteins in *Staphylococcus aureus*. The complete nucleotide sequence and characterization of the second gene. *Eur. J. Biochem.* **202,** 1041–1048.

35. Greene, C., Vaudaux, P. E., Francois, P., Proctor, R. A., McDevitt, D., and Foster, T. J. (1996) A low-fibronectin-binding mutant of *Staphylococcus aureus* 879R4S has Tn918 inserted into its single *fnb* gene. *Microbiology* **142 (Pt 8),** 2153–2160.

36. Rice, K., Huesca, M., Vaz, D., and McGavin, M. J. (2001) Variance in fibronectin binding and *fnb* locus polymorphisms in *Staphylococcus aureus*: identification of antigenic variation in a fibronectin binding protein adhesin of the epidemic CMRSA-1 strain of methicillin-resistant *S. aureus*. *Infect. Immun.* **69,** 3791–3799.

37. Cheung, A. L. and Zhang, G. (2002) Global regulation of virulence determinants in *Staphylococcus aureus* by the SarA protein family. *Front. Biosci.* **7,** d1825–d1842.

38. Schmidt, K. A., Manna, A. C., Gill, S., and Cheung, A. L. (2001) SarT, a repressor of alpha-hemolysin in Staphylococcus aureus. *Infect. Immun.* **69,** 4749–4758.

39. Manna, A. C. and Cheung, A. L. (2003) *sar*U, a *sar*A homolog, is repressed by SarT and regulates virulence genes in *Staphylococcus aureus*. *Infect. Immun.* **71,** 343–353.

40. Schmidt, K. A., Manna, A. C., and Cheung, A. L. (2003) SarT influences *sar*S expression in *Staphylococcus aureus*. *Infect. Immun.* **71,** 5139–5148.

41. Tegmark, K., Karlsson, A., and Arvidson, S. (2000) Identification and characterization of SarH1, a new global regulator of virulence gene expression in *Staphylococcus aureus*. *Mol. Microbiol.* **37,** 398–409.

42. Cheung, A. L. and Projan, S. J. (1994) Cloning and sequencing of *sar*A of *Staphylococcus aureus*, a gene required for the expression of *agr*. *J. Bacteriol.* **176,** 4168–4172.

43. Peng, H. L., Novick, R. P., Kreiswirth, B., Kornblum, J., and Schlievert, P. (1988) Cloning, characterization, and sequencing of an accessory gene regulator (*agr*) in *Staphylococcus aureus*. *J. Bacteriol.* **170,** 4365–4372.

44. Roche, F. M., Meehan, M., and Foster, T. J. (2003) The *Staphylococcus aureus* surface protein SasG and its homologues promote bacterial adherence to human desquamated nasal epithelial cells. *Microbiology* **149,** 2759–2767.

45. Hussain, M., Herrmann, M., von Eiff, C., Perdreau-Remington, F., and Peters, G. (1997) A 140-kilodalton extracellular protein is essential for the accumulation of *Staphylococcus epidermidis* strains on surfaces. *Infect. Immun.* **65,** 519–524.

46. Savolainen, K., Paulin, L., Westerlund-Wikstrom, B., Foster, T. J., Korhonen, T. K., and Kuusela, P. (2001) Expression of *pls*, a gene closely associated with the *mec*A gene of methicillin-resistant *Staphylococcus aureus*, prevents bacterial adhesion in vitro. *Infect. Immun.* **69,** 3013–3020.

47. Juuti, K. M., Sinha, B., Werbick, C., Peters, G., and Kuusela, P. I. (2004) Reduced adherence and host cell invasion by methicillin-resistant *Staphylococcus aureus* expressing the surface protein Pls. *J. Infect. Dis.* **189,** 1574–1584.

48. Huesca, M., Peralta, R., Sauder, D. N., Simor, A. E., and McGavin, M. J. (2002) Adhesion and virulence properties of epidemic Canadian methicillin-resistant *Staphylococcus aureus* strain 1: identification of novel adhesion functions associated with plasmin-sensitive surface protein. *J. Infect. Dis.* **185,** 1285–1296.

49. Josefsson, E., McCrea, K. W., Ni Eidhin, D., et al. (1998) Three new members of the serine-aspartate repeat protein multigene family of *Staphylococcus aureus*. *Microbiology* **144,** 3387–3395.

50. Tung, H., Guss, B., Hellman, U., Persson, L., Rubin, K., and Ryden, C. (2000) A bone sialoprotein-binding protein from *Staphylococcus aureus*: a member of the Staphylococcal Sdr family. *Biochem. J.* **345,** 611–619.

51. Cassat, J. E., Dunman, P. M., McAleese, F., Murphy, E., Projan, S. J., and Smeltzer, M. S. (2005) Comparative genomics of *Staphylococcus aureus* musculoskeletal isolates. *J. Bacteriol.* **187,** 576–592.

52. Blevins, J. S., Beenken, K. E., Elasri, M. O., Hurlburt, B. K., and Smeltzer, M. S. (2002) Strain-dependent differences in the regulatory roles of *sar*A and *agr* in *Staphylococcus aureus*. *Infect. Immun.* **70,** 470–480.

53. Yamaguchi, T., Hayashi, T., Takami, H., et al. (2000) Phage conversion of exfoliative toxin A production in *Staphylococcus aureus*. *Mol. Microbiol.* **38,** 694–705.

54. Kaneko, J., Kimura, T., Narita, S., Tomita, T., and Kamio, Y. (1998) Complete nucleotide sequence and molecular characterization of the temperate staphylococcal bacteriophage phiPVL carrying Panton-Valentine leukocidin genes. *Gene* **215,** 57–67.

55. Pragman, A. A., Yarwood, J. M., Tripp, T. J., and Schlievert, P. M. (2004) Characterization of virulence factor regulation by SrrAB, a two-component system in *Staphylococcus aureus*. *J. Bacteriol.* **186,** 2430–2438.

56. Yarwood, J. M., McCormick, J. K., and Schlievert, P. M. (2001) Identification of a novel two-component regulatory system that acts in global regulation of virulence factors of *Staphylococcus aureus*. *J. Bacteriol.* **183,** 1113–1123.

57. Throup, J. P., Zappacosta, F., Lunsford, R. D., et al. (2001) The *srh*SR gene pair from *Staphylococcus aureus*: genomic and proteomic approaches to the identification and characterization of gene function. *Biochemistry* **40,** 10,392–10,401.

58. Katayama, Y., Ito, T., and Hiramatsu, K. (2000) A new class of genetic element, staphylo- coccus cassette chromosome *mec*, encodes methicillin resistance in *Staphylococcus aureus*. *Antimicrob. Agents Chemother.* **44**, 1549–1555.

59. Ito, T., Ma, X. X., Takeuchi, F., Okuma, K., Yuzawa, H., and Hiramatsu, K. (2004) Novel type V staphylococcal cassette chromosome *mec* driven by a novel cassette chromosome recombinase, *ccr*C. *Antimicrob. Agents Chemother.* **48**, 2637–2651.

60. Ito, T., Katayama, Y., Asada, K., et al. (2001) Structural comparison of three types of staphylococcal cassette chromosome *mec* integrated in the chromosome in methicillin-resis- tant *Staphylococcus aureus*. *Antimicrob. Agents Chemother.* **45**, 1323–1336.

61. Daum, R. S., Ito, T., Hiramatsu, K., et al. (2002) A novel methicillin-resistance cassette in community-acquired methicillin-resistant *Staphylococcus aureus* isolates of diverse gene- tic backgrounds. *J. Infect. Dis.* **186**, 1344–1347.

62. O'Brien, F. G., Price, C., Grubb, W. B., and Gustafson, J. E. (2002) Genetic characteriza- tion of the fusidic acid and cadmium resistance determinants of *Staphylococcus aureus* plasmid pUB101. *J. Antimicrob. Chemother.* **50**, 313–321.

63. Tveten, Y., Jenkins, A., and Kristiansen, B. E. (2002) A fusidic acid-resistant clone of Staphylococcus aureus associated with impetigo bullosa is spreading in Norway. *J. Anti- microb. Chemother.* **50**, 873–876.

64. Osterlund, A., Eden, T., Olsson-Liljequist, B., Haeggman, S., and Kahlmeter, G. (2002) Clonal spread among Swedish children of a *Staphylococcus aureus* strain resistant to fusi- dic acid. *Scand. J. Infect. Dis.* **34**, 729–734.

65. O'Neill, A. J., Larsen, A. R., Henriksen, A. S., and Chopra, I. (2004) A fusidic acid-resis- tant epidemic strain of *Staphylococcus aureus* carries the *fus*B determinant, whereas *fus*A mutations are prevalent in other resistant isolates. *Antimicrob. Agents Chemother.* **48**, 3594–3597.

66. Gilot, P., Jossin, Y., and Content, J. (2000) Cloning, sequencing and characterisation of a *Listeria monocytogenes* gene encoding a fibronectin-binding protein. *J. Med. Microbiol.* **49**, 887–896.

67. Gilot, P. and Content, J. (2002) Specific identification of *Listeria welshimeri* and *Listeria monocytogenes* by PCR assays targeting a gene encoding a fibronectin-binding protein. *J. Clin. Microbiol.* **40**, 698–703.

68. Luong, T. T., Ouyang, S., Bush, K., and Lee, C. Y. (2002) Type 1 capsule genes of *Staphy- lococcus aureus* are carried in a staphylococcal cassette chromosome genetic element. *J. Bacteriol.* **184**, 3623–3629.

69. Mongkolrattanothai, K., Boyle, S., Murphy, T. V., and Daum, R. S. (2004) Novel non-*mec*A-containing staphylococcal chromosomal cassette composite island containing *pbp*4 and *tag*F genes in a commensal staphylococcal species: a possible reservoir for anti- biotic resistance islands in *Staphylococcus aureus*. *Antimicrob. Agents Chemother.* **48**, 1823– 1836.

70. Schertzer, J. W. and Brown, E. D. (2003) Purified, recombinant TagF protein from *Bacil- lus subtilis* 168 catalyzes the polymerization of glycerol phosphate onto a membrane accep- tor in vitro. *J. Biol. Chem.* **278**, 18,002–18,007.

71. Fitzgerald, S. N. and Foster, T. J. (2000) Molecular analysis of the *tag*F gene, encoding CDP-Glycerol:Poly(glycerophosphate) glycerophosphotransferase of *Staphylococcus epi- dermidis* ATCC 14990. *J. Bacteriol.* **182**, 1046–1052.

72. Hilden, P., Savolainen, K., Tyynela, J., Vuento, M., and Kuusela, P. (1996) Purification and characterisation of a plasmin-sensitive surface protein of *Staphylococcus aureus*. *Eur. J. Biochem.* **236**, 904–910.

73. Zhang, Y. Q., Ren, S. X., Li, H. L., et al. (2003) Genome-based analysis of virulence genes in a non-biofilm-forming *Staphylococcus epidermidis* strain (ATCC 12228). *Mol. Micro- biol.* **49**, 1577–1593.

74. Lindsay, J. A., Ruzin, A., Ross, H. F., Kurepina, N., and Novick, R. P. (1998) The gene for toxic shock toxin is carried by a family of mobile pathogenicity islands in *Staphylococcus aureus*. *Mol. Microbiol.* **29,** 527–543.

75. Fitzgerald, J. R., Monday, S. R., Foster, T. J., et al. (2001) Characterization of a putative pathogenicity island from bovine *Staphylococcus aureus* encoding multiple superantigens. *J. Bacteriol.* **183,** 63–70.

76. Yarwood, J. M., McCormick, J. K., Paustian, M. L., Orwin, P. M., Kapur, V., and Schlievert, P. M. (2002) Characterization and expression analysis of *Staphylococcus aureus* pathogenicity island 3. Implications for the evolution of staphylococcal pathogenicity islands. *J. Biol. Chem.* **277,** 13,138–13,147.

77. Ruzin, A., Lindsay, J., and Novick, R. P. (2001) Molecular genetics of SaPI1—a mobile pathogenicity island in *Staphylococcus aureus*. *Mol. Microbiol.* **41,** 365–377.

78. Sebulsky, M. T., Shilton, B. H., Speziali, C. D., and Heinrichs, D. E. (2003) The role of FhuD2 in iron(III)-hydroxamate transport in *Staphylococcus aureus*. Demonstration that FhuD2 binds iron(III)-hydroxamates but with minimal conformational change and implication of mutations on transport. *J. Biol. Chem.* **278,** 49,890–49,900.

79. Sebulsky, M. T. and Heinrichs, D. E. (2001) Identification and characterization of *fhu*D1 and *fhu*D2, two genes involved in iron-hydroxamate uptake in *Staphylococcus aureus*. *J. Bacteriol.* **183,** 4994–5000.

80. Moore, P. C. and Lindsay, J. A. (2002) Molecular characterisation of the dominant UK methicillin-resistant *Staphylococcus aureus* strains, EMRSA-15 and EMRSA-16. *J. Med. Microbiol.* **51,** 516–521.

81. Rice, K., Peralta, R., Bast, D., de Azavedo, J., and McGavin, M. J. (2001) Description of staphylococcus serine protease (*ssp*) operon in *Staphylococcus aureus* and nonpolar inactivation of *ssp*A-encoded serine protease. *Infect. Immun.* **69,** 159–169.

82. Reed, S. B., Wesson, C. A., Liou, L. E., et al. (2001) Molecular characterization of a novel *Staphylococcus aureus* serine protease operon. *Infect. Immun.* **69,** 1521–1527.

12

Type III Secretion Systems in *Yersinia pestis* and *Yersinia pseudotuberculosis*

James B. Bliska, Michelle B. Ryndak, and Jens P. Grabenstein

Summary

Three species of bacteria in the genus *Yersinia* are pathogenic for humans. *Yersinia entero-colitica* and *Yersinia pseudotuberculosis* cause enteric diseases. *Yersinia pestis* causes the disease known as plague. Studies utilizing DNA hybridization and multilocus DNA sequencing show that *Y. pestis* and *Y. pseudotuberculosis* are closely related at the genetic level, while *Y. enterocolitica* represents a distinct evolutionary lineage. It has been known for some time that *Y. pestis*, *Y. pseudo-tuberculosis*, and *Y. enterocolitica* encode homologous type III secretion system (TTSS) gene clusters on a common virulence plasmid. More recently, genome scale sequence analysis has revealed that *Y. pestis* and *Y. pseudotuberculosis* also encode homologous TTSS gene clusters on their chromosomes. In this chapter, we describe the genetic organization of TTSSs in *Y. pestis* and *Y. pseudo-tuberculosis*. We also describe several genetic changes in these TTSSs that have occurred during the evolution of *Y. pseudotuberculosis* and *Y. pestis,* and discuss the implications of these changes on our understanding of TTSS function during *Yersinia* pathogenesis.

Key Words: *Yersinia*; plague; type III secretion; Yop protein; pathogenicity island.

1. Overview of the Pathogenic *Yersinia*

1.1. Classification of the Pathogenic *Yersinia*

The genus *Yersinia* contains three species that are frequent pathogens of humans *(1)*. *Yersinia pseudotuberculosis* and *Yersinia enterocolitica* are enteropathogenic bacteria that cause a variety of self-limiting intestinal infections. These bacteria invade and replicate within Peyer's patches and mesenteric lymph nodes. These infections can result in enterocolitis, mesenteric lymphadenitis, and (rarely) septicemia. *Yersinia pestis* is the agent of plague, which is an acute, often fatal disease. Three major types of plague infections, bubonic, septicemic, and pneumonic, are defined by the nature of the organs colonized by the bacteria. Bubonic plague is an infection of regional lymph nodes, septicemic plague is an infection of the blood stream, and pneumonic plague is an infec-tion of the lungs.

Twenty-one different serological groups of *Y. pseudotuberculosis* have been identi-fied *(2)*. *Y. pestis* lacks O-antigen and has traditionally been classified into three biovars (Antiqua, Mediaevalis, and Orientalis) based on several biochemical differences *(3)*. More recently, a fourth biovar designation (Microtus) has been proposed for a group of low-virulence *Y. pestis* strains indigenous to China *(4)*.

From: *Bacterial Genomes and Infectious Diseases*
Edited by: V. L. Chan, P. M. Sherman, and B. Bourke © Humana Press Inc., Totowa, NJ

Y. enterocolitica is classified into five biogroups and approx 60 *O*-antigen serogroups *(5)*. Strains of biogroup 1B (serotypes O4, O8, O13, and O21) form a geographically distinct group known as the "New World" strains, and are considered highly pathogenic.

1.2. Pathogenesis of Y. pseudotuberculosis and Y. enterocolitica Infections

Y. pseudotuberculosis infections are typically initiated following consumption of contaminated foodstuffs. The ingested bacteria transit through the intestinal tract until they reach the small intestine. Specialized phagocytic cells (microfold or M cells) that reside on the surface of the follicle-associated epithelia of the small intestine are specifically targeted by these pathogens *(6,7)*. *Y. pseudotuberculosis* carries the *inv* gene, which encodes a specialized surface protein called invasin. Invasin promotes efficient bacterial uptake into M cells *(7,8)*. The internalized bacteria are transported through the M cell, exit from the basolateral side, and gain access to a specialized lymphoid follicle called the Peyer's patch. Bacterial colonization of Peyer's patches results in local inflammation that causes enterocolitis. Dissemination from Peyer's patches to mesenteric lymph nodes can also occur, resulting in mesenteric lymphadenitis and abdominal pain that can mimic appendicitis. In rare cases, *Y. pseudotuberculosis* will enter the blood stream and cause septicemia in humans.

The pathogenic strategy of *Y. enterocolitica* is similar to that of *Y. pseudotuberculosis*. However, there are subtle differences in the pathologies of their infections. *Y. pseudotuberculosis* appears to preferentially colonize mesenteric lymph nodes, causes a characteristic mesenteric lymphadenitis, and is more likely to cause septicemia. On the other hand, *Y. enterocolitica* infections are typically restricted to the Peyer's patches and surrounding tissues (lamina propria). In addition, *Y. enterocolitica* causes a more severe enterocolitis than *Y. pseudotuberculosis*.

1.3. Pathogenesis of Y. pestis Infections

The pathogenesis of *Y. pestis* infections is distinct from that of the enteropathogenic *Yersinia* in several important aspects. Plague is usually transmitted from a rodent host to humans by the bite of an infected flea. When an infected flea takes a blood meal on a human, the bacteria are introduced into the capillary system of the dermis *(3)*. The bacteria may be internalized by blood-derived monocytes, in which they can proliferate *(9)*. Subsequently, the bacteria colonize regional lymph nodes resulting in bubonic plague, or disseminate to the blood stream resulting in primary septicemic plague. If the infection is not eliminated in either case, the bacteria will ultimately spread via the circulatory system to the spleen, liver, and lungs. Following extensive bacterial replication in these organs, the host dies of septic shock and multiple organ failure *(3)*. A second mode of *Y. pestis* transmission that can occur is by direct inhalation of aerosolized bacteria into lungs, resulting in primary pneumonic plague *(3)*.

2. Evolution of the Pathogenic Yersinia Species

Early studies utilizing DNA–DNA hybridization demonstrated that *Y. pseudotuberculosis* and *Y. pestis* are closely related at the genetic level, while *Y. enterocolitica* represents a separate evolutionary lineage *(1)*. More recently, multilocus sequence typing of housekeeping genes has indicated that *Y. pestis* is a clone of a *Y. pseudotuberculosis* O1b strain. *Y. pestis* appears to have evolved as recently as 1500–20,000 yr ago *(2,10)*.

Although *Y. enterocolitica* and *Y. pseudotuberculosis/Y. pestis* strains can be grouped into two lineages, these bacteria share a number of homologous virulence genes presumably inherited from a common ancestor. Determinants such as these represent "core" virulence factors of the pathogenic *Yersinia* species. Of special importance to this discussion is an approx 70 kb plasmid, referred to generically as the virulence plasmid, which encodes a TTSS (Fig. 1, and *see* **Subheading 3.3.**). Additional factors that may fall into this group include genes that promote survival and replication in macrophages *(11)*, and a set of genes involved in acquisition of iron, e.g., the high pathogenicity island (HPI). *(12)*. Unique sets of pathogenicity genes were acquired by the different *Y. enterocolitica* and *Y. pseudotuberculosis/Y. pestis* lineages during evolution. For example, all enteropathogenic *Y. enterocolitica* carry the chromasomally encoded *yst* gene, which encodes a heat stable enterotoxin (Yst). The *yst* gene is absent from *Y. pseudotuberculosis/Y. pestis*; such a factor could explain why *Y. enterocolitica* infections are associated with more severe gastroenteritis *(13)*. On the other hand, all *Y. pseudotuberculosis* and *Y. pestis* strains carry a pathogenicity island (PI) referred to as the pigmentation (*pgm*) segment, which is absent from *Y. enterocolitica (14)*. The hemin storage (*hms*) genes within the *pgm* segment have been linked genetically to the ability of *Y. pseudotuberculosis* and *Y. pestis* strains to form biofilms on the mouth of the nematode worm *(13)*. Biofilm formation prevents worm feeding and may have evolved in *Y. pseudotuberculosis* as a mechanism to avoid predation by nematodes.

Two plasmids, pMT1 (~101 kb) and pPCP1 (9.6 kb), are uniquely found in *Y. pestis (3)*. pMT1 and pPCP1 are associated with the increased virulence (pMT1 and pPCP1) and vector-borne transmissibility (pMT1) of *Y. pestis (3)*. The acquisition of these two plasmids was likely a key event in the evolution of *Y. pestis* from a *Y. pseudotuberculosis* ancestor. What additional genetic changes occurred in *Y. pestis* that can explain its dramatic increase in virulence, compared with *Y. pseudotuberculosis*? Some answers to this question are being gleaned from analysis of the chromosome sequences of three *Y. pestis* strains (CO92, biovar Orientalis, KIM, biovar Mediaevalis, and 91001, biovar Microtus) and a *Y. pseudotuberculosis* strain (IP32953, serotype 01) that have recently been published *(15–18)*. Soon to be published is the genome sequence of a *Y. enterocolitica* strain (8081, serotype 08, NC_003222). The overall structure of the *Y. pseudotuberculosis/Y. pestis* chromosome is similar to that of other *Enterobacteriaceae*, with a shared genetic backbone and high colinear arrangements of housekeeping genes and operons. However, large segments of the chromosome have apparently undergone re-arrangement in some *Y. pestis* strains (e.g., CO92), either by inversion or translocation mechanisms, indicating that there is the potential for genomic instability in *Y. pestis* *(18)*. The *Y. pseudotuberculosis/Y. pestis* genome also contains numerous islands of *Yersinia*-specific genes. For example, analysis of the *Y. pestis* chromosome for regions of distinct GC content suggest that approx 21 PI are present *(13)*. These chromosomal PIs harbor a number of intriguing genes, including two insecticidal toxin gene clusters, a TTSS gene cluster, a flagellar gene cluster, and a type II secretion gene cluster. Interestingly, comparison of the *Y. pestis* and *Y. pseudotuberculosis* genomes shows that the vast majority of these islands (~18) are present in *Y. pseudotuberculosis (13,16,19)*. The major genetic differences between *Y. pestis* and *Y. pseudotuberculosis* correspond to the presence of phage-related sequences and a higher number of insertion sequence elements in the *Y. pestis* chromosome. The expansion of these insertion sequence elements, combined with other types of mutations (e.g., insertions, deletions), has resulted in the inactivation of a large number of nonessential pathogenicity and metabolic genes in *Y. pestis*. Examples of pathogenicity genes inherited from *Y. pseudotuberculosis* that have been inactivated in *Y. pestis* include *inv*, O antigen biosynthetic genes, and one of the flagellar gene clusters.

A view that is emerging is that most of the genes required for pathogenicity in *Y. pestis* were inherited from *Y. pseudotuberculosis (13)*. For example, *Y. pestis* forms biofilms in the throats of fleas as part of its strategy for efficient vector transmission, and the *hms* genes are required for this activity. Thus, *Y. pestis* did not have to acquire a new set of genes for this purpose, but rather genes that promote biofilm formation on nematodes were inherited from *Y. pseudotuberculosis* and adapted for efficient flea transmission. In addition, it has been suggested that the dramatic increase in virulence of *Y. pestis* is owing to the inactivation of genes encoding metabolical functions *(13)*. Thus, increased virulence could result from energy conservation as a consequence of inactivating dispensable metabolical pathways. However, it is important to point out that *Y. pseudotuberculosis* is nearly as virulent as *Y. pestis* when the bacteria are injected directly into the blood stream of mice *(20)*. Therefore, if metabolic streamlining does play a role in the increased virulence of *Y. pestis*, this phenotype would need to be expressed during the early stages of infection from peripheral routes.

3. Type III Secretion Gene Clusters

3.1. Overview of the TTSS

TTSSs are essential determinants of pathogenicity in a large number of Gram-negative bacteria *(21)*. TTSSs function during host cell contact to introduce virulence proteins into host cells. Secretion systems involved in the export of flagella will also be discussed here because they are structurally and functionally related to TTSSs. For example, the flagellar export apparatus can secrete virulence proteins in addition to flagella *(22)*. In fact, under certain conditions, the flagellar export pathway can secrete TTSS substrates *(23)*. TTSSs and flagella export pathways are distinguished from other secretion systems by the following characteristics: (1) the secretion apparatus is a highly organized structure of greater than 20 proteins and this structure spans the inner membrane, periplasm, and outer membrane of the bacteria; (2) secretion of substrates is independent of the general secretory *(sec)* pathway; (3) there is no periplasmic intermediate; and (4) the secreted proteins are not processed in any way, such as by removal of a signal sequence. Topologically, these multi-protein systems span the bacterial envelope and are able to secrete components across this barrier in one step without processing. Both TTSSs and flagella export pathways require ATP hydrolysis for this activity. Genes encoding the basal bodies are highly similar *(24,25)*. However, the protruding appendages of the two systems are quite different. The flagellum is comprised of a rigid hook and flexible filament, which is used for bacterial motility. The TTSS appendage is comprised of a rigid, hollow needle, which is used for the transport of substrates to the surface of the host cell and beyond. Only TTSSs appear to be able to transfer virulence proteins into the host cell cytosol *(25)*. It is commonly believed that the TTSS is evolutionarily related to the flagellar export pathway, but how these two systems are related remains controversial. It is generally accepted that the flagellar system did not evolve from the TTSS. Which of the two alternative scenarios (evolution of the TTSS from the flagellar system, or coevolution from a common ancestor) is correct, remains to be determined *(24,26–28)*.

3.2. Flagellar Gene Clusters

Y. pseudotuberculosis and *Y. enterocolitica* strains are typically motile. Motility is important in *Y. enterocolitica* for efficient invasion into host cells in vitro *(29)*, and therefore motility may be important during intestinal infection by both pathogens. Interestingly, *Y. pseudotuberculosis* and *Y. pestis* contain two homologous flagellar gene clusters. One cluster (cluster II in Fig. 1) is similar to the systems found in other *Enterobacteriaceae*. This gene cluster is presumably responsible for motility in *Y. pseudotuberculosis*, since it is inactivated by multiple mutations in *Y. pestis (17,18)*, and *Y. pestis* is known to be nonmotile. Comparative DNA hybridization by microarray of several *Y. pseudotuberculosis* strains indicates that this gene cluster has undergone marked divergence within the species, suggesting it is under selective pressure from the host immune system to vary *(30)*. The second flagellar gene cluster (cluster I in Fig. 1) represents a mystery. This gene cluster is less similar to well-characterized flagellar systems *(17,18)*. It is presently not clear if it represents a complete, functional gene cluster in either species. In addition, comparative DNA hybridization by microarray of this gene cluster in several *Y. pseudotuberculosis* strains shows no divergence from *Y. pestis*. A lack of divergence may mean that this cluster is under no selective pressure to vary, and therefore may not be expressed in either species during infection *(30)*.

3.3. Virulence Plasmid-Encoded TTSS

The virulence plasmid-encoded TTSS has been extensively studied and has been the subject of several recent reviews *(31–33)*. The plasmid carrying this TTSS is designated pYV in *Y. pseudotuberculosis* and pCD1 in *Y. pestis*. The virulence plasmid carried by *Y. enterocolitica* strains is also referred to as pYV. Complete sequences have been determined for plasmids from three *Y. pestis* strains (NC_003131, NC_004836, NC_005813), one *Y. pseudotuberculosis* strain (NC_006153) and three *Y. enterocolitica* strains (NC_002120, NC_004564, NC_005017). The *Y. pseudotuberculosis* and *Y. pestis* plasmids are nearly identical at the DNA sequence level *(16)*. In addition, the TTSS genes are highly conserved between all the plasmids, with approx 98% sequence identity in the genes comprising the major operons of the TTSS (between *yopD* and *yscM* in Fig. 1). The contiguous arrangement of the major TTSS operons is also conserved between *Y. pseudotuberculosis, Y. pestis*, and *Y. enterocolitica* plasmids. However, the overall structure of the *Y. enterocolitica* plasmid differs from the plasmids in *Y. pseudotuberculosis* and *Y. pestis (34)*. These differences appear to be because of rearrangements caused by inversions in the *Y. enterocolitica* plasmid *(34)*.

The following scenario is consistent with the current understanding of how the TTSS functions during *Yersinia*–host cell interactions. Growth of the bacteria at 37°C results in production of the TTSS components and assembly of the apparatus in the cell envelope. Most of the secretion substrates (*Yersinia* outer proteins [Yops] and low calcium response protein V [LcrV]) are also produced and held within the bacterium until activation of secretion by contact with host cells. Type III secretion can be elicited in vitro by culturing *Yersinia* at 37°C in medium that is low in free calcium (the low calcium

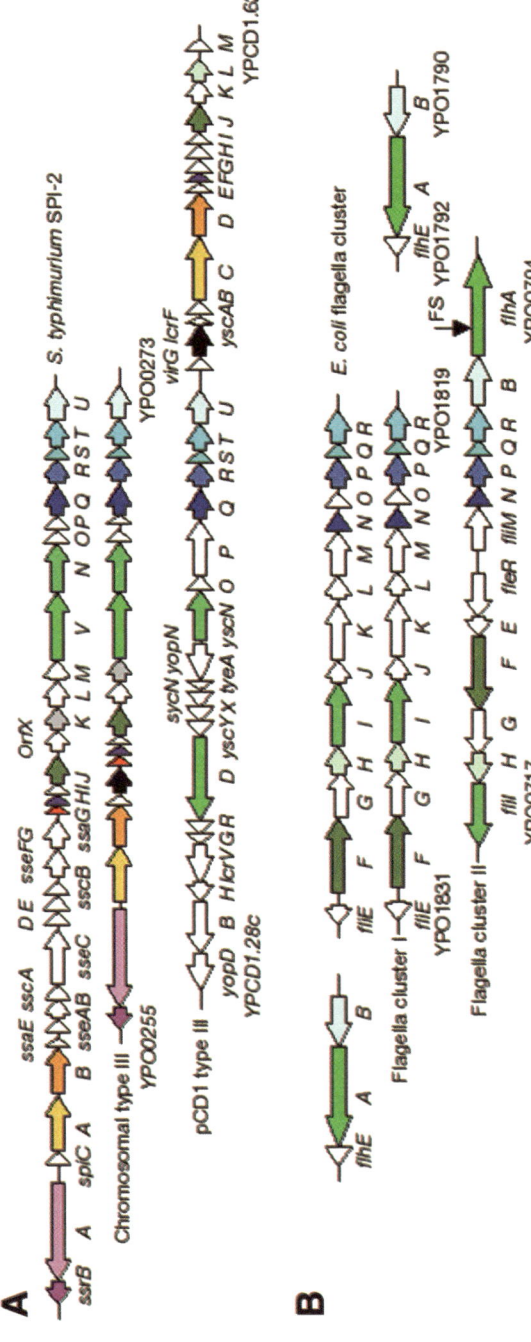

Fig. 1. Type III secretion system and flagellar gene clusters in *Yersinia pseudotuberculosis* and *Yersinia pestis*. Arrangements of the gene clusters are shown in comparison with the *Salmonella enterica* serotype typhimurium TTSS encoded in SPI-2 (**A**), and the *Escherichia coli* flagella gene cluster (**B**). Genes are arbitrarily color coded to show related genes. In (**A**), the TTSS gene clusters encoded on the chromosome (YPO0255-YPO0273), and on the pCD1 virulence plasmid (YPCD1.28c-YPCD1.62), are illustrated. In (**B**), the flagellar gene clusters that are of unknown function (cluster 1, YPO1831-YPO1790), or responsible for motility in *Y. pseudotuberculosis* (cluster II, YPO0717-YPO0704), are illustrated. The position of a frameshift mutation in flagellar gene cluster II is indicated by the arrow. (Adapted with permission from ref. *18*.)

response). However, it remains unclear how intimate contact with the host cell triggers activation of type III secretion. At the same time that the bacterium is receiving signals to activate type III secretion, the host cell initiates a response to the pathogen. For example, a macrophage encodes integrin receptors that recognize proteins on the bacterial surface (e.g., invasin in the case of enteropathogenic *Yersinia* species), and this interaction can stimulate actin-dependent bacterial phagocytosis *(35)*. Macrophages also encode Toll-like receptors, which recognize components of the bacterial envelope (e.g., lipopolysaccharide), and trigger production of cytokines or apoptosis *(36,37)*. Probably the first substrates to be secreted through the tip of the TTSS needle are YopB, YopD and LcrV, as these proteins are thought to form a channel which mediates delivery of six additional substrates (YopO [YpkA], YopH, YopM, YopT, YopJ [YopP], and YopE) across the host cell membrane *(31,32)*. After the Yops are delivered into the host cytoplasm, they act to disrupt key signal transduction pathways. For example, several Yops antagonize signaling pathways important for actin polymerization and cytokine production *(38)*. LcrV is also secreted into the extracellular milieu, where it functions as a diffusible factor to downmodulate the inflammatory response *(39)*. The combined actions of the Yops and LcrV cause a localized suppression of the innate immune response, allowing the bacteria to replicate in host tissues.

3.4. Loss of yopT *in* Y. pseudotuberculosis O3 *Strains*

Although the virulence plasmids are in general highly similar among *Y. pseudotuberculosis* and *Y. pestis* strains, a notable difference is found in the *yopT-yopD* region of virulence plasmids carried by *Y. pseudotuberculosis* O3 strains. The virulence plasmid carried by several *Y. pseudotuberculosis* O3 strains contains a deletion of approx 1790 base pairs that removes most of the *yopT* coding region, as well as the entire coding regions for *sycT* (the YopT chaperone), and an uncharacterized ORF YPCD1.22c (Fig. 2). The mechanism of this deletion is not clear. Only a very short repetitive sequence (CAG) is found at the site of the deletion (Fig. 2). This is the only known example of a genetic change on the virulence plasmid that ablates the expression of a TTSS substrate. The genes encoding the secreted effectors are generally extremely well conserved, with greater than 92% identity in protein coding sequence between *Y. pseudotuberculosis*, *Y. pestis*, and *Y. enterocolitica (34)*, arguing that there is strong selective pressure to maintain these genes. Thus, it is unusual that *Y. pseudotuberculosis* O3 strains lack functional *yopT* and *sycT* genes. It should be noted that only a small number of O3 strains have been examined thus far. Additional strains need to be tested before it can be concluded that this mutation is specific for O3.

A possible explanation for this genetic alteration in *Y. pseudotuberculosis* O3 strains comes from molecular studies of YopT and another Yop protein, YopE. These studies indicate that YopT is dispensable for pathogenesis because it has functional redundancy with YopE. To understand the basis for this functional redundancy, it is necessary to first describe the molecular functions of YopT and YopE. YopT and YopE act on a subfamily of small guanosine triphosphate (GTP)-binding proteins that have GTP hydrolyzing activity and are referred to as GTPases *(40)*. Small GTPases act as molecular switches to turn key signaling pathways on and off. When bound to GTP, the GTPase is switched on allowing it to interact with downstream effectors; hydrolysis of GTP to GDP switches the GTPase off (Fig. 3). The three small GTPases that are targeted

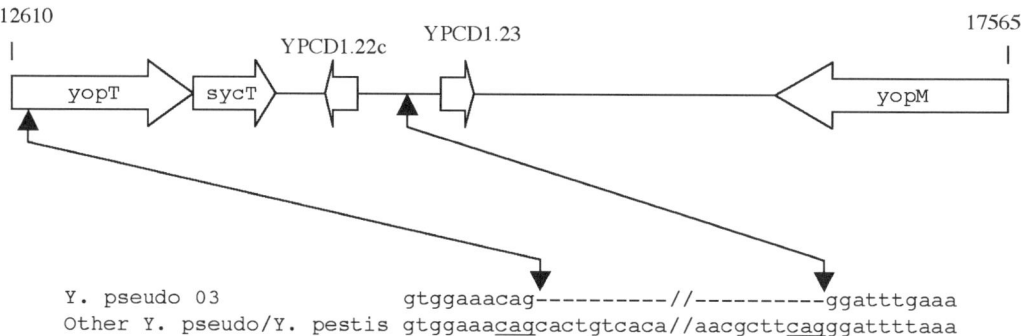

Fig. 2. Representation of the *yopT-yopM* region in the pCD1 virulence plasmid. The region shown at the top corresponds to nucleotides 12610-17565 of pCD1 (NC_003131) and contains three genes (*yopT, sycT*, and *yopM*) and two open reading frames (YPCD1.22c and YPCD1.23) represented by thick arrows. Double headed thin arrows point to end points of the deletion in the virulence plasmid of *Y. pseudotuberculosis* O3 strains that removes most of *yopT* and all of *sycT* and YPCD1.22c. Nucleotide sequences at the deletion end point are shown at the bottom. The top line corresponds to the sequence in the *Yersinia pseudotuberculosis* O3 virulence plasmid (U18804). The bottom line corresponds to the sequence in *Y. pseudotuberculosis* O1 virulence plasmid (NC_006153) and the *Yersinia pestis* pCD1 virulence plasmid (NC_003131). A tri-nucleotide repeat at the site of the deletion is underlined.

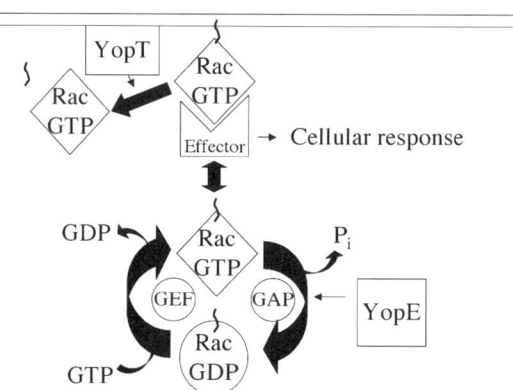

Fig. 3. Model of Rho GTPase inactivation by YopT and YopE. The model depicts Rac1, but is representative for all three Rho GTPases (Rac1, Cdc42, and RhoA). Rho GTPases cycle between an active GTP-bound form (diamond) and an inactive GDP-bound form (oval). Guanine nucleotide exchange factors facilitate the release of GDP and the binding of GTP by Rho GTPases. GTPase activating proteins accelerate hydrolysis of GTP. Rho GTPases modified by prenylation (squiggle) localize to the membrane, where the GTP-bound forms interact with effector proteins to stimulate cellular responses. YopE functions as a GAP for Rho GTPases. YopT cleaves near the C-terminus of Rho GTPases, removing the prenyl group, and preventing interaction with the membrane. GTP, guanosine triphosphate; GDP, guanosine diphosphate; GEF, guanine nucleotide exchange factor; GAP, GTPase activating protein.

by YopT and YopE, (RhoA, Cdc42, and Rac1) are referred to as the Rho GTPase subfamily. Rho GTPases regulate signaling pathways that are important for controlling actin polymerization and cytokine production *(41)*.

YopE inactivates Cdc42, RhoA, and Rac1 by accelerating GTP hydrolysis (Fig. 3). YopE thus functions as a GTPase activating protein (GAP) *(42)*. YopT, on the other hand, is a cysteine protease that functions to remove a key post-translational modification (prenylation) from the C-terminus of Rho GTPases *(43)*. Prenylation of Rho GTPases promotes their association with the inner side of the plasma membrane, which is necessary to bring them into proximity with their effectors. YopT-induced cleavage of the prenyl group prevents the GTPase from anchoring to the membrane, and from interacting productively with effectors (Fig. 3). Thus, YopT and YopE inactivate the same targets, but by completely distinct mechanisms.

Although YopT and YopE can inactivate all three Rho GTPases in vitro, studies carried out using cultured cell infection models suggest that YopT acts selectively on RhoA *(44)*, whereas YopE acts selectively on Rac1 *(45)*. However, in experimental mouse infection assays, *yopE* mutant *Y. enterocolitica* strains are significantly attenuated for virulence, whereas *yopT* mutant strains are not attenuated for virulence *(46,47)*. This finding coincides with the observation that *Y. pseudotuberculosis* O3 strains (naturally *yopE*$^+$ *yopT*$^-$) appear to be roughly as virulent as other *Y. pseudotuberculosis* serotypes that are *yopE*$^+$ *yopT*$^+$. These results argue that YopE can compensate for the loss of YopT in vivo, but that YopT cannot fully compensate for the loss of YopE. One possibility is that YopT is more selective in its action; for example, YopT may preferentially inactivate RhoA, whereas YopE has broader substrate specificity, which encompasses all three Rho GTPases.

Although YopT may be dispensable for pathogenesis owing to overlapping function with YopE, what selective advantage results from the inactivation of *yopT* in *Y. pseudotuberculosis* O3 strains? In a recent study, a *Y. enterocolitica yopT* mutant was found to be slightly more pathogenic than the isogenic wild-type strain in a mouse infection assay *(47)*. Specifically, the mutant grew to higher numbers in certain tissues (e.g., liver) than its parent. Therefore, *yopT* appears to function as an "avirulence" gene to downmodulate the pathogenicity of *Yersinia*. Homologs of YopT that are found in plant pathogens also function as avirulence proteins *(48)*. It is conceivable that inactivation of *yopT* provided a selective advantage to *Y. pseudotuberculosis* O3 strains by increasing their virulence in a specific host.

3.5. Chromosome-Encoded TTSS

The TTSS encoded on the chromosome of *Y. pseudotuberculosis* and *Y. pestis* (referred to here as TTSS-2) shares a high degree of similarity with respect to gene order and content to the TTSS encoded in SPI-2 of *Salmonella enterica* (Fig. 1). Biotype 1B strains of *Y. enterocolitica* encode a different chromosomal TTSS (*ysa* TTSS) that is related to the *mxi-spa* TTSS of *Shigella* species and to the SPI-1 encoded TTSS of *S. enterica (49)*. TTSS-2 is encoded within a PI of approx 60 kb along with genes involved in hemin uptake (*hmu*), tellurite resistance, and a chaperone-usher system. An interesting difference between the SPI-2 TTSS and TTSS-2 is that a block of genes that encode substrates of the *S. enterica* system are absent from TTSS-2, and in their place is a single gene encoding a transcriptional activator of the arabinose utilization protein C (AraC) family (YPO0260 in Fig. 1). The transcription factor encoded by this gene may regulate the expression of TTSS-2 substrates encoded at different locations in the genome. Genes encoding homologs of substrates secreted by the SPI-2 TTSS have been identified at several different locations in the genome of *Y. pestis (50)*. Whether these genes encode substrates of TTSS-2 remains to be determined.

The SPI-2 TTSS plays an important role in the ability of *S. enterica* to replicate in macrophages *(51)*. The TTSS encoded in SPI-2 is selectively expressed during the intracellular phase of *S. enterica* growth. *Y. pseudotuberculosis* and *Y. pestis* can replicate in macrophages, but, surprisingly, the TTSS-2 does not appear to be required for this activity *(52)*. Thus, the function of TTSS-2 in *Yersinia* pathogenesis remains a mystery.

3.6. Inactivation of TTSS-2 Genes in Y. pestis

Deng et al. *(17)* compared TTSS-2 sequences from *Y. pestis* KIM and CO92 and noted that a TTSS-2 gene in KIM (y0521) contains a frame shift mutation. The gene inactivated in KIM corresponds to YPO0263 in CO92 (Fig. 1). This gene encodes a lipoprotein homologous to SsaJ in *S. enterica*.

We have compared TTSS-2 sequences between *Y. pseudotuberculosis* IP32953 and several *Y. pestis* strains and identified a second mutation present in KIM and CO92 strains. In KIM and CO92 the gene encoding the putative AraC family transcription factor (YPO0260 in Fig. 1) contains a 20-bp deletion located downstream of the predicted ATG initiation codon (Fig. 4). A GTG codon located 3' to the 20-bp deletion was assigned as the initiation codon of this gene in KIM and CO92 (Fig. 4). Interestingly, this mutation is not present in the low-virulence *Y. pestis* 91001 strain (biovar Microtus), suggesting that the mutation arose in a common ancestor of KIM and CO92 after it diverged from 91001. Given that there are one or more mutations in TTSS-2 in highly virulent *Y. pestis* strains, we suggest that TTSS-2 is a genetic "remnant" inherited from *Y. pseudotuberculosis* that is not required for plague pathogenesis, and is, therefore, undergoing genetic decay in *Y. pestis*. Therefore, a possible role for TTSS-2 in the pathogenesis of enteric infections caused by *Y. pseudotuberculosis* should be further investigated.

4. Conclusion

Studies of the pathogenic *Yersinia* species have yielded many important insights into the mechanisms of bacterial pathogenesis. For example, pioneering investigations of the LcrV and Yop proteins were instrumental in the discovery of the TTSS. The pathogenic *Yersinia* species also represents a model genus for the study of pathogen evolution. Sequence analysis at the genomic level, made possible by the completion of sequencing projects on all three pathogenic *Yersinia* species, confirms that *Y. pestis* is a clone of *Y. pseudotuberculosis*, whereas *Y. enterocolitica* is on a distinct evolutionary pathway. A striking conclusion of these genomic studies is that loss of genetic information in *Y. pestis* likely contributes to its increased virulence relative to *Y. pseudotuberculosis*. A number of nonessential metabolic and pathogenicity genes have been inactivated in *Y. pestis* by genetic decay. Pathogenicity genes inactivated in *Y. pestis* may be important for intestinal pathogenesis by *Y. pseudotuberculosis*. Included in this group are a flagellar export pathway and a second TTSS encoded on the chromosome. Future studies of these specialized secretions systems and their roles during infection will likely yield additional insights into bacterial pathogenesis.

Acknowledgments

The authors thank Céline Pujol, Gloria Viboud, and Yue Zhang for editorial corrections to the manuscript. Research on type III secretion in the laboratory was supported by Public Health Service grant AI43389.

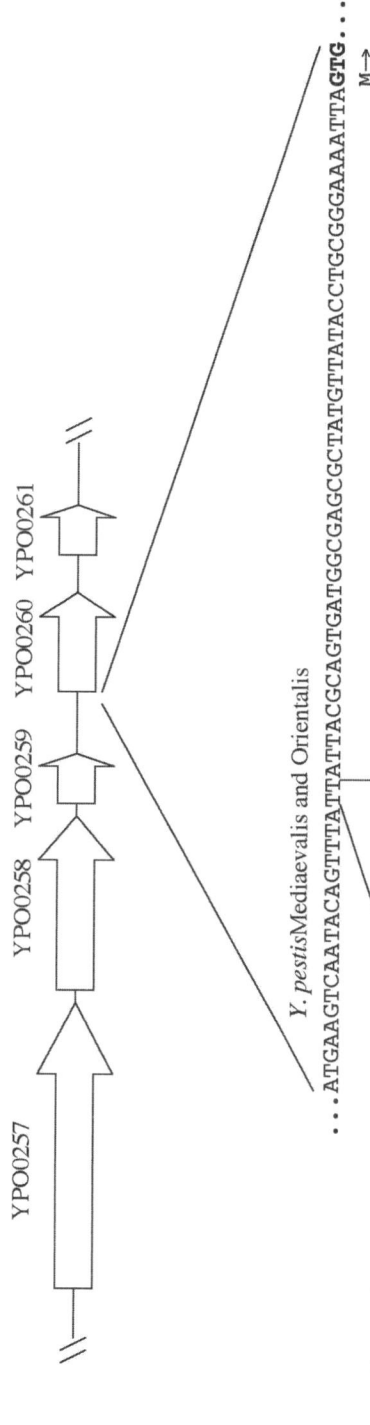

Y. pestis Mediaevalis and Orientalis
...ATGAAGTCAATACAGTTTATTATTACGCAGTGATGGCGAGCGCTATGTTATACCTGCGGGAAAAT**TAGTG**...
                                            M→

Y. pseudotuberculosis
Y. pestis Microtus
...**ATGA**AGTCAATACAGTTTATT<u>TATTTCGGCAAGAGGGCG</u>TATTACGCAGTGATGGCGAGCGCTATGTTATACCTGCGGGAAAATTAGTG...
M K S I Q F I L F R Q E G V L R S D G E R Y V I P A G K L V→

Fig. 4. Representation of the YPO0257-YPO0261 region of TTSS-2. Open reading frames (ORFs) are shown at the top as thick arrows and are labeled with the nomenclature used for *Yersinia pestis* CO92 (*18*). ORF YPO0260 is predicted to encode an AraC family transcription factor. Nucleotide sequences corresponding to the 5' end of ORF YPO0260 are shown at the bottom. The top sequence shown is from *Y. pestis* KIM (biovar Mediaevalis; NC_004088) and CO92 (biovar Orientalis; NC_003143). The bottom sequence shown is from *Y. pestis* 91001 (biovar Microtus; NC_005810) and *Yersinia pseudotuberculosis* IP32953 (serogroup O1; NC_006155). Thin lines between the sequences mark the site of the 20-bp deletion in the KIM and CO92 sequences. The region deleted is underlined in the 91001 and IP32953 sequences. Predicted initiation codons are in bold. The predicted translation product of YPO0260, in single letter code, is shown below the DNA sequence.

References

1. Brubaker, R. R. (1991) Factors promoting acute and chronic diseases caused by yersiniae. *Clin. Microbiol. Rev.* **4,** 309–324.
2. Skurnik, M., Peippo, A., and Ervela, E. (2000) Characterization of the O-antigen gene clusters of *Yersinia pseudotuberculosis* and the cryptic O-antigen gene cluster of *Yersinia pestis* shows that the plague bacillus is most closely related to and has evolved from *Y. pseudotuberculosis* serotype O:1b. *Mol. Microbiol.* **37,** 316–330.
3. Perry, R. D. and Fetherston, J. D. (1997) *Yersinia pestis*-etiologic agent of plague. *Clin. Microbiol. Rev.* **10,** 35–66.
4. Zhou, D., Tong, Z., Song, Y., et al. (2004) Genetics of metabolic variations between *Yersinia pestis* biovars and the proposal of a new biovar, Microtus. *J. Bacteriol.* **186,** 5147–5152.
5. Bottone, E. J. (1997) *Yersinia enterocolitica*: the charisma continues. *Clin. Microbiol. Rev.* **10,** 257–276.
6. Clark, M. A., Hirst, B. H., and Jepson, M. A. (1998) M-cell surface beta1 integrin expression and invasin-mediated targeting of *Yersinia pseudotuberculosis* to mouse Peyer's patch M cells. *Infect. Immun.* **66,** 1237–1243.
7. Marra, A. and Isberg, R. R. (1997) Invasin-dependent and invasin-independent pathways for translocation of *Yersinia pseudotuberculosis* across the Peyer's patch intestinal epithelium. *Infect. Immun.* **65,** 3412–3421.
8. Pepe, J. and Miller, V. L. (1993) *Yersinia enterocolitica* invasin: a primary role in the initiation of infection. *Proc. Natl. Acad. Sci. USA* **90,** 6473–6477.
9. Cavanaugh, D. C. and Randall, R. (1959) The role of multiplication of *Pasteurella pestis* in mononuclear phagocytes in the pathogenesis of fleaborne plague. *J. Immunol.* **85,** 348–363.
10. Achtman, M., Zurth, K., Morelli, G., Torrea, G., Guiyoule, A., and Carniel, E. (1999) *Yersinia pestis,* the cause of plague, is a recently emerged clone of *Yersinia pseudotuberculosis. Proc. Natl. Acad. Sci. USA* **96,** 14,043–14,048.
11. Pujol, C. and Bliska, J. B. (2005) Turning *Yersinia* pathogenesis outside in: subversion of macrophage function by intracellular yersiniae. *Clin. Immunol.* **114,** 216–226.
12. Carniel, E. (1999) The *Yersinia* high-pathogenicity island. *Int. Microbiol.* **2,** 161–167.
13. Wren, B. W. (2003) The yersiniae—a model genus to study the rapid evolution of bacterial pathogens. *Nat. Rev. Microbiol.* **1,** 55–64.
14. Carniel, E. (2002) Plasmids and pathogenicity islands of *Yersinia. Curr. Top. Microbiol. Immunol.* **264,** 89–108.
15. Song, Y., Tong, Z., Wang, J., et al. (2004) Complete genome sequence of *Yersinia pestis* strain 91001, an isolate avirulent to humans. *DNA Res.* **11,** 179–197.
16. Chain, P. S., Carniel, E., Larimer, F. W., et al. (2004) Insights into the evolution of *Yersinia pestis* through whole-genome comparison with *Yersinia pseudotuberculosis. Proc. Natl. Acad. Sci. USA* **101,** 13,826–13,831.
17. Deng, W., Burland, V., Plunkett, G. 3rd, et al. (2002) Genome sequence of *Yersinia pestis* KIM. *J. Bacteriol.* **184,** 4601–4611.
18. Parkhill, J., Wren, B. W., Thomson, N. R., et al. (2001) Genome sequence of *Yersinia pestis*, the causative agent of plague. *Nature* **413,** 523–527.
19. Zhou, D., Han, Y., Song, Y., et al. (2004) DNA microarray analysis of genome dynamics in *Yersinia pestis*: insights into bacterial genome microevolution and niche adaptation. *J. Bacteriol.* **186,** 5138–5146.
20. Une, T. and Brubaker, R. R. (1984) In vivo comparison of avirulent Vwa$^-$ and Pgm$^-$ or Pst^r phenotypes of yersiniae. *Infect. Immun.* **43,** 895–900.

21. Hueck, C. J. (1998) Type III protein secretion systems in bacterial pathogens of animals and plants. *Microbiol. Mol. Biol. Rev.* **62,** 379–433.

22. Young, G. M., Schmiel, D. H., and Miller, V. L. (1999) A new pathway for the secretion of virulence factors by bacteria: the flagellar export apparatus functions as a protein-secretion system. *Proc. Natl. Acad. Sci. USA* **96,** 6456–6461.

23. Lee, S. H. and Galan, J. E. (2004) *Salmonella* type III secretion-associated chaperones confer secretion-pathway specificity. *Mol. Microbiol.* **51,** 483–495.

24. Macnab, R. M. (1999) The bacterial flagellum: reversible rotary propellor and type III export apparatus. *J. Bacteriol.* **181,** 7149–7153.

25. Blocker, A., Komoriya, K., and Aizawa, S. (2003) Type III secretion systems and bacterial flagella: insights into their function from structural similarities. *Proc. Natl. Acad. Sci. USA* **100,** 3027–3030.

26. Gophna, U., Ron, E. Z., and Graur, D. (2003) Bacterial type III secretion systems are ancient and evolved by multiple horizontal-transfer events. *Gene* **312,** 151–163.

27. Aizawa, S. I. (2001) Bacterial flagella and type III secretion systems. *FEMS Microbiol. Lett.* **202,** 157–164.

28. Nguyen, L., Paulsen, I. T., Tchieu, J., Hueck, C. J., and Saier, M. H. Jr. (2000) Phylogenetic analyses of the constituents of Type III protein secretion systems. *J. Mol. Microbiol. Biotechnol.* **2,** 125–144.

29. Young, G. M., Badger, J. L., and Miller, V. L. (2000) Motility is required to initiate host cell invasion by *Yersinia enterocolitica. Infect. Immun.* **68,** 4323–4326.

30. Hinchliffe, S. J., Isherwood, K. E., Stabler, R. A., et al. (2003) Application of DNA microarrays to study the evolutionary genomics of *Yersinia pestis* and *Yersinia pseudotuberculosis. Genome Res.* **13,** 2018–2029.

31. Ramamurthi, K. S. and Schneewind, O. (2002) Type III protein secretion in *Yersinia* species. *Annu. Rev. Cell. Dev. Biol.* **18,** 107–133.

32. Cornelis, G. R. (2002) *Yersinia* type III secretion: send in the effectors. *J. Cell. Biol.* **158,** 401–408.

33. Plano, G. V., Day, J. B., and Ferracci, F. (2001) Type III export: new uses for an old pathway. *Mol. Microbiol.* **40,** 284–293.

34. Snellings, N. J., Popek, M., and Lindler, L. E. (2001) Complete DNA sequence of *Yersinia enterocolitica* serotype O:8 low-calcium-response plasmid reveals a new virulence plasmid-associated replicon. *Infect. Immun.* **69,** 4627–4638.

35. Isberg, R. R. and Van Nhieu, G. T. (1995) The mechanism of phagocytic uptake promoted by invasin-integrin interaction. *Trends Cell Biol.* **5,** 120–124.

36. Zhang, Y. and Bliska, J. B. (2003) Role of Toll-like receptor signaling in the apoptotic response of macrophages to *Yersinia* infection. *Infect. Immun.* **71,** 1513–1519.

37. Haase, R., Kirschning, C. J., Sing, A., et al. (2003) A dominant role of Toll-like receptor 4 in the signaling of apoptosis in bacteria-faced macrophages. *J. Immunol.* **171,** 4294–4303.

38. Viboud, G. I., So, S. S., Ryndak, M. B., and Bliska, J. B. (2003) Proinflammatory signalling stimulated by the type III translocation factor YopB is counteracted by multiple effectors in epithelial cells infected with *Yersinia pseudotuberculosis. Mol. Microbiol.* **47,** 1305–1315.

39. Brubaker, R. R. (2003) Interleukin-10 and inhibition of innate immunity to Yersiniae: roles of Yops and LcrV (V antigen). *Infect. Immun.* **71,** 3673–3681.

40. Boquet, P. (2000) Small GTP binding proteins and bacterial virulence. *Microbes Infect.* **2,** 837–844.

41. Bishop, A. L. and Hall, A. (2000) Rho GTPases and their effector proteins. *Biochem. J.* **348(Pt 2),** 241–255.

42. Black, D. S. and Bliska, J. B. (2000) The RhoGAP activity of the *Yersinia pseudotuberculosis* cytotoxin YopE is required for antiphagocytic function and virulence. *Mol. Microbiol.* **37,** 515–527.

43. Juris, S. J., Shao, F., and Dixon, J. E. (2002) *Yersinia* effectors target mammalian signalling pathways. *Cell. Microbiol.* **4,** 201–211.

44. Aepfelbacher, M., Trasak, C., Wilharm, G., et al. (2003) Characterization of YopT effects on Rho GTPases in *Yersinia enterocolitica*-infected cells. *J. Biol. Chem.* **278,** 33,217–33,223.

45. Andor, A., Trulzsch, K., Essler, M., et al. (2001) YopE of *Yersinia*, a GAP for Rho GTPases, selectively modulates Rac-dependent actin structures in endothelial cells. *Cell Microbiol.* **3,** 301–310.

46. Iriarte, M. and Cornelis, G. (1998) YopT, a new *Yersinia* Yop effector protein, affects the cytoskeleton of host cells. *Mol. Microbiol.* **29,** 915–929.

47. Trulzsch, K., et al. (2004) Contribution of the major secreted Yops of *Yersinia enterocolitica* O:8 to pathogenicity in the mouse infection model. *Infect. Immun.* **72,** 5227–5234.

48. Shao, F., Merritt, P. M., Bao, Z., Innes, R. W., and Dixon, J. E. (2002) A *Yersinia* effector and a *Pseudomonas* avirulence protein define a family of cysteine proteases functioning in bacterial pathogenesis. *Cell* **109,** 575–588.

49. Foultier, B., Troisfontaines, P., Muller, S., Opperdoes, F. R., and Cornelis, G. R. (2002) Characterization of the *ysa* pathogenicity locus in the chromosome of *Yersinia enterocolitica* and phylogeny analysis of type III secretion systems. *J. Mol. Evol.* **55,** 37–51.

50. Miao, E. A., Scherer, C. A., Tsolis, R. M., et al. (1999) *Salmonella typhimurium* leucine-rich repeat proteins are targeted to the SPI1 and SPI2 type III secretion systems. *Mol. Microbiol.* **34,** 850–864.

51. Waterman, S. R. and Holden, D. W. (2003) Functions and effectors of the *Salmonella* pathogenicity island 2 type III secretion system. *Cell Microbiol.* **5,** 501–511.

52. Pujol, C. and Bliska, J. B. (2003) The ability to replicate in macrophages is conserved between *Yersinia pestis* and *Yersinia pseudotuberculosis*. *Infect. Immun.* **71,** 5892–5899.

13

Genomics and the Evolution of Pathogenic *Vibrio cholerae*

William S. Jermyn, Yvonne A. O'Shea,
Anne Marie Quirke, and E. Fidelma Boyd

Summary

In this chapter, the complete genome sequence of the human pathogen *Vibrio cholerae* is examined. We discuss, in particular, the level of gene acquisition in the form of pathogenicity and genomic islands within the species, and the role of these elements in the various lifestyles of the organism. This chapter will highlight the significant role horizontal gene transfer plays in the evolution, ecology, and virulence of *V. cholerae*-specific traits.

Key Words: *Vibrio cholerae*; evolution; horizontal gene transfer.

1. Introduction

Vibrio cholerae is the causative agent of the severe diarrheal disease cholera (Table 1). Cholera is a devastating disease that is prevalent in Southern Asia, where the disease is endemic in many areas. Cholera is unusual in that it causes severe explosive outbreaks, particularly in areas where social upheaval has occurred owing to flooding or war. Pathogenic *V. cholerae* isolates belonging to the O1 and O139 serogroups are the predominant cause of epidemics and pandemics of cholera *(1,2)*. *V. cholerae* isolates belonging to other serogroups, collectively referred to as non-O1/non-O139, are implicated only in moderate to severe forms of gastroenteritis *(2–5)*. Since 1817, seven pandemics of cholera have been record, all of which are associated with two biotypes of the O1 serogroup named classical and El Tor biotypes. The first six pandemics were caused by the classical biotype, whereas the seventh and current pandemic, which began in 1961, is caused by the El Tor biotype. In 1992, a newly recognized serogroup, O139, was identified as the cause of a cholera outbreak in South Asia *(6)*. The O139 serogroup initially replaced the El Tor biotype as the dominant cause of cholera in India and Bangladesh *(7)*. Recently, El Tor strains have re-established themselves as the main etiological agent of cholera *(7)*. The emergence of the O139 serogroup in 1992 and the re-emergence of cholera in South America in 1991 led to renewed interest in *V. cholerae* evolution and pathogenesis.

In 1996, it was discovered that the *ctxAB* genes, which encode cholera toxin (CT), the main cause of the profuse, watery diarrhea characteristic of cholera, were part of the genome of a filamentous phage CTXϕ *(8)*. Subsequently, it was shown that the receptor for CTXϕ on the *V. cholerae* cell surface, the toxin coregulated pilus (TCP), was

From: *Bacterial Genomes and Infectious Diseases*
Edited by: V. L. Chan, P. M. Sherman, and B. Bourke © Humana Press Inc., Totowa, NJ

Table 1
Human Disease Syndromes Caused by *Vibrio* Species

Species	Disease	Source	Symptoms
Vibrio cholerae	Cholera	Poor sanitation resulting in contaminated water supplies	Vary from a mild, watery diarrhea to an acute diarrhea, with characteristic rice water stools. Abdominal cramps, nausea, vomiting, dehydration, and shock; after severe fluid and electrolyte loss, death may occur.
Vibrio parahaemolyticus	Gastroenteritis	Consumption of raw, improperly cooked, or cooked, recontaminated fish and shellfish	Diarrhea, abdominal cramps, nausea, vomiting, headache, fever, and chills
Vibrio vulnificus	Gastroenteritis	Consumption of contaminated seafood especially oysters, clams, and crabs	Vomiting, diarrhea, and abdominal pain
	Primary septicemia	Consumption of raw contaminated seafood by individuals with underlying chronic disease, particularly liver disease	In these individuals, the microorganism enters the bloodstream, causing fever and chills, decreased blood pressure (septic shock), resulting in septic shock, rapidly followed by death in many cases (about 50%)
	Wound infection	Contamination of an open wound with sea water harboring the organism, or by lacerating part of the body on coral, fish, and others, followed by contamination with the organism	Skin breakdown and ulceration
Vibrio mimicus	Gastroenteritis	Consumption of raw, improperly cooked, or cooked, recontaminated fish and shellfish	Diarrhea, abdominal cramps, nausea, and vomiting

Table 2
Size and Percent G+C Content of the *Vibrio*-Sequenced Genomes

Species strain	chr 1 Size (Mb)	chr 2 Size (Mb)	Overall Size (Mb)	Overall Percent G + C content
Vibrio cholerae N16961	2.96	1.07	4.03	47%
Vibrio parahaemolyticus RIM2210633	3.29	1.88	5.17	45%
Vibrio vulnificus YJ016	3.35	1.86	5.26	45%
Vibrio vulnificus CMCP6	3.28	1.84	5.12	46%
Vibrio fischeri ES114	2.91	1.33	4.24	38%

chr, chromosome; nk, not known.

Table 3
Location and Size of Super Integrons Among *Vibrio* Species

Vibrio species	Integron size	Location	Number of ORFs	Reference
Vibrio cholerae N16961 chromosome 2	125 kb	VCA0282–VCA0498	216	*(11)*
Vibrio parahaemolyticus RIMD 2210633 chromosome 1	48 kb	VP1787–VP1865	79	*(91)*
Vibrio vulnificus YJ016 chromosome 1	138 kb	VV1740–VV1941	202	*(99)*
Vibrio vulnificus CMCP6 chromosome 1	151 kb	VV12401–VV12550	150	*(100)*

ORF, open reading frame.

encoded on a pathogenicity island (PAI) named *Vibrio* pathogenicity island (VPI) *(9)*. In 1998, pulse-field gel electrophoresis analysis demonstrated that the genome of *V. cholerae* strain N16961 consisted of two circular chromosomes of unequal size *(10)*. The 4.1-Mb genome sequence of *V. cholerae* strain N16961 was published in 2000, and confirmed the existence of two circular chromosomes: a large 3.0-Mb chromosome 1 (encoding 2770 open reading frames [ORFs]) and a smaller 1.1-Mb chromosome 2 (encoding 1115 ORFs) (Table 2) *(11)*. Analysis of gene content on both chromosomes indicated the vast majority of genes for essential cell functioning, such as DNA replication, transcription, translation, cell wall biosynthesis, and pathogenicity, for example adhesins, surface antigens, and toxins, are located on chromosome 1, whereas chromosome 2 contained a higher proportion of hypothetical genes (58%) than chromosome 1 (42%) *(11)*. The presence of a larger number of hypothetical genes on chromosome 2, however, is highly localized to a large super integron (SI) of 125 kb encoding 216 ORFs (Figs. 1 and 2; Table 3). In addition, chromosome 2 does contain several essential metabolic and regulatory pathways *(11)*.

In this chapter, we first describe the evolutionary genetic relationships among *V. cholerae* natural isolates based on multilocus enzyme electrophoresis (MLEE), comparative

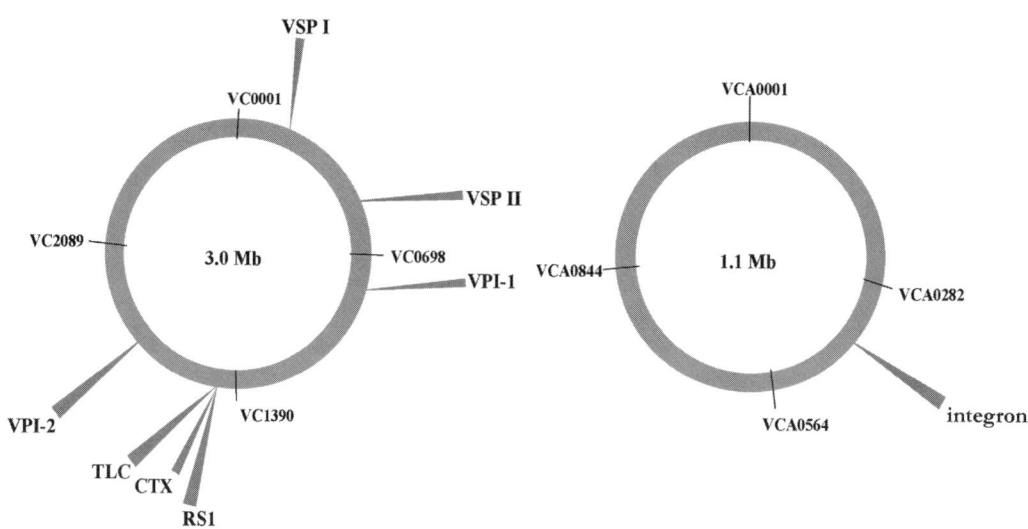

Fig. 1. Circular map of *Vibrio cholerae* chromosomes 1 and 2 showing the location of various mobile genetic elements.

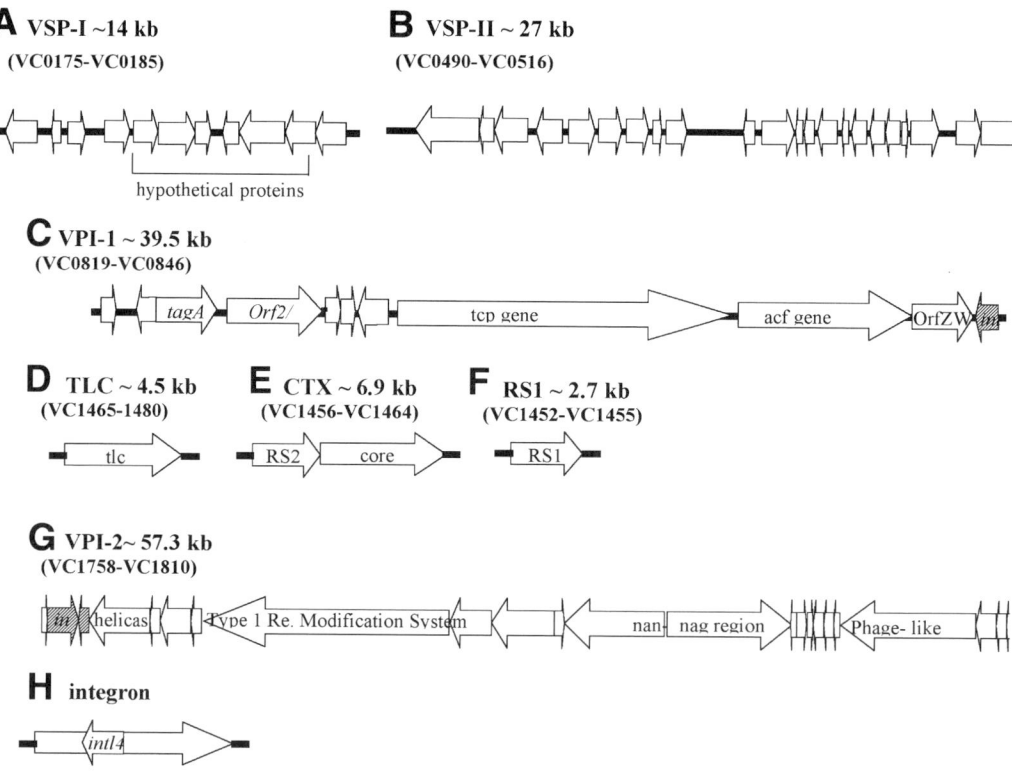

Fig. 2. Schematic representation of eight regions that are associated with pathogenesis in *Vibrio cholerae*. The position and direction of transcription of the open reading frames are indicated by the direction of the arrows. The numbers within the arrows refer to the genetic organization of the genes, e.g., 175 is VC0175.

single-locus and multilocus sequence analyses (SLSA and MLSA), as well as whole genome microarray analysis. We report on the role mobile genetic elements (MGEs) play in *V. cholerae* diversity and evolution, and then we describe the global regulation of virulence genes encoded on MGEs and the core genome. Finally, we report on the four additional *Vibrio* whole-genome sequences available and what comparative genomic analysis between *Vibrio* species has revealed.

2. Evolutionary Genetic Relationships Among *V. cholera* Natural Isolates

Until relatively recently, it was assumed that the *V. cholerae* O1 antigen was an essential component of epidemic isolates since non-O1 serogroup strains were only associated with sporadic cases of cholera. This assumption changed when a *V. cholerae* O139 serogroup emerged in Bengal in 1992 to cause epidemic outbreaks of cholera. Subsequently, it has been shown that *V. cholerae* O139 serogroup isolates evolved from a *V. cholerae* O1 El Tor strain by horizontal gene transfer (HGT) of the O antigen genes *(12,13)*. This finding limited the value of serotyping for predicting the epidemiological potential of a strain, and established the importance of understanding the genetic relationships among isolates. A precise account of the determinants that differentiate nonpathogenic *V. cholerae* strains from those causing disease in humans is essential. Numerous studies have illustrated that at least three essential elements are required in the emergence of pathogenic *V. cholerae* isolates: (1) the presence of VPI (named VPI-1 in this chapter), which encodes the type IV pilus TCP, an essential colonization factor, and the receptor for CTXϕ, (2) the presence of CTXϕ, which encodes CT, and (3) the presence of ToxR, an essential regulatory protein *(14,15)*. Environmental strains of *V. cholerae* may acquire these virulence genes to become pathogenic to humans, but most of these strains are unlikely to attain pandemic potential by acquisition of VPI and CTXϕ alone *(7,15)*. In order to better understand the defining properties of epidemic isolates, their underlying genetic relationships must be elucidated.

Several genotypic studies have endeavoured to examine the evolutionary genetic relationship between *V. cholerae* pathogenic and nonpathogenic isolates using methods that survey the entire genome, such as MLEE *(16–20)* and MLSA of housekeeping genes *(21–23)*, or methods that examine a locus representative of the genome, such as SLSA of housekeeping genes *(24–28)*. In this section we will report on the findings of these studies and their impact on our understanding of *V. cholerae* evolution.

2.1. Multilocus Enzyme Electrophoresis

MLEE analyses electrophoretic mobility differences in multiple enzymes (20–35 enzymes) distributed around the genome among large collections of isolates (50–500 isolates typically). Electrophoretic mobility differences are converted into genetic distances, which are used to construct phylogenetic trees indicating overall evolutionary relationships among isolates. Several groups have carried out MLEE analysis of epidemic *V. cholerae* classical, El Tor sixth and seventh isolates, and nonepidemic environmental isolates *(16–20, 29)*. All of these MLEE studies found that the epidemic *V. cholerae* isolates formed a separate lineage from nonepidemic isolates. Therefore, epidemic *V. cholerae* O1 serogroup isolates were all found to exhibit similar electrophoretic profiles, differing at only 1–3 of the 12–16 loci analyzed, implying a closer relationship among these pathogenic isolates than the nonpathogenic environmental isolates, which display diverse electrophoretic patterns *(16–20,29)*. However, differences among the studies were found regarding the relationship between sixth (classical) and seventh (El Tor) pandemic isolates. Some studies showed that limited genetic diversity exists between these biotypes, hence pathogenic isolates from various sources were shown to have identical electrophoretic profiles, or differed very slightly in their profiles *(16,17, 29)*. Other studies indicated that the sixth pandemic and seventh pandemic isolates were independent clones, whereas others have postulated that the two clones are very closely related *(18,19)*. More recently, Farfan and colleagues applied MLEE methods to a large collection of isolates of diverse origin, including serogroup O1/O139 and non-O1/non-O139 strains of *V. cholerae* and the population showed a high degree of genetic variation *(20)*.

2.2. Single-Locus Sequence Analysis

The evolutionary genetic relationships among *V. cholerae* have also been assessed using comparative SLSA. In 1995, *asd*, a chromosomal housekeeping gene, which encodes aspartate-semialdehyde dehydrogenase, was sequenced from 45 *V. cholerae* isolates *(24)*. Comparative sequence analysis of *asd* found no nucleotide polymorphisms within either the sixth or seventh pandemic isolates; however, nucleotide variation was found between the two clones and between clinical US Gulf strains isolated in the 1980s, which formed a third independent clone. It was reported that all three clones evolved independently from environmental, nontoxigenic non-O1 El Tor organisms *(24)*. Subsequently, it was shown that the variation at the *asd* locus was owing to recombination, and is, therefore, not a suitable indicator of evolutionary relationships *(21)*.

A SLSA study carried out by Stine and colleagues *(25)* examined a 705-bp fragment of the *recA* gene, which encodes RecA protein involved in homologous recombination, from 113 *V. cholerae* isolates, and found the sixth (classical) pandemic and seventh pandemic strains tended to cluster separately from one another. Although *V. cholerae* O139 strains generally fell within the biotype El Tor cluster as expected, there were several strains, which were not within this clade.

Comparative sequence analysis of a 648-bp region of the malate dehydrogenase (*mdh*) locus among 64 *V. cholerae* isolates showed that clinical O1 (classical and El Tor biotypes) and O139 epidemic isolates formed a highly uniform clone (an epidemic clonal complex), whereas non-O1/non-O139 environmental isolates formed several more diverse lineages *(28)*. In addition, multilocus virulence gene profiling among these isolates found that the 12 virulence regions examined were associated predominantly with the epidemic clonal complex of sixth and seventh pandemic isolates *(28)*. These data taken together indicated that the emergence of *V. cholerae* sixth and seventh pandemic strains resulted from the successive acquisition of virulence regions by an O1 progenitor. Interestingly, O'Shea and colleagues identified a number of O37 and O8 serogroup isolates that clustered within the epidemic clonal complex and which had similar multilocus virulence gene profiles, which suggests that these isolates may have arose through the acquisition of a novel O antigen by an O1 strain similar to O139 isolates *(28)*. This is consistent with a study by Bik and colleagues *(30)*, in which they determined by IS1004 fingerprinting that a *V. cholerae* O37 serogroup strain from Sudan was closely related to O1 strains, and may have acquired the O37 biosynthesis genes via lateral gene transfer, similar to the emergence of the O139 serogroup strains from India.

A recent study using polymerase chain reaction single-stranded conformational polymorphism analysis at the housekeeping genes *groEL*-I on chromosome 1 and *groEL*-II on chromosome 2 gave profile patterns, which differentiated O1 classical and El Tor isolates from non-O1/non-O139 isolates *(31)*. This study also demonstrates that both chromosomes have a similar evolutionary history, and that *groEL*-I, *groEL*-II, and *mdh* are good indicators of overall genetic diversity *(31)*.

2.3. Multilocus Sequence Analysis

Because MLEE analysis enables one to examine regions throughout the genome, and given some of the discordance between MLEE and SLSA studies, MLSA has been posited as a better method to determine overall genetic relationships. Byun and co-workers *(21)* sequenced five chromosomal genes; *asd*, *mdh*, *hlyA* (hemolysin), *recA*, and *dnaE*, from 32 *V. cholerae* isolates representing sixth pandemic, seventh pandemic, and US Gulf coast isolates. They showed that for each of the five genes examined, no variation was found within the sixth pandemic, seventh pandemic, or US Gulf isolates. In addition, they found the *mdh* and *hlyA* sequences from these three pathogenic clones to be identical, except for the previously reported 11-bp deletion in *hly*A in sixth pandemic isolates. Both sixth pandemic and seventh pandemic isolates had identical *dnaE* sequences *(21)*. Differences between clones were identified at the *asd* and *recA* loci, however, further analysis indicated that the variation at both these loci was the result of recombination rather than random point mutations *(21)*. This study suggests that the rate of recombination in *V. cholerae* is high and highlights the importance of choosing the right loci (recombination-free) for evolutionary studies.

2.4. Comparative Genomic Microarray Analysis

Dziejman et al. *(15)* constructed a *V. cholerae* genomic microarray that contained over 93% of the predicted genes of the seventh pandemic El Tor strain N16961. They used this *V. cholerae* DNA chip to examine the gene content of O1 (El Tor, classical, US Gulf Coast) and O139 serogroup isolates. Their comparative genomic analysis indicated that the O1 and O139 strains examined showed a high degree of conservation with the N16961 genome, which suggested that only a small group of *V. cholerae* strains might be capable of evolving to become epidemic through the acquisition of CTXφ and VPI-1 *(15)*. As well as supporting previous studies that showed that the sixth and seventh pandemic strains are highly clonal, microarray analysis identified genes specific to seventh pandemic isolates. Most of these seventh pandemic specific genes were located in three clusters: the 2.7-kb repeat sequence-1 (RS1) prophage, and two newly identified genomic islands (GEIs), the 16-kb *Vibrio* seventh pandemic island-1 (VSP-I), and the 7.5-kb VSP-II. The RS1 prophage encodes an anti-repressor, and is required for the production of infectious CTXφ, which explains why classical sixth pandemic isolates cannot produce infectious CTXφ particles *(32)*. The role(s) that the VSP islands play in *V. cholerae* survival remains unknown, but potentially they could be responsible for more efficient infection of the human host or improved environmental survival *(15)*. This study, along with previous studies examining the evolutionary genetic relationships among *V. cholerae*, indicates that the differences between pathogenic isolates have arisen predominantly by the acquisition of regions via HGT and recombination, rather than mutation.

Table 4
Genome Location of Mobile Genetic Elements Identified in *Vibrio cholerae*

MGE	Location	Size	ORFs	Function/virulence factor	Reference
VSP-I	VC0175–VC0185	14 kb	11	Unknown	*15*
VSP-II	VC0490–VC0516	27 kb	24	Unknown	*15*
VPI-1	VC0819–VC0846	39.5 kb	28	TCP, ACF, ToxT, TcpPH	*9*
RS1ϕ	VC1452–VC1455	2.7 kb	4	CT anti-repressor	*39*
CTXϕ	VC1456–VC1464	6.9 kb	9	CT, Zot, Ace	*8*
TLC	VC1465–VC1480	4.5 kb	16	Unknown	*59*
VPI-2	VC1758–VC1809	57.3 kb	52	Neuraminidase	*55*
Integron	VCA0282–VCA0498	125.3 kb	216	Drug resistance	*11*

MGE, mobile genetic element; ORFs, open reading frames; TCP, toxin coregulatory pilus; ACF, accessory colonization factor; Zot, zonula occludens toxin; Ace, accessory cholera enterotoxin.

3. Impact of Mobile Genetic Elements on *V. cholerae* Genome Diversity

Gene acquisition via HGT results in genomic changes that can rapidly and radically alter the lifestyle of a bacterium in "quantum leaps" *(33)*. *V. cholerae* is an aquatic organism that exists in a number of different states: (1) in association with crustacean, (2) as a free-living organism, (3) in a viable but nonculturable state, and (4) as a pathogen *(34)*. Pathogenic *V. cholerae* probably evolved from an aquatic form that acquired the ability to colonize the human intestine. Many essential virulence factors of bacteria are encoded on MGEs, such as bacteriophages, pathogenicity and GEIs, integrons, and integrative and conjugative elements. In this next section, we will describe a number of these MGEs and their role in *V. cholerae* virulence (Figs. 1 and 2; Table 4).

3.1. Bacteriophages in *V. cholerae*

The *ctxAB* genes, which encode CT, are located on a filamentous phage CTXϕ. The 6.9-kb CTXϕ genome contains two functionally distinct regions, the core and the RS2 (Figs. 1 and 2; Table 4) *(8)*. The core region encodes CT and the genes involved in bacteriophage morphogenesis, including genes that are thought to encode the major and minor bacteriophage coat proteins (Psh, Cep, OrfU, and Ace) and a protein required for CTXϕ assembly (Zot) *(8)*. The repeat sequence-2 (RS2) region encodes genes required for replication (*rstA*), integration (*rstB*), and regulation (*rstR*) of CTXϕ *(35)*. In pathogenic *V. cholerae* El Tor and O139 isolates, a 2.7-kb RS1 prophage is integrated upstream of the CTX prophage, which is closely related to the RS2 region of CTXϕ *(35–39)*. The RS1 element contains an additional ORF encoding RstC *(36,40)*. RstC is an anti-repressor that controls CTXϕ lysogeny, production of CTXϕ particles, and expression of CT *(32)*.

It was recently reported that the RS1 element (Figs. 1 and 2; Table 4) is in fact the genome of a satellite bacteriophage that utilizes CTXϕ morphogenesis genes to produce RS1ϕ particles *(39)*. An additional filamentous phage named KSF-1ϕ has also recently been identified in *V. cholerae* isolates, and this phage, which is similar to CTXϕ, plays a role in production of RS1ϕ particles *(32,39)*. Additionally, a *V. cholerae* filamentous bacteriophage called VGJϕ has been isolated that transmits CTXϕ or RS1ϕ by a TCP-independent mechanism *(41)*. It was recently found that VGJϕ infects host cells through the mannose sensitive hemagglutinin pilus and that it uses the same integration site as CTXϕ, that is, the *attB* chromosomal site *(41)*. These recent studies indicate that an array of filamentous bacteriophages have the ability to alternatively package different DNA elements in *V. cholerae (39)*.

In addition to filamentous phages, generalized transducing phage has also been identified in *V. cholerae*. One such phage, CP-T1, has the ability to transduce both CTX and VPI to nonpathogenic strains *(42,43)*. Similar to a range of pathogenic bacteria, prophages in *V. cholerae* play a very important role in the emergence of pathogenic isolates, and also in the regulation of virulence *(44)*.

3.2. PAIs and GEIs in V. cholerae

PAIs are large chromosomal regions typically 10–200 kb that encode one or more virulence genes. Their guanine and cytosine (G+C) content varies from the core genome, and are very often found adjacent to transfer RNA (tRNA) genes and are associated with direct repeats, integrases, or transposases *(45)*. Typically, PAIs are present in the genomes of pathogenic bacteria but are absent from the genomes of nonpathogenic isolates of the same species. GEIs are similar to PAIs except they encode genes such as iron uptake systems, which may increase the fitness of the bacterium GEIs in pathogenic and environmental micro-organisms *(45)*. A number of PAIs and GEIs have been identified in *V. cholerae* and in the following sections some of these chromosomal regions are described.

3.2.1. Vibrio *Pathogenicity Island-1*

The 39.5-kb VPI-1 encodes approx 28 ORFs, including TCP, an essential colonization factor and receptor for CTXφ (Figs. 1 and 2; Table 4) *(8,9,46,47)*. VPI-1 has a sporadic distribution among *V. cholerae* natural isolates and has also been identified in *Vibrio mimicus (9,28,48)*. The species *V. mimicus* is closely related to *V. cholerae*; however, *V. mimicus* is phenotypically and genotypically distinct from *V. cholerae* and can be readily differentiated from *V. cholerae (21,28,31,48,49)*. The natural habitat of *V. mimicus*, similar to *V. cholerae*, is the aquatic ecosystem, where it has been found both as a free-living bacterium and in association with phytoplankton and crustaceans *(50,51)*. To date, the mode and mechanism of VPI-1 transfer among isolates remains unknown, however several pathways have been proposed. It was suggested in 1999 that the VPI-1 was a filamentous bacteriophage *(52)*, however, this was greeted with skepticism resulting from a lack of gene sequence similarity between VPI-1 and canonical filamentous bacteriophages, and the inability to independently verify the existence of VPIφ *(7)*. It was shown that VPI-1 could be transferred to a recipient strain by the generalized transducing bacteriophage CP-T1 *(43)*. This led to the suggestion that VPI-1 corresponds to a satellite element, which can be efficiently packaged by a helper bacteriophage but are not viable bacteriophages themselves *(7,43,53)*. VPI-1 can precisely excise from the chromosome to form extra-chromosomal circular excision products (pVPI), which is mediated by the VPI-encoded recombinases, *int* and *vpiT*, which may suggest a possible transfer route *(54)*.

3.2.2. Vibrio *Pathogenicity Island-2*

VPI-2 is a57.3-kb chromosomal region that has all the characteristics of a PAI (Figs. 1 and 2; Table 4) *(55)*. All pathogenic *V. cholerae* isolates contain VPI-2, whereas nonpathogenic isolates lacked the region. VPI-2 encodes several gene clusters; the *nan-nag* region involved in the utilization of sialic acid, which may be an alternative nutrient source; a type-1 restriction modification system, which may protect the bacteria from viral infection; neuraminidase (*nanH*), which acts on higher order gangliosides in the intestines converting them to GM_1 gangliosides with the release of sialic acids, and a gene cluster that shows homology to Mu phage. Among *V. cholerae* O139 serogroup isolates, the *nan-nag* and *hsd* type-1 restriction modification regions are deleted, indicating the instability of the region *(55)*. In addition, the loss of this region among *V. cholerae* O139 isolates may have resulted in their decline as a serious cause of epidemic cholera. Recently, a 14.1-kb region of VPI-2 comprised of ORFs VC1773 to VC1787 was identified in *V. mimicus* isolates *(56)*. Interestingly, the VPI-2 region was found in all 17 *V. mimicus* strains examined, suggesting that this region is essential for survival. The VPI-2 region in *V. mimicus* was inserted adjacent to a serine tRNA similar to VPI-2 in *V. cholerae*. In 11 of the 17 *V. mimicus* isolates examined, an additional 5.3-kb region encoding VC1758 and VC1804 toVC1809 was present, which suggests that additional regions of VPI-2 may have been deleted. The evolutionary history of VPI-2 was reconstructed by comparative analysis of the *nanH* (VC1784) gene tree with the species gene tree, deduced from the housekeeping gene *mdh* among *V. cholerae* and *V. mimicus* isolates. Both gene trees showed an overall congruence; on both gene trees *V. cholerae* O1 and O139 serogroup isolates clustered together, whereas non-O1/non-O139 serogroup isolates formed separate divergent branches with similar clustering of strains within the branches. One exception was noted. On the *mdh* gene tree *V. mimicus* sequences formed a distinct divergent lineage from *V. cholerae* sequences as expected for a distinct species; however, on the *nanH* gene tree, *V. mimicus* clustered with *V. cholerae* non-O1/ non-O139 isolates suggesting horizontal transfer of this region between these species *(56)*. In addition, analysis of the genomes of *Vibrio vulnificus* strains YJ016 and CMCP6 demonstrated the presence of homologs of the *nan-nag* region on chromosome 2 of these strains (Fig. 4) *(57)*. Twenty-nine of the 52 ORFs of VPI-2 showed homology to ORFs in the genome of *V. vulnificus* strain YJ016. These ORFs were clustered into three distinct regions on the *V. vulnificus* YJ016 genome at ORFs VV2151 to VV2162, VV2250 to VV2262 on chromosome 1, and VVA1196 to VVA1206 on chromosome 2. Only homologs of *V. cholerae* ORFs VC1773 to VC1783 encoding a sialic acid metabolism gene cluster (*nan-nag*) were identified in both *V. vulnificus* YJ016 (VVA1196 to VVA1206) and CMCP6 (VV20726 to VV20736) genomes (Fig. 4). Additional homologs of VPI-2 were identified in only *V. vulnificus* strain YJ016. Comparative sequence analysis suggests that the *nan-nag* region maybe ancestral to *V. vulnificus*.

3.2.3. Vibrio *Seventh Pandemic Island-I and* Vibrio *Seventh Pandemic Island-II*

As stated previously in **Subheading 2.4.**, comparative genomic analysis using a *V. cholerae* genome microarray identified two genomic regions designated VSP-I and VSP-II that were present only in seventh pandemic strains (Figs. 1 and 2; Table 4) *(15)*. These regions showed several characteristics of PAIs. VSP-I spans a 16-kb region encoding 11 ORFs designated VC0175–VC0185, which are mostly hypothetical proteins, with a G+C content of 40%, in contrast to 47% for the entire genome *(15)*. The VSP-II region encompassed eight ORFs VC0490–VC0502, but the boundaries were not well defined. Subsequently, it has been shown that VSP-II is a 26.9-kb region that shows homology to a 43.4-kb GEI from *V. vulnificus* (Fig. 3) *(58)*. The VSP-II region from ORFs VC0490 to VC0516 encoded transcriptional regulators, a putative ribonuclease, a putative type

IV pilin, and a number of methyl accepting chemotaxis proteins, as well as a large number of hypothetical proteins. VC0516 encodes an integrase and inserts adjacent to a tRNA gene (Fig. 3) *(58)*. Interestingly, *V. cholerae* ORFs VC0493 to VC0498, VC0504 to VC0510, and VC0516 were homologous to *V. vulnificus* strain YJ016 ORFs VV0510 to VV0516, VV0518 toVV0525, and VV0560, respectively (Fig. 3). Some ORFs showed amino acid identities greater than 90% between the two species in these regions. In *V. vulnificus* strain YJ016, a 43.4-kb low GC-content (43%) GEI encompassing ORFs VV0509 to VV0560, designated *V. vulnificus* island-I (VVI-I), was identified *(58)*. The VVI-Is 52 ORFs included a phosphotransferase system gene cluster, genes required for sugar metabolism, as well as two transposase genes and two insertion sequence elements. There was synteny and homology between the 5' region of *V. cholerae* VSP-II and the 5' region of *V. vulnificus* VVI-I; however, VVI-I contained an additional 31.5 kb of DNA between VV0526 and VV0560 in strain YJ016 *(58)*. Comparative genomic analysis between *V. vulnificus* strain CMCP6 and YJ016 identified only two ORFs between the 5' and the 3' flanking ORFs VV10636 and VV10632, which show 100% identity to the VVI-I flanking ORFs VV0508 and VV0561, respectively. This indicated that the 43.4-kb VVI-I region is absent from strain YJ016, and that this site is empty in this strain (Fig. 3) *(58)*.

It was suggested by Dziejman and colleagues that the genes encoded on the VSP-I and VSP-II islands are likely to be responsible for the unique characteristics of the seventh pandemic strains *(15)*. For example, they might allow the seventh pandemic strains to survive more efficiently than preseventh pandemic strains either in the aquatic ecosystem or in the human host.

3.3. Plasmids in V. cholerae

Upstream of the CTX prophage is an integrated 4.7-kb plasmid named pTLC (toxin-linked cryptic element) (Figs. 1 and 2; Table 4) *(59)*. The plasmid also exists as an extrachromosomal circular double-stranded DNA form of a tandemly duplicated chromosomal element. The size and low copy number of pTLC suggest that it is identical to the three megadalton plasmid identified in many classical *V. cholerae* strains and in the small cryptic plasmid identified in classical strain V58 *(60,61)*.

As the TLC element is tandemly duplicated on the *V. cholerae* chromosome, the extrachromosomal circular pTLC may arise from homologous recombination between directly repeated TLC element copies. Once excised as an extrachromosomal circle, pTLC may replicate under the control of its own replication functions *(59)*. As of yet, the function of pTLC remains undiscovered.

Another plasmid identified in *V. cholerae* is the P factor, which is capable of immobilizing chromosomal genes. Its presence in wild-type strains is extremely rare *(62)* and results in partial attenuation, with a decrease in intestinal colonization ability *(63)*. In some *V. cholerae* non-O1 strains, a plasmid carrying genes that encode a thermostable direct hemolysin-like toxin has been reported *(64)*.

3.4. Super Integrons

Reports of toxigenic *V. cholerae* strains resistant to the antibiotics commonly used for treatment are appearing with increasing frequency, although the genetic mechanisms for the resistance are often not determined *(65–67)*. The 1996–1997 cholera epidemic in Guinea-Bissau, which involved a reported 26,967 cases, was caused by a *V. cholerae*

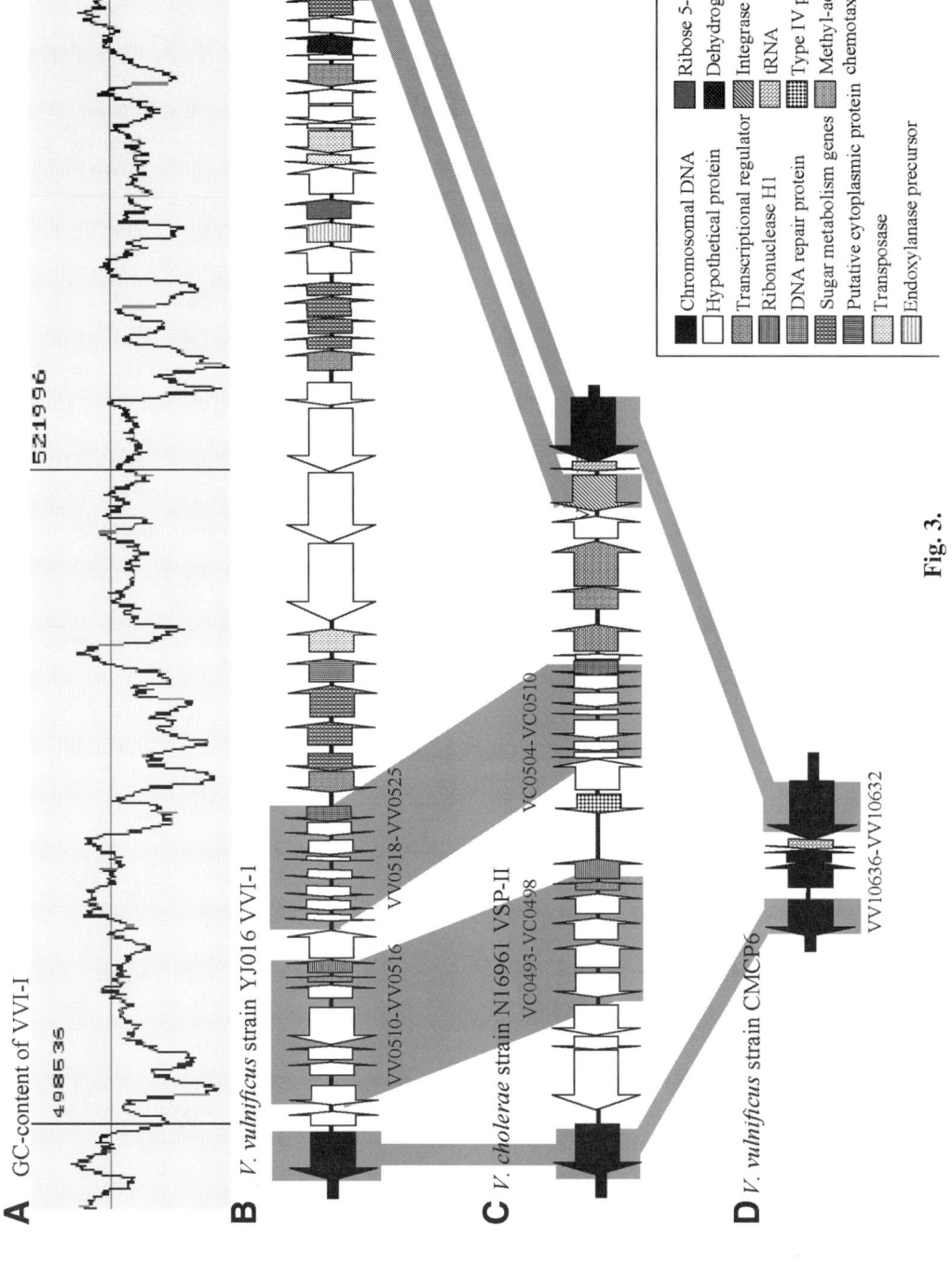

A GC-content of VVI-I

498535   521996

B *V. vulnificus* strain YJO16 VVI-I

VV0510-VV0516   VV0518-VV0525

C *V. cholerae* strain N16961 VSP-II

VC0493-VC0498   VC0504-VC0510

D *V. vulnificus* strain CMCP6

VV10636-VV10632

Chromosomal DNA   Ribose 5-
Hypothetical protein   Dehydrog
Transcriptional regulator   Integrase
Ribonuclease H1   tRNA
DNA repair protein   Type IV ɪ
Sugar metabolism genes   Methyl-ac
Putative cytoplasmic protein   chemotax
Transposase
Endoxylanase precursor

Fig. 3.

238

O1 strain that had a genotype identical to that of the 1994–1995 epidemic strain, but that contained a 150-kb conjugative multiple-antibiotic-resistance plasmid with class 1 integron-borne gene cassettes encoding resistance to trimethoprim and aminoglycosides *(68)*. Studies indicate that super integrons are prevalent among *Vibrio* species and can be highly variable even within the same species *(69)*. In *V. cholerae* strain N16961, the SI encompasses VCA0331–VCA420 on chromosome 2 and encodes 216 genes (Figs. 1 and 2; Table 3) *(11)*.

3.5. Integrative and Conjugative Elements

The 99.5-kb SXT element (SXT) is an integrative and conjugative element that was originally identified in a 1993 *V. cholerae* O139 serogroup isolate called MO10. The MO10-derived SXT encodes resistance to sulfamethoxazole, trimethoprim, chloramphenicol, and streptomycin. The transfer of the SXT element has features reminiscent of both temperate bacteriophages and conjugative plasmids. The element encodes a λ family recombinase (Int) that is required for excision from the chromosome and circularization by recombination between the left and right ends of the integrated element. Generation of this extrachromosomal intermediate is an essential step in the successful transfer of the SXT element; it must precede conjugative transfer to recipient cells. Once transferred, the SXT element integrates specifically into the chromosome in an *int*-dependent, *recA*-independent fashion via recombination between element (*attP*) and chromosomal (*attB*) sequences that are nearly identical *(70)*.

4. Virulence Gene Global Regulation and Expression in *V. cholerae*

The *ctxAB* and *tcp* genes belong to a network of genes called the ToxR regulon, whose expression is modulated by transcriptional regulators encoded on the core chromosome (*toxRS*) and on the VPI-1 (*toxT* and *tcpPH*). In addition to CT, ToxR controls expression of at least 17 genes that make up the ToxR regulon, these include the TCP colonization factor, the *acf* colonization factor, and the outer membrane proteins OmpT and OmpU *(71)*. ToxRS are not sufficient to activate transcription of *ctxAB*, and an additional factor, ToxT, is required *(72,73)*. In addition, TcpP and TcpH function as positive regulators of *toxT (74–79)*.

4.1. In Vivo Virulence Gene Expression Analysis

A number of molecular techniques have been devised to examine the regulatory networks of virulence gene expression of *V. cholerae* in vivo. Many of these technologies used a whole genome approach to identify and elucidate virulence gene expression in vivo; these included recombination-based in vivo expression technology (RIVET), signature-tagged mutagenesis (STM), and microarray analysis using a *V. cholerae* DNA chip.

Fig. 3. *(Opposite page)* (**A**) Regional variation in the mean proportion of GC-content of *Vibrio vulnificus* island-I (VVI-I) from *V. vulnificus* strain YJ016 based on a sliding window of 500 bp. (**B**) Schematic representation of the organization of the VVI-1 from *V. vulnificus* strain YJ016. (**C**) Schematic representation of the organization of the *Vibrio* seventh pandemic island-II from *Vibrio cholerae* strain N16961. (**D**) Schematic representation of the organization of the VVI-1 in *V. vulnificus* strain CMCP6. The position and direction of transcription of the open reading frames (ORFs) are indicated by the direction of the arrows. The numbers refer to the genetic organization of the genes. Gray bars link homologous ORFs. Genes are pattern coded according to their function.

The RIVET system is based on recombinase gene fusions, which, on induction during infection, mediate a site-specific recombination, the product of which can be screened for after recovery of bacteria from host tissues *(80)*. Lee and colleagues *(81)* using RIVET demonstrated that CT and TCP were expressed sequentially during infection, and that full CT expression was dependent on prior TCP expression. A biphasical TCP induction was noted, in which early expression occurred in the lumen of the gastrointestinal tract, this is followed by full induction in the small intestine. Therefore, expression of CT only occurred after, and was dependent upon colonization; so CT was only expressed after coming into close proximity to its target cells. They also found that there were differences in the requirements of particular regulators in vivo as opposed to in vitro. For example, the TcpP regulator and its accessory protein, TcpH, which are essential for TCP expression in vitro, are not required during infection in vivo *(82)*.

STM is a whole genome approach that uses insertional mutagenesis with in vivo negative selection of attenuated strains in an animal host model, enabling many different mutants to be screened in a single host to identify factors that are essential for colonization *(83)*. STM analysis showed that mutations in a number of genes involved in metabolism, adaptation to stress, transport, regulation of cellular processes, and others of unknown function were found to have an adverse effect on colonization and survival *(84,85)*.

More recently, a whole *V. cholerae* genome microarray, consisting of 3890 full-length polymerase chain reaction products representing the *V. cholerae* strain N16961 genome sequence, was used to elucidate virulence gene expression in vivo by determining changes in gene expression on exposure of *V. cholerae* to various environments. A number of studies have been conducted using different environments: (1) *V. cholerae* derived from stool sample vs in vitro grown *V. cholerae*, (2) *V. cholerae* grown to midexponential phase derived from the rabbit ileal loop model vs in vitro grown *V. cholerae*, (3) *V. cholerae* derived from stool sample vs wild-type and *toxRS*, *tcpPH*, and *toxT* mutant *V. cholerae* grown under optimal conditions for CT and TCP expression *(86–88)*. Microarray analysis revealed that under both in vivo and in vitro conditions, the genes showing the highest levels of expression resided primarily on chromosome 1. However, a shift in expression occurred in vivo resulting in more genes on chromosome 2 being expressed. These genes were involved in the production of adhesions, amino acids, type IV pili, chemotaxis, iron transport, anaerobic metabolism, and the formation of a periplasmic nitrate reductase complex that may allow for respiration under low oxygen tension *(86–88)*. Expression of genes involved in motility was down regulated, possibly to aid shedding of *V. cholerae* from the human host *(86)*. The *ace* and *cep* genes involved in CTXϕ morphogenesis, and genes involved in repeat in toxin (RTX) toxin production, were also highly expressed. Bina and workers examined the transcriptional profile of *toxT*, *tcpPH*, or *toxR* mutant cells against a wild-type background under optimal CT-producing conditions. They revealed that the *tcpPH* mutant showed TcpPH being required for expression of CT, TCP, and the regulator *arcA*, whereas the OmpT and OmpW porins and genes involved in glycerol metabolism exhibited a decrease in expression *(87)*.

5. Whole-Genome Comparisons Between *Vibrio* Species

Vibrios are Gram-negative γ-proteobacteria that is ubiquitous in marine and estuarine environments. A common genomic feature among *Vibrio* species and probably all *Vibrionaceae* is the existence of two chromosomes of unequal size *(10,89,90)*. Pulse-field gel

Table 5
Comparison of the Genome Composition of *Vibrio* Species

Vibrio species	ORF	Conserved hypo. protein	Hypo. protein	tRNA	rRNA	Integron
Vibrio cholerae N16961						
chromosome 1	2770	478	515	94	8	0
chromosome 2	1115	165	419	4	0	1
Vibrio parahaemolyticus RIMD2210633						
chromosome 1	3080	377	714	112	10	1
chromosome 2	1752	122	636	14	1	0
Vibrio vulnificus YJ016						
chromosome 1	3262	495	569	100	8	1
chromosome 2	1697	240	377	12	1	0
Vibrio vulnificus CMCP6						
chromosome 1	3205	348	732	98	8	1
chromosome 2	689	136	479	13	1	0

Hypo, hypothetical.

electrophoresis analysis of a *V. cholerae* O1 classical biotype isolate by Trucksis and colleagues, estimated a genome size of 2.41 Mb for chromosome 1 and 1.58 Mb for chromosome 2 *(10,11)*. The complete sequenced genome of *V. cholerae* El Tor stain N16961 was 2.96 and 1.07 Mb, respectively, for chromosome 1 and 2 *(11)*. Since the completion of *V. cholerae* O1 strain N16961 genome, a further four complete genome sequences of members of the family *Vibrionaceae* are available; *Vibrio parahaemolyticus* strain RIMD2210633, two *V. vulnificus* strains, YJO16 and CMCP6, and *Vibrio fischeri* strain E116 (Table 2) *(91)*. A related deep sea bacterium *Photobacterium profundum* SS9 whole genome is also available *(92)*. *V. cholerae*, a human pathogen, and *V. fischeri*, an endosymbiont, have the smallest *Vibrio* genomes of 4.1 and 4.3 Mb, respectively, whereas the other three *Vibrio* genomes are all approx 5 Mb in size. *P. profundum* SS9 has the largest genomes at 6.2 Mb. All sequenced *Vibrio* genomes have several features in common; the presence of two chromosomes, a similar number of ORFs of known and unknown function, and all genomes contain an integron (Table 5). The relative position of the integron, its gene content and its size, differs among the *Vibrio* species (Table 3). The integron is present on chromosome 1 of *V. vulnificus* and *V. parahaemolyticus*, but in *V. cholerae* N16961 the SI is present on chromosome 2. The ToxRS regulatory proteins are also a common feature of *Vibrio* species. Whole-genome comparisons of the relative positions of conserved genes among *V. cholerae*, *V. parahaemolyticus*, and *V. vulnificus* indicate that the gene order is more highly conserved between *V. vulnificus* and *V. parahaemolyticus*, than between *V. cholerae* and *V. vulnificus* or *V. parahaemolyticus* (Fig. 5). Whole-genome comparison analysis also indicates that the gene order and content is more highly conserved on chromosome 1 than on chromosome 2 (Fig. 5).

A striking difference among the six *Vibrionaceae* genomes is the G+C content of the DNA. *V. fischeri* has the lowest, with a genome-wide value of 38% G+C, next is *P. profundum* SS9 with 42% G+C, whereas V. *cholerae*, *V. parahaemolyticus*, and *V.*

vulnificus vary from 45 to 47% G+C. In keeping with their size, the genome of *P. profundum* SS9 has 5640 annotated ORFs, whereas *V. fischeri* ES114 has only 3747. The number of tRNA operons is similar among the species with the larger *P. profundum* SS9 genome having 169, more than the average of 110 found in the other genomes. The distribution of *rrn* operons (most are present on chromosome 1) is similar amongst all six species examined, with the exception of the absence of a rrn operon on chromosome 2 of *V. cholerae*. Many of the metabolic processes that are critical to the normal life cycle of the bacterial cell are conserved among the six genomes examined. The genome percentage for metabolic processes is maintained across the *V. vulnificus* YJ016 and CMCP6, *V. parahaemolyticus*, and *V. cholerae* species. Transport and metabolism of carbohydrates, amino acids, nucleotides, coenzymes, lipids, and inorganic ions accounts for approx 19% of protein-coding genes. Similarly, functions such as signal transduction, cell wall biogenesis, and transcription are encoded for by 15.5% of the genome. On the other hand, a higher percentage of genes in *V. vulnificus* are involved in replication, recombination, and repair (5.8% for both strains) than in *V. parahaemolyticus* RIMD2210633 (4.6%) and *V. cholerae* N16961 (4.5%). Conversely, more genes in *V. cholerae* N16961 and *V. parahaemolyticus* RIMD2210633 are involved in energy metabolism (6.3%) than is the case for *V. vulnificus* (average of 5.4% for both strains) (Table 2). *P. profundum* SS9 has the highest number (216 in total) of transposase genes and *V. fischeri* the least number (1) of transposase genes among the six genomes examined, which may be a reflection of the very different lifestyles of these bacteria. *P. profundum* is a free-living deep sea organism, and *V. fischeri* is an endosymbiont of squid, which may reduce its opportunities for the acquisition of MGEs by HGT from other organisms. However, the *V. fischeri* ES114 genome did show evidence of the presence of a retron, an integron and a filamentous prophage with homology to the CT-encoding CTX prophage from *V. cholerae (93)*. *V. fischeri* ES114 also encoded a region that shows homology to a type IV pilus from *V. cholerae* that is encoded on a PAI. Interestingly, both *V. vulnificus* strains contain a large number of transposase genes compared with *V. parahaemolyticus* strain RIMD2210633. *V. vulnificus* YJ016 and CMCP6 contain 48 and 43 transposase genes each, whereas *V. parahaemolyticus* RIMD2210633 only contains two transposase genes. The difference in the number of transposase or remnant transposase genes between the two species has not altered the genome size, however, many of the transposase genes in *V. vulnificus* YJ016 and CMCP6 are associated with novel GEIs unique to each strain *(57)*. The reason for the abundance or lack of transposase genes is not apparent from the lifestyles of these two species, which are very similar. Within *V. cholerae* N16961 21 transposase genes are annotated, many of which are associated with PAIs.

5.1. V. parahaemolyticus *Strain RIMD2210633 Genome Sequence*

V. parahaemolyticus is a major causative agent of gastroenteritis, particularly in areas of high seafood consumption (Table 1) and is an emerging pathogen in North America *(94)*. A characteristic of pathogenic *V. parahaemolyticus* strains is the production of a thermostable direct hemolysin on Wagatsuma's blood agar resulting in the haemolytic Kanagawa phenomenon (KP) *(95)*. Similar to *V. cholerae*, most strains of *V. parahaemolyticus* are not pathogenic to humans. The complete genome sequence of *V. parahaemolyticus* strain RIMD2210633, a 1996 pandemic KP-positive serotype

O3:K6 isolate from Japan, is available *(91)*. The *V. parahaemolyticus* strain RIMD2210633 genome is 5.2 Mb, comprised of a 3.29 Mb chromosome 1 and a 1.88 Mb chromosome 2, and contains 4832 genes, 40% of which were annotated as hypothetical proteins (Tables 2 and 5) *(91)*. A high number of rRNA operons, 11 in all, were identified (Table 5). Examination of the genome sequence revealed that, as in *V. cholerae*, most of the genes essential for growth and viability are located on chromosome 1 *(91)*. Chromosome 2, however, contained several genes involved in important metabolic pathways and encoded more genes required for transcriptional regulation and transport of various substances than chromosome 1. It was suggested that these types of genes have a role in response to environmental changes and that chromosome 2, therefore, may have a role in adaptation to environmental changes *(91)*. For example, *V. parahaemolyticus* produces two types of flagella, polar and lateral. The polar flagella, encoded on chromosome 1, are constitutively expressed, whereas the lateral flagella, encoded on chromosome 2, are induced for movement through viscous surroundings or swarming on solid surfaces. An approx 80 kb PAI encoding the *tdh* gene (VPA1313) was identified on chromosome 2, along with homologs to a cytotoxic necrotising factor (VPA1321) and an exoenzyme T (VPA1327). A type III secretion system (TTSS) was also identified on this PAI *(91)*. The TTSS region is only found in KP-positive isolates and the PAI in which it is embedded has a GC content of 40% compared with 45% for the entire genome. The TTSS is an essential virulence factor for a number of enteric pathogens, such as *Shigella* and *Salmonella*. The presence of a TTSS is probably responsible for the characteristic inflammatory diarrhea associated with *V. parahaemolyticus* gastroenteritis. *V. cholerae* isolates do not contain a TTSS and cause a noninflammatory diarrhea. A second TTSS was identified in *V. parahaemolyticus* strain RIMD2210633 on chromosome 1, encompassing ORFs VP1654 to VP1702. This second TTSS was found to be present in all *V. parahaemolyticus* examined and had a percent GC content similar to the entire genome, suggesting that this region is ancestral to the species *(91)*. Many of the ORFs within this second TTSS (VP1654 to VP1702) showed homology to a TTSS described in *V. harveyi* (ORFs VP1657 to VP1675) and *Yersinia entercolitica* (VP1687 to VP1700) *(91)*. A bacteriophage named f237 was also present on chromosome 1 between ORFs VP1547 and VP1588 in strain RIMD2210633, which showed strong homology to Vf33 *(96)*. The replication origin (*ori*) of chromosome 1 had several feature in common to that of prokaryotic genomes, however, the *ori* of chromosome 2 has no features in common to known bacterial or plasmid *ori*s. The *ori* on chromosome 2 of *V. parahaemolyticus* was similar in sequence to that identified in *V. cholerae (91)*.

5.1.1. V. parahaemolyticus *vs* V. cholerae *Genomic Comparison*

The 5.2-Mb *V. parahaemolyticus* genome is 1.1 Mb larger than the *V. cholerae* genome, with chromosome 2 showing the greatest size difference between the species; 1.1 Mb vs 1.9 Mb. The reason for the size difference is unclear but could be accounted for by gene acquisition in *V. parahaemolyticus* and/or gene loss in *V. cholerae*. Most of the ORFs present on chromosome 1 of *V. cholerae* had homologs on chromosome 1 of *V. parahaemolyticus* and similarly with chromosome 2 of both species. As mentioned earlier in **Subheading 5.**, whole genome comparative analysis between *V. parahaemolyticus* and *V. cholerae* N16961 revealed multiple intrachromosomal rearrangements, with the most divergence in size and gene content on chromosome 2 (Fig. 5) *(91)*.

5.2. V. vulnificus *Strains YJ016 and CMCP6 Genome Sequences*

V. vulnificus is the major cause of mortality associated with food-borne disease resulting in the highest death rate of any causative agent (Table 1) *(97)*. *V. vulnificus* causes a severe human infection acquired through wounds or contaminated seafood *(98,99)*. *V. vulnificus* invades connective tissue, where it releases extracellular proteins, causing blistering and hemorrhagic necrosis of the tissue *(100)*. *V. vulnificus* is halophilic and is present in the aquatic ecosystem. The *V. vulnificus* strain YJ016 genome was recently completed and consists of two circular chromosomes of 3.35 Mb (3262 ORFs) and 1.86 Mb (1697 ORFs), respectively, and a 49-kb plasmid (Table 2) *(101)*. *V. vulnificus* strain YJ016 is a biotype 1 hospital isolate from Taiwan. Of the 4959 genes identified, 1688 (34%) encoded hypothetical proteins, which account for most of the genes that are unique to the *V. vulnificus* genome (Table 5). A SI was identified on chromosome 1 spanning 138 kb and containing 202 ORFs of which 160 are hypothetical (Table 3). The *V. vulnificus* strain CMCP6, a biotype 1 hospital isolate from South Korea, complete genome sequence is also available in the databases (accession no. AE016795) *(102)*. The genome of strain CMCP6 is similar in size (5,126,798 bp) and sequence to strain *V. vulnificus* strain YJ016. *V. vulnificus* contains a number of virulence genes, such as genes involved in type IV pilus formation, capsular polysaccharide biosynthesis, iron acquisition, extracellular enzyme, toxin production, and RTX toxin production. The extracellular enzymes metalloprotease, phospholipase, and cytolysin, and the RTX toxin, which have been implicated in causing tissue damage and subsequent bacterial invasion into the bloodstream, are all present on chromosome 2 *(101)*.

5.2.1. V. vulnificus *Strain YJ016 vs* V. cholerae *and* V. parahaemolyticus *Genomic Comparisons*

The 5.13-Mb genome of *V. vulnificus* strain CMCP6 is similar in size to *V. parahaemolyticus*, but is larger than the 4.07-Mb of *V. cholerae* strain N16961 owing to a larger chromosome 2 (1.84 vs 1.88 vs 1.07 Mb for *V. vulnificus*, *V. parahaemolyticus*, and *V. cholerae*, respectively) *(101)*. *V. vulnificus* shows a higher degree of conservation in gene organization in the two chromosomes to *V. parahaemolyticus* than to *V. cholerae* (Fig. 5) *(101)*. Comparative genome analysis between *V. cholerae* strain N16961 and *V. vulnificus* strain YJ016 revealed that multiple intra- and interchromosomal rearrangements have occurred. Between chromosomes 1 and 2, the gene content and position is better conserved among chromosome 1 from the three species, unlike chromosome 2, which is divergent in gene content between *V. cholerae*, *V. parahaemolyticus*, and *V. vulnificus* (Fig. 5). Overall, the distribution of functionally known genes is similar to that of *V. cholerae* and *V. parahaemolyticus*. *V. vulnificus* contains more genes involved in transcription, carbohydrate transport, metabolism, and secondary metabolism biosynthesis than *V. cholerae*. *V. cholerae* lacks the operon involved in capsular polysaccharide biosynthesis, whereas *V. vulnificus* strains exhibit different capsular types *(103)*. The iron metabolism gene clusters in *V. vulnificus* are located on the chromosome 1, whereas in *V. cholerae* the genes are dispersed between the two chromosomes. Chen and colleagues *(101)* compared the number, distribution, and position of gene family members in the *V. vulnificus* and *V. cholerae* genomes and concluded that duplication and transposition events occurred more frequently in *V. vulnificus* (260 duplications added 495 genes compared with 113 duplications adding 147 genes in *V. cholerae*). This suggested that these events could account for genome size differences between the species.

Homologs of *V. cholerae nan-nag* region of VPI-2 were identified in both *V. vulnificus* strains YJ016 and CMCP6 (Fig. 4). Three regions were identified on the genome of *V. vulnificus* strain YJ016 that showed homology to ORFs VC1758 to VC1809 of VPI-2 from *V. cholerae*. Homologs of VC1773 to VC1784 of VPI-2 were identified on chromosome 2 of both *V. vulnificus* strains. The two other regions of homology were found only in strain YJ016 dispersed on chromosome 1 (Fig. 4) *(57)*. Homologs of the *V. cholerae* VSP-II region were also identified in *V. vulnificus* strain YJ016 but were absent from strain CMCP6 (Fig. 3) *(58)*. The region of homology to VSP-II in strain YJ016 was embedded in a 43 kb GEI named VVI-1 *(58)*. In addition to VSP-II homologs, VVI-I encoded a fructose/mannose phosphotransferase system (PTS) system and a sugar metabolism gene cluster, and was integrated in *V. vulnificus* YJ016 at the same site as VSP-II in *V. cholerae* N16961 (Fig. 3) *(58)*. The metalloprotease, phospholipase, and the RTX cluster of *V. vulnificus* YJ016 exhibited homologies to those in *V. cholerae*.

5.2.2. V. vulnificus *Strain YJ016* vs V. vulnificus *Strain CMCP6 Genomic Comparisons*

Genomic comparison between the two *V. vulnificus* strains indicates that a few small intrachromosomal rearrangements have taken place on the CMCP6 genome *(57)*. A total of 17 GEIs ranging in size from 10 to 117 kb were identified between the two strains, 9 unique GEIs were identified in strain YJ016, and 5 unique GEIs in CMCP6. These GEIs encoded a range of genes involved in transport, sugar metabolism, and restriction modification systems *(57)*.

5.3. V. fischeri *Strain ES114* *and* Photobacterium profundum *Strain SS9 Genome Sequences*

V. fischeri is found worldwide, mainly in temperate and subtropical waters. *V. fischeri* exists as a light-organ symbiont of several species of squids and fishes, and as a pathogen of certain invertebrates *(104)*. The ability to emit visible levels of light has been shown to be caused primarily by the activity of a small cluster of five genes, *luxCDABE*, and the regulatory genes, *luxR* and *luxI (105)*. The *V. fischeri* genome is complete and information is available at the web address (http://ergo.integratedgenomics.com/Genomes/VFI/vibrio_fischeri.html). The *V. fischeri* strain ES114 genome is 4.3 Mb encoding 3802 genes, of which 2810 have been assigned function, and has an overall GC content of 38% (Table 2) *(93)*. Homologs of CTXφ and TCP are found on the genome. The whole genome sequence of *Photobacterium profundum* strain SS9, a member of the family *Vibrionaceae*, is also available in the databases *(92)*. *Photobacterium profundum* strain SS9 is a moderately piezophilic ("pressure loving") psychrotolerant marine bacterium. Chromosome 1 consists of 4,085,304 bp (accession number CR354531) and chromosome 2 consists of 2,237,943 bp (accession number CR354532) and an 80,033-bp circular plasmid is also sequenced from this strain *(92)*.

6. Conclusion

Examination of the genomic content and organization of the *Vibrio* species gives an indication of how these species have evolved to cause disease in different ways and to

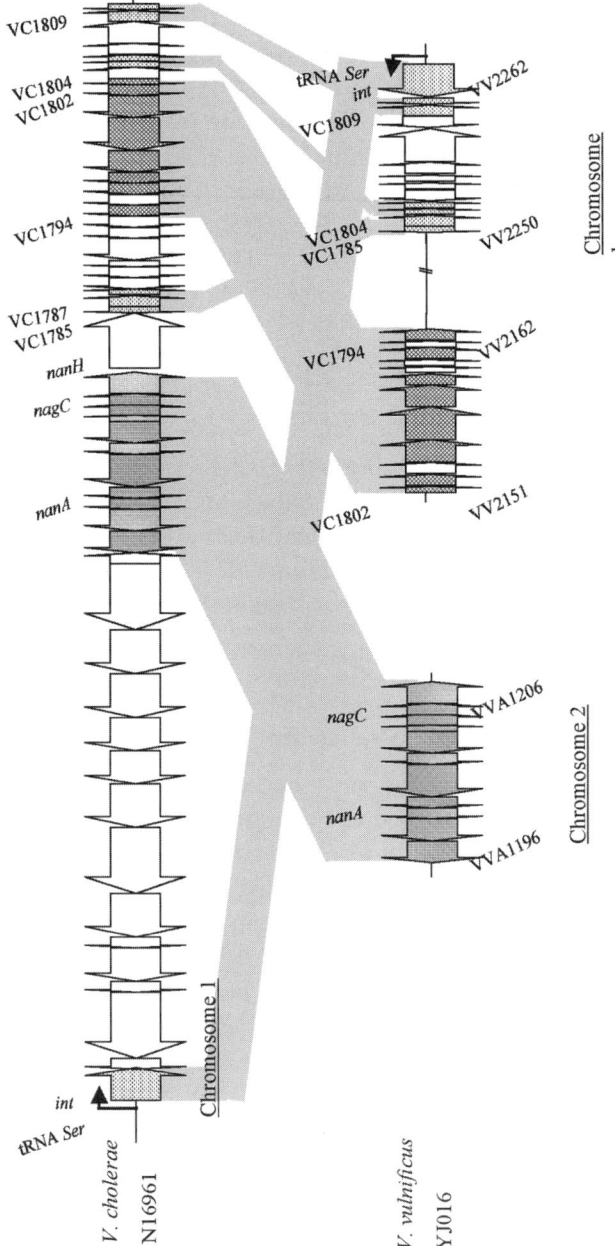

Fig. 4. Schematic representation and genetic organization of the *Vibrio* pathogenicity island (VPI)-2 in *Vibrio cholerae* and *Vibrio vulnificus*. *V. cholerae* toxigenic isolates, such as the sequenced strain N16961, contained the entire 57.3-kb VPI-2 on chromosome 1. The sequenced *V. vulnificus* strain YJ016 contained elements of the VPI-2 dispersed into three regions between the two chromosomes. The position and direction of transcription of the open reading frames (ORFs) are indicated by the direction of the arrows. The numbers refer to the genetic organization of the genes. Gray bars link homologous ORFs between the two species. The transfer RNA serine is represented by a black arrow.

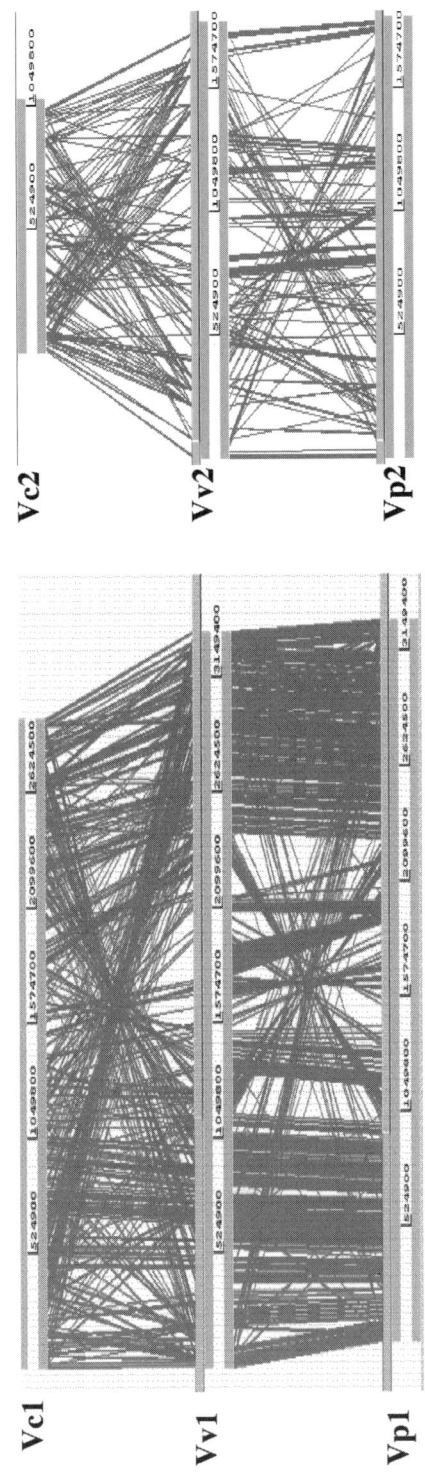

Fig. 5. Comparison of the relative position of conserved genes between *Vibrio vulnificus* and *Vibrio cholerae*, and *V. vulnificus* and *Vibrio parahaemolyticus* on chromosome 1 and chromosome 2. Gene pairs were generated by Blastn analysis of predicted genes from each of the genomes. Red lines connect both conserved genes between the organisms. A line between chromosomes indicates a homologous block of genomic sequences; crossed lines represent intrachromosomal rearrangements.

occupy different niches. The voluminous diarrhea associated with *V. cholerae* is caused by CT, which causes electrolyte imbalance leading to massive water loss; *V. cholerae* is the prototypical extracellular pathogen. In *V. parahaemolyticus*, the TTSS is believed to be responsible for the inflammatory diarrhea. TTSS are required to inject bacterial proteins directly into host cells. In *V. vulnificus*, the extensive tissue damage and septicemia are the result of extracellular proteins and toxin production. A number of the virulence factors found in *Vibrio* species, and in particular *V. cholerae*, are encoded on MGEs. In addition, among the six *Vibrionaceae* species sequenced to date, three tRNA sites, a tmRNA, a tRNA-met, and a tRNA-ser are all hotspots for the insertions of MGEs among the six species *(57,106)*. For example, in *V. parahaemolyticus*, at a tRNA-met locus a 23-kb GEI was identified only among strains isolated after 1996 including all pandemic O3:K6 isolates *(106)*. Elucidation of these MGEs, and the virulence factors that they encode, could help our understanding of how pathogenic strains originate and evolve. Through the use of comparative and functional genomic analysis, we may also be able to identify potential antibacterial targets that may inhibit the disease causing ability of these bacteria or their transmission from aquatic reservoir to human host.

Acknowledgments

We thank our colleagues who have provided ideas and suggestions that have contributed greatly to this chapter. We thank members of the Department of Microbiology, UCC, National University of Ireland, Cork for continued support. Owing to space constraints, literature citations have been limited in some cases to recent relevant reviews. Therefore, we apologize to those authors whose important work has not been included or cited. Research in EFB's laboratory is supported by Enterprise Ireland basic research grants, EMBARK IRCSET postdoctoral and postgraduate fellowship grants, a Science Foundation Ireland (SFI) Research Frontiers Program grant 05/RFP/Gen0006, a SFI Investigator Program grant 04/IN3/B651, and a Higher Education Authority PRTLI-3 grant.

References

1. Faruque, S. M., Albert, M. J., and Mekalanos, J. J. (1998) Epidemiology, genetics, and ecology of toxigenic *Vibrio cholerae. Microbiol. Mol. Biol. Rev.* **62,** 1301–1314.
2. Kaper, J. B., Morris, J. G. Jr., and Levine, M. M. (1995) Cholera. *Clin. Microbiol. Rev.* **8,** 48–86.
3. Janda, J. M., Powers, C., Bryant, R. G., and Abbott, S. L. (1988) Current perspectives on the epidemiology and pathogenesis of clinically significant *Vibrio* spp. *Clin. Microbiol. Rev.* **1,** 245–267.
4. Mukhopadhyay, A. K., Saha, P. K., Garg, S., et al. (1995) Distribution and virulence of *Vibrio cholerae* belonging to serogroups other than O1 and O139: a nationwide survey. *Epidemiol. Infect.* **114,** 65–70.
5. Ramamurthy, T., Garg, S., Sharma, R., et al. (1993) Emergence of novel strain of *Vibrio cholerae* with epidemic potential in southern and eastern India. *Lancet* **341,** 703–704.
6. Cholera Working Group. (1993) Large epidemic of cholera-like disease in Bangladesh caused by *Vibrio cholerae* O139 synonym Bengal. *Lancet* **342,** 387–390.
7. Faruque, S. M., Zhu, J., Asadulghani, Kamruzzaman, M., and Mekalanos, J. J. (2003) Examination of diverse toxin-coregulated pilus-positive Vibrio cholerae strains fails to demonstrate evidence for Vibrio pathogenicity island phage. *Infect. Immun.* **71,** 2993–2999.

8. Waldor, M. K. and Mekalanos, J. J. (1996) Lysogenic conversion by a filamentous phage encoding cholera toxin. *Science* **272,** 1910–1914.

9. Karaolis, D. K., Johnson, J. A., Bailey, C. C., Boedeker, E. C., Kaper, J. B., and Reeves, P. R. (1998) A *Vibrio cholerae* pathogenicity island associated with epidemic and pandemic strains. *Proc. Natl. Acad. Sci. USA* **95,** 3134–3139.

10. Trucksis, M., Michalski, J., Deng, Y. K., and Kaper, J. B. (1998) The *Vibrio cholerae* genome contains two unique circular chromosomes. *Proc. Natl. Acad. Sci. USA* **95,** 14,464–14,469.

11. Heidelberg, J. F., Eisen, J. A., Nelson, W. C., et al. (2000) DNA sequence of both chromosomes of the cholera pathogen *Vibrio cholerae. Nature* **406,** 477–483.

12. Bik, E. M., Bunschoten, A. E., Gouw, R. D., and Mooi, F. R. (1995) Genesis of the novel epidemic *Vibrio cholerae* O139 strain: evidence for horizontal transfer of genes involved in polysaccharide synthesis. *EMBO J.* **14,** 209–216.

13. Comstock, L. E., Maneval, D. Jr., Panigrahi, P., et al. (1995) The capsule and O antigen in *Vibrio cholerae* O139 Bengal are associated with a genetic region not present in *Vibrio cholerae* O1. *Infect. Immun.* **63,** 317–323.

14. Faruque, S. M., Asadulghani, Alim, A. R., Albert, M. J., Islam, K. M., and Mekalanos, J. J. (1998) Induction of the lysogenic phage encoding cholera toxin in naturally occurring strains of toxigenic *Vibrio cholerae* O1 and O139. *Infect. Immun.* **66,** 3752–3757.

15. Dziejman, M., Balon, E., Boyd, D., Fraser, C. M., Heidelberg, J. F., and Mekalanos, J. J. (2002) Comparative genomic analysis of *Vibrio cholerae*: genes that correlate with cholera endemic and pandemic disease. *Proc. Natl. Acad. Sci. USA* **99,** 1556–1561.

16. Salles, C. A. and Momen, H. (1991) Identification of *Vibrio cholerae* by enzyme electrophoresis. *Trans. Soc. Trop. Med. Hyg.* **85,** 544–547.

17. Evins, G. M., Cameron, D. N., Wells, J. G., et al. (1995) The emerging diversity of the electrophoretic types of *Vibrio cholerae* in the Western Hemisphere. *J. Infect. Dis.* **172,** 173–179.

18. Wachsmuth, I. K., Evins, G. M., Fields, P. I., et al. (1993) The molecular epidemiology of cholera in Latin America. *J. Infect. Dis.* **167,** 621–626.

19. Beltran, P., Delgado, G., Navarro, A., Trujillo, F., Selander, R. K., and Cravioto, A. (1999) Genetic diversity and population structure of *Vibrio cholerae. J. Clin. Microbiol.* **37,** 581–590.

20. Farfan, M., Minana, D., Fuste, M. C., and Loren, J. G. (2000) Genetic relationships between clinical and environmental *Vibrio cholerae* isolates based on multilocus enzyme electrophoresis. *Microbiology* **146,** 2613–2626.

21. Byun, R., Elbourne, L. D., Lan, R., and Reeves, P. R. (1999) Evolutionary relationships of pathogenic clones of *Vibrio cholerae* by sequence analysis of four housekeeping genes. *Infect. Immun.* **67,** 1116–1124.

22. Karaolis, D. K., Lan, R., Kaper, J. B., and Reeves, P. R. (2001) Comparison of *Vibrio cholerae* pathogenicity islands in sixth and seventh pandemic strains. *Infect. Immun.* **69,** 1947–1952.

23. Li, M., Kotetishvili, M., Chen, Y., and Sozhamannan, S. (2003) Comparative genomic analyses of the vibrio pathogenicity island and cholera toxin prophage regions in nonepidemic serogroup strains of *Vibrio cholerae. Appl. Environ. Microbiol.* **69,** 1728–1738.

24. Karaolis, D. K., Lan, R., and Reeves, P. R. (1995) The sixth and seventh cholera pandemics are due to independent clones separately derived from environmental, nontoxigenic, non-O1 *Vibrio cholerae. J. Bacteriol.* **177,** 3191–3198.

25. Stine, O. C., Sozhamannan, S., Gou, Q., Zheng, S., Morris, J. G. Jr., and Johnson, J. A. (2000) Phylogeny of *Vibrio cholerae* based on *recA* sequence. *Infect. Immun.* **68,** 7180–7185.

26. Boyd, E. F., Heilpern, A. J., and Waldor, M. K. (2000) Molecular analysis of a putative CTXφ precursor and evidence for independent acquistion of distinct CTXφs by toxigenic *Vibrio cholerae*. *J. Bacteriol.* **182,** 5530–5538.

27. Kotetishvili, M., Stine, O. C., Chen, Y., et al. (2003) Multilocus sequence typing has better discriminatory ability for typing *Vibrio cholerae* than does pulsed-field gel electrophoresis and provides a measure of phylogenetic relatedness. *J. Clin. Microbiol.* **41,** 2191–2196.

28. O'Shea, Y. A., Reen, F. J., Quirke, A. M., and Boyd, E. F. (2004) Evolutionary genetic analysis of the emergence of epidemic *Vibrio cholerae* isolates based on comparative nucleotide sequence analysis and multilocus virulence gene profiles. *J. Clin. Microbiol.* **42,** 4657–4671.

29. Chen, F., Evins, G. M., Cook, W. L., Almeida, R., Hargrett-Bean, N., and Wachsmuth, K. (1991) Genetic diversity among toxigenic and nontoxigenic *Vibrio cholerae* O1 isolated from the Western Hemisphere. *Epidemiol. Infect.* **107,** 225–233.

30. Bik, E., Gouw, R., and Mooi, F. (1996) DNA fingerprinting of *Vibrio cholerae* strains with a novel insertion sequence element: a tool to identify epidemic strains. *J. Clin. Microbiol.* **34,** 1453–1461.

31. Reen, F. J. and Boyd, E. F. (2005) Molecular typing of epidemic and nonepidemic *Vibrio cholerae* isolates and differentiation of *V. cholerae* and *V. mimicus* isolates by PCR-single-strand conformation polymorphism analysis. *J. Appl. Microbiol.* **98,** 544–555.

32. Davis, B. M., Kimsey, H. H., Kane, A. V., and Waldor, M. K. (2002) A satellite phage-encoded antirepressor induces repressor aggregation and cholera toxin gene transfer. *EMBO J.* **21,** 4240–4249.

33. Groisman, E. A. and Ochman, H. (1996) Pathogenicity islands: bacterial evolution in quantum leaps. *Cell* **87,** 791–794.

34. Colwell, R. R. and Haq, A. (1994) Vibrios in the environment: Viable but nonculturable *Vibrio cholerae,* in Vibrio cholerae *and Cholera: Molecular to Global Perspectives* (Wachsmuth, I. K., Blake, P. A., and Olsvik, O., eds.). American Society for Microbiology, Washington, DC, pp. 117–133.

35. Waldor, M. K., Rubin, E. J., Pearson, G. D., Kimsey, H., and Mekalanos, J. J. (1997) Regulation, replication, and integration functions of the *Vibrio cholerae* CTXΦ are encoded by region RS2. *Mol. Microbiol.* **24,** 917–926.

36. Pearson, G. D., Woods, A., Chiang, S. L., and Mekalanos, J. J. (1993) CTX genetic element encodes a site-specific recombination system and an intestinal colonization factor. *Proc. Natl. Acad. Sci. USA* **90,** 3750–3754.

37. Campos, J., Fando, R., Silva, A., Rodriguez, B. L., and Benitez, J. A. (1998) Replicating function of the RS1 element associated with *Vibrio cholerae* CTX phi prophage. *FEMS Microbiol. Lett.* **164,** 141–147.

38. Davis, B. M., Moyer, K. E., Boyd, E. F., and Waldor, M. K. (2000) CTX prophages in classical biotype *Vibrio cholerae*: functional phage genesbut dysfunctional phage genomes. *J. Bacteriol.* **182,** 6992–6998.

39. Faruque, S. M., Asadulghani, Kamruzzaman, M., et al. (2002) RS1 element of Vibrio cholerae can propagate horizontally as a filamentous phage exploiting the morphogenesis genes of CTXphi. *Infect. Immun.* **70,** 163–170.

40. Mekalanos, J. J. (1983) Duplication and amplification of toxin genes in *Vibrio cholerae*. *Cell* **35,** 253–263.

41. Campos, J., Martinez, E., Suzarte, E., et al. (2003) VGJ phi, a novel filamentous phage of *Vibrio cholerae*, integrates into the same chromosomal site as CTX phi. *J. Bacteriol.* **185,** 5685–5696.

42. Boyd, E. F. and Waldor, M. K. (1999) Alternative mechanism of cholera toxin acquisition by *Vibrio cholerae*: Generalized transduction of CTXΦ by bBacteriophage CP-T1 *Infect. Immun.* **67,** 5898–5905.

43. O'Shea, Y. A. and Boyd, E. F. (2002) Mobilization of the Vibrio pathogenicity island between *Vibrio cholerae* isolates mediated by CP-T1 generalized transduction. *FEMS Microbiol. Lett.* **214,** 153–157.

44. Wagner, P. L. and Waldor, M. K. (2002) Bacteriophage control of bacterial virulence. *Infect. Immun.* **70,** 3985–3993.

45. Dobrindt, U., Hochhut, B., Hentschel, U., and Hacker, J. (2004) Genomic islands in pathogenic and environmental microorganisms. *Nat. Rev. Microbiol.* **2,** 414–424.

46. Taylor, R. K., Miller, V. L., Furlong, D. B., and Mekalanos, J. J. (1987) Use of *phoA* gene fusions to identify a pilus colonization factor coordinately regulated with cholera toxin. *Proc. Natl. Acad. Sci. USA* **84,** 2833–2837.

47. Herrington, D. A., Hall, R. H., Losonsky, G., Mekalanos, J. J., Taylor, R. K., and Levine, M. M. (1988) Toxin, toxin-coregulated pili, and the toxR regulon are essential for *Vibrio cholerae* pathogenesis in humans. *J. Exp. Med.* **168,** 1487–1492.

48. Boyd, E. F., Moyer, K. L., Shi, L., and Waldor, M. K. (2000) Infectious CTXϕ and the *Vibrio* pathogenicity island prophage in *Vibrio mimicus*: evidence for recent horizontal transfer between *V. mimicus* and *V. cholerae. Infect. Immun.* **68,** 1507–1513.

49. Davis, B. R., Fanning, G. R., Madden, J. M., et al. (1981) Characterization of biochemically atypical *Vibrio cholerae* strains and designation of a new pathogenic species, *Vibrio mimicus. J. Clin. Microbiol.* **14,** 631–639.

50. Campos, E., Bolanos, H., Acuna, M. T., et al. (1996) *Vibrio mimicus* diarrhea following ingestion of raw turtle eggs. *Appl. Environ. Microbiol.* **62,** 1141–1144.

51. Acuna, M. T., Diaz, G., Bolanos, H., et al. (1999) Sources of *Vibrio mimicus* contamination of turtle eggs. *Appl. Environ. Microbiol.* **65,** 336–338.

52. Karaolis, D. K., Somara, S., Maneval, D. R. Jr., Johnson, J. A., and Kaper, J. B. (1999) A bacteriophage encoding a pathogenicity island, a type-IV pilus and a phage receptor in cholera bacteria. *Nature* **399,** 375–379.

53. Boyd, E. F. and Brussow, H. (2002) Common themes among bacteriophage-encoded virulence factors and diversity among the bacteriophages involved. *Trends Microbiol.* **10,** 521–529.

54. Rajanna, C., Wang, J., Zhang, D., et al. (2003) The vibrio pathogenicity island of epidemic *Vibrio cholerae* forms precise extrachromosomal circular excision products. *J. Bacteriol.* **185,** 6893–6901.

55. Jermyn, W. S. and Boyd, E. F. (2002) Characterization of a novel *Vibrio* pathogenicity island (VPI-2) encoding neuraminidase (*nanH*) among toxigenic *Vibrio cholerae* isolates. *Microbiology* **148,** 3681–3693.

56. Jermyn, W. S. and Boyd, E. F. (2005) Molecular evolution of *Vibrio* pathogenicity island-2 (VPI-2): mosaic structure among *Vibrio cholerae* and *Vibrio mimicus* natural isolates *Microbiology* **151,** 311–322.

57. Quirke, A. M., Reen, F. J., Claussen, M., and Boyd, E. F. (2006) Genomic island identification in *Vibrio vulnificus* reveals high genome plasticity in this emerging pathogen. *Bioinformatics* (in press).

58. O'Shea, Y. A., Finnan, S., Reen, F. J., Morrissey, J. P., O'Gara, F., and Boyd, E. F. (2004) The *Vibrio* seventh pandemic island-II is a 26.9 kb genomic island present in *Vibrio cholerae* El Tor and O139 serogroup isolates that shows homology to a 43.4 kb island in *V. vulnificus. Microbiology* **150,** 4053–4063.

59. Rubin, E. J., Lin, W., Mekalanos, J. J., and Waldor, M. K. (1998) Replication and integration of a *Vibrio cholerae* cryptic plasmid linked to the CTX prophage. *Mol. Microbiol.* **28,** 1247–1254.

60. Cook, W. L., Wachsmuth, K., Johnson, S. R., Birkness, K. A., and Samadi, A. R. (1984) Persistence of plasmids, cholera toxin genes, and prophage DNA in classical *Vibrio cholerae* O1. *Infect. Immun.* **45,** 222–226.

61. Bartowsky, E. J., Morelli, G., Kamke, M., and Manning, P. A. (1987) Characterization and restriction analysis of the P sex factor and the cryptic plasmid of *Vibrio cholerae* strain V58. *Plasmid* **18,** 1–7.

62. Kaper, J. B., Michalski, J., Ketley, J. M., and Levine, M. M. (1994) Potential for reacquistion of cholera entertoxin genes by attenuated *Vibrio cholerae* vaccine strain CVD103-HgR. *Infect. Immun.* **62,** 1480–1483.

63. Bartowsky, E. J., Attridge, S. R., Thomas, C. J., Matyrhofer, G., and Manning, P. A. (1990) Role of the P plasmid in attenuation of *Vibrio cholerae* O1. *Infect. Immun.* **58,** 3129–3134.

64. Honda, T., Nishibuchi, M., Miwatani, T., and Kaper, J. B. (1986) Demonstration of a plasmid-borne gene encoding a thermostable direct hemolysin in *Vibrio cholerae* non-O1 strains. *Appl. Environ. Microbiol.* **52,** 1218–1220.

65. Bag, P. K., Matiti, S., Sharma, C., et al. (1998) Rapid spread of the new clone of *Vibrio cholerae* O1 El Tor in cholera epidemic areas in India. *Epidemiol. Infect.* **121,** 245–251.

66. Dubon, J. M., Palmer, C. J., Ager, A. L., Shor-Posner, G., and Baum, M. K. (1997) Emergence of multiple drug-resistant *Vibrio cholerae* O1 in San Pedro Sula, Honduras. *Lancet* **349,** 924.

67. Glass, R. I., Huq, A., Alim, A. M. R., and Yunus, M. (1980) Emergence of multiple antibiotic-resistant *Vibrio cholerae* in Bangladesh. *J. Infect. Dis.* **142,** 939–943.

68. Dalsgaard, A., Forslund, A., Tam, N. V., Vinh, D. X., and Cam, P. D. (1999) Cholera in Vietnam: changes in genotypes and emergence of class I integrons containing aminoglycoside resistance gene cassettes in *Vibrio cholerae* O1 strains isolated from 1979 to 1996. *J. Clin. Microbiol.* **37,** 734–741.

69. Mazel, D., Dychinco, B., Webb, V. A., and Davis, J. (1998) A distinctive class of integron in the *Vibrio cholerae* genome. *Science* **280,** 605–608.

70. Hochhut, B. and Waldor, M. K. (1999) Site-Specific integration of the conjugal *Vibrio cholerae* SXT element into *prfC. Mol. Microbiol.* **32,** 99–110.

71. Cotter, P. A. and DiRita, V. J. (2000) Bacterial virulence gene regulation: an evolutionary perspective. *Annu. Rev. Microbiol.* **54,** 519–565.

72. Miller, V. L. and Mekalanos, J. J. (1984) Synthesis of cholera toxin is positively regulated at the transcriptional level by toxR. *Proc. Natl. Acad. Sci. USA* **81,** 3471–3475.

73. Champion, G. A., Neely, M. N., Brennan, M. A., and DiRita, V. J. (1997) A branch in the ToxR regulatory cascade of *Vibrio cholerae* revealed by characterization of *toxT* mutant strains. *Mol. Microbiol.* **23,** 323–331.

74. Carroll, P. A., Tashima, K. T., Rogers, M. B., DiRita, V. J., and Calderwood, S. B. (1997) Phase variation in *tcpH* modulates expression of the ToxR regulon in *Vibrio cholerae. Mol. Microbiol.* **25,** 1099–1111.

75. Hase, C. C. and Mekalanos, J. J. (1998) TcpP protein is a positive regulator of virulence gene expression in *Vibrio cholerae. Proc. Natl. Acad. Sci. USA* **95,** 730–734.

76. Hase, C. C. and Mekalanos, J. J. (1999) Effects of changes in membrane sodium flux on virulence gene expression in *Vibrio cholerae. Proc. Natl. Acad. Sci. USA* **96,** 3183–3187.

77. Higgins, D. E. and DiRita, V. J. (1994) Transcriptional control of toxT, a regulatory gene in the ToxR regulon of *Vibrio cholerae. Mol. Microbiol.* **14,** 17–29.

78. Medrano, A. I., DiRita, V. J., Castillo, G., and Sanchez, J. (1999) Transient transcriptional activation of the *Vibrio cholerae* El Tor virulence regulator toxT in response to culture conditions. *Infect. Immun.* **67,** 2178–2183.

79. Murley, Y. M., Carroll, P. A., Skorupski, K., Taylor, R. K., and Calderwood, S. B. (1999) Differential transcription of the tcpPH operon confers biotype-specific control of the *Vibrio cholerae* ToxR virulence regulon *Infect. Immun.* **67,** 5117–5123.

80. Camilli, A., Beattie, D. T., and Mekalanos, J. J. (1994) Use of genetic recombination as a reporter of gene expression. *Proc. Natl. Acad. Sci. USA* **91,** 2634–2638.

81. Lee, S. H., Hava, D. L., Waldor, M. K., and Camilli, A. (1999) Regulation and temporal expression patterns of *Vibrio cholerae* virulence genes during infection. *Cell* **99,** 625–634.

82. Lee, S. H., Butler, S. M., and Camilli, A. (2001) Selection for in vivo regulators of bacterial virulence. *Proc. Natl. Acad. Sci. USA* **98,** 6889–6894.

83. Hensel, M., Shea, J. E., Gleeson, C., Jones, M. D., Dalton, E., and Holden, D. W. (1995) Simultaneous identification of bacterial virulence genes by negative selection. *Science* **269,** 400–403.

84. Chiang, S. L. and Mekalanos, J. J. (1998) Use of signature-tagged transposon mutagenesis to identify *Vibrio cholerae* genes critical for colonization. *Mol. Microbiol.* **27,** 797–805.

85. Merrell, D. S., Hava, D. L., and Camilli, A. (2002) Identification of novel factors involved in colonization and acid tolerance of *Vibrio cholerae*. *Mol. Microbiol.* **43,** 1471–1491.

86. Merrell, D. S., Butler, S. M., Qadri, F., et al. (2002) Host-induced epidemic spread of the cholera bacterium. *Nature* **417,** 642–645.

87. Bina, J., Zhu, J., Dziejman, M., Faruque, S., Calderwood, S., and Mekalanos, J. (2003) ToxR regulon of *Vibrio cholerae* and its expression in vibrios shed by cholera patients. *Proc. Natl. Acad. Sci. USA* **100,** 2801–2806.

88. Xu, Q., Dziejman, M., and Mekalanos, J. J. (2003) Determination of the transcriptome of *Vibrio cholerae* during intraintestinal growth and midexponential phase in vitro. *Proc. Natl. Acad. Sci. USA* **100,** 1286–1291.

89. Yamaichi, Y., Iida, T., Park, K. S., Yamamoto, K., and Honda, T. (1999) Physical and genetic map of the genome of Vibrio parahaemolyticus: presence of two chromosomes in *Vibrio* species. *Mol. Microbiol.* **31,** 1513–1521.

90. Tagomori, K., Iida, T., and Honda, T. (2002) Comparison of genome structures of vibrios, bacteria possessing two chromosomes. *J. Bacteriol.* **184,** 4351–4358.

91. Makino, K., Oshima, K., Kurokawa, K., et al. (2003) Genome sequence of *Vibrio parahaemolyticus*: a pathogenic mechanism distinct from that of *V. cholerae*. *Lancet* **361,** 743–749.

92. Vezzi, A., Campanaro, S., D'Angelo, M., et al. (2005) Life at depth: *Photobacterium profundum* genome sequence and expression analysis. *Science* **307,** 1459–1461.

93. Ruby, E. G., Urbanowski, M., Campbell J., et al. (2005) Complete genome sequence of *Vibrio fischeri*: a symbiotic bacterium with pathogenic congeners. *Proc. Natl. Acad. Sci. USA* **102,** 3004–3009.

94. Daniels, N. A., MacKinnon, L., Bishop, R., et al. (2000) *Vibrio parahaemolyticus* infections in the United States, 1973-1998. *J. Infect. Dis.* **181,** 1661–1666.

95. Sakazaki, R., Tamura, K., Kato, T., Obara, Y., and Yamai, S. (1968) Studies on the enteropathogenic, facultatively halophilic bacterium, *Vibrio parahaemolyticus*. 3. Enteropathogenicity. *Jpn. J. Med. Sci. Biol.* **21,** 325–331.

96. Nasu, H., Iida, T., Sugahara, T., et al. (2000) A filamentous phage associated with recent pandemic *Vibrio parahaemolyticus* O3:K6 strains. *J. Clin. Microbiol.* **38,** 2156–2161.

97. Todd, E. C. (1989) Costs of acute bacterial foodborne disease in Canada and the United States. *Int. J. Food Microbiol.* **9,** 313–326.

98. Blake, P. A., Merson, M. H., Weaver, R. E., Hollis, D. G., and Heublein, P. C. (1979) Disease caused by a marine *Vibrio*. Clinical characteristics and epidemiology. *N. Engl. J. Med.* **300,** 1–5.

99. Hlady, W. G. and Klontz, K. C. (1996) The epidemiology of *Vibrio* infections in Florida, 1981–1993. *J. Infect. Dis.* **173,** 1176–1183.

100. Chuang, Y. C., Yuan, C. Y., Liu, C. Y., Lan, C. K., and Huang, A. H. (1992) Vibrio vulnificus infection in Taiwan: report of 28 cases and review of clinical manifestations and treatment. *Clin. Infect. Dis.* **15,** 271–276.

101. Chen, C. Y., Wu, K. M., Chang, Y. C., et al. (2003) Comparative genome analysis of *Vibrio vulnificus*, a marine pathogen. *Genome Res.* **13,** 2577–2587.

102. Kim, Y. R., Lee, S. E., Kim, C. M., et al. (2003) Characterization and pathogenic significance of *Vibrio vulnificus* antigens preferentially expressed in septicemic patients. *Infect. Immun.* **71**, 5461–5471.

103. Hayat, U., Reddy, G. P., Bush, C. A., Johnson, J. A., Wright, A. C., and Morris, J. G. Jr. (1993) Capsular types of *Vibrio vulnificus*: an analysis of strains from clinical and environmental sources. *J. Infect. Dis.* **168**, 758–762.

104. Boettcher, K. J. and Ruby, E. G. (1990) Depressed light emission by symbiotic *Vibrio fischeri* of the sepiolid squid *Euprymna scolopes*. *J. Bacteriol.* **172**, 3701–3706.

105. Meighen, E. A. and Dunlap, P. V. (1993) Physiological, biochemical and genetic control of bacterial bioluminescence. *Adv. Microb. Physiol.* **34**, 1–67.

106. Hurley, C., Quirke, A. M., Reen, F. J., and Boyd, E. F. (2005) Four novel genomic islands mark O3:K6 and post-1995 *Vibrio parahaemolyticus* isolates. (In review).

14

Future Directions of Infectious Disease Research

Philip M. Sherman, Billy Bourke, and Voon Loong Chan

Summary

Sequencing of bacterial genomes and comparative genomics provide novel approaches for the identification of previously unrecognized microbial pathogens that likely cause a variety of infectious diseases in humans and animals. In addition, the genetic approaches probably will discover new virulence determinants that can be used as targets in the development of novel intervention strategies for both the prevention and treatment of infectious diseases. This chapter cites specific examples in support of these contentions, with particular reference to recent advances of selected infectious and chronic inflammatory diseases involving the gastrointestinal tract.

Key Words: Microbiota; Crohn's disease; Whipples disease; *Helicobacter*; probiotics; commensals.

1. Introduction

Less than a decade ago, prognosticators forecast the importance of whole genome sequencing for the field of microbial pathogenesis in particular, and for microbiology as a whole *(1,2)*. Since then, there has been incredible progress in the sequencing of bacterial genomes *(3,4)*, including about 95 of the roughly 200 or so associated with human diseases (December 20, 2004; www.genomes.org).

Genome sequencing has also identified virulence determinants in microbial pathogenesis, including the acquisition of large blocks of DNA, such as pathogenicity islands and molecular syringes encoded by type III and type IV secretion systems, through horizontal gene transfer *(3)*. Loss of genes from microbes adapted for life in specific microenvironments *(5)* and phase variation in gene expression also both affect the virulence properties of pathogens.

2. Identification of Nonculturable Organisms

2.1. Unidentified Microbial Pathogens

It is well known that not all microbes can be grown in the laboratory using standard culture techniques. A recent phylogenetic analysis, based on 16S ribosomal RNA (rRNA) sequences available in the database, identified 52 phyla of which only 26 have cultivated representatives *(6)*. For example, Whipple's disease is a chronic enteropathy, which has been known for decades to be caused by an infectious agent that could be visualized by histology and electron microscopy. There is a symptomatic response to empirical antimicrobial therapy. However, it is only in the past decade, using novel methodological

From: *Bacterial Genomes and Infectious Diseases*
Edited by: V. L. Chan, P. M. Sherman, and B. Bourke © Humana Press Inc., Totowa, NJ

approaches, that the causative agent of Whipple's disease was cultured successfully *(7)*. Successful culture from the cerebrospinal fluid of an affected patient then allowed sequencing of the entire genome of *Tropheryma whipplei (8)*. Nevertheless, much was known about the organism in advance of its being cultured, based on a genetic analysis of prokaryotic DNA extracted from intestinal biopsies of affected patients *(9)*. For this approach, conservation of portions of 16S rDNA amongst prokaryotes can be employed as the basis for developing templates for polymerase chain reaction (PCR) to identify novel and unculturable microorganisms.

Peptic ulcer disease in humans was, for decades, considered to be a disorder of excess acid production and reduced cytoprotection that causes recurrent gastrointestinal hemorrhage, intestinal perforation, and bowel obstruction. In the past, such recurrences meant that affected patients underwent surgical interventions, including antrectomy and vagotomy with pyloroplasty.

With the advent in the late 1970s of microaerobic culture conditions, it became possible to culture intestinal *Campylobacters*, including *C. jejuni* and *C. coli*. Using microaerophilic conditions, Marshall and Warren (reviewed in ref. *10*) were the first investigators to successfully culture a related bacterium, now referred to as *Helicobacter pylori*, from the antrum (i.e., the portion of the stomach proximal to the pylorus) of patients with chronic-active gastritis and associated duodenal ulcers.

Remarkably, many groups confirmed this original observation and then showed that the natural history for peptic ulcers to recur is completely abolished if the gastric infection is eradicated. This dramatic change in the conceptualization of peptic ulcer disease from a disorder of acid production, to that of an infectious disease, has raised the possibility that other chronic, recurrent disorders have an infectious etiology that has yet to be identified.

For instance, there is increasing evidence that bacterial infections are promoters of the development of cancers. *H. pylori* was the first bacterium defined as a class 1 (that is, a definitive) carcinogen based on compelling epidemiological evidence associated with chronic infection in humans with development of gastric adenocarcinomas and MALTomas (mucosa-associated lymphoid tumors) of the stomach *(11)*. *Helicobacter felis* infection of mice is associated with the development of stomach cancer, which appears to be related to the honing of malignant precursor cells of bone marrow origin to sites of inflammation in the gastric mucosa *(12)*. *C. jejuni* infection has been associated with the development of a small bowel lymphoma variously referred to as immunoproliferative small intestinal disease and α chain disease *(13)*.

As another example, there is great interest in the potential for chronic inflammatory bowel diseases in humans (Crohn's disease and ulcerative colitis) to have an infectious etiology *(14)*. Although standard culture techniques have not proven successful, the unexpected observations by Marshall and Warren for peptic ulcer disease have fueled the hope that newer methodological approaches might prove more fruitful. As summarized in Table 1, there are a number of observations supporting the hypothesis that at least some cases of inflammatory bowel disease in humans are infectious in origin.

Animal models of inflammatory bowel disease support the consideration that Crohn's disease might have an infectious etiology. For instance, Johne's disease of ruminants has some features in common with Crohn's disease affecting humans. Because *Mycobacterium paratuberculosis* is the causative agent of Johne's disease, there has been at least

Table 1
Observations Which Suggest That Crohn's Disease Has an Infectious Etiology

Positive response to antibiotic therapy.
Positive response to surgical diversion of the fecal stream.
High prevalence of antibodies to microbial antigens.
Absence of inflammation in susceptible animals raised germ-free.
Susceptibility gene (*Nod2*) involved in recognition of bacterium-derived peptidoglycan.

two decades of studies assessing this possibility, with inconsistent results between research groups *(15)*.

Novel helicobacters that colonize the bowel, rather than the stomach, have been associated with the spontaneous development of colitis in experimental animals, including the cotton topped tamarin *(16)* and mice *(17)*. Eradication of the infection reduces the severity of colitis and will prevent the development of colitis if therapy is provided before the onset of mucosal inflammation *(18)*. Isolation of *Campylobacter* species in fecal specimens of Rhesus macaques with chronic colitis also has been reported *(19)*. However, studies in the human setting have, to date, provided inconsistent results.

The possibility has also been proposed that virulent strains of *Escherichia coli* may be associated with either Crohn's disease or ulcerative colitis *(20)*. Some of the studies support the concept that there are bacteria present adherent to the surface mucosa in affected patients that are not observed in asymptomatic controls. Fluorescent *in situ* hybridization analysis provides a complementary experimental approach to support these observations, which were initially provided using standard culture techniques of luminal materials and intestinal biopsy specimens.

It is possible that other nonculturable organisms are associated with the onset of chronic inflammatory bowel diseases in genetically susceptible persons. As a result, several groups are now undertaking broad range PCR with 16S rDNA as template using tissues obtained from either new-onset disease or surgical resection specimens. The potential for such an approach bearing fruit is provided by other successes using this methodological approach to identify the hepatitis C virus. Such an approach also has been used to identify at least 50 novel, nonculturable microbial species in the oral cavity of humans *(21)*. Whether the organisms are simply nonpathogenic inhabitants of a complex microflora or are opportunistic pathogens, at least in certain circumstances, associated with clinical disease (such as periodontitis), remains to be determined.

16S rRNA templates for PCR have also been used to demonstrate that archaea can inhabit the large intestine of humans. Whether the number and composition of archaea are altered in the intestinal microflora of subjects with chronic inflammatory bowel diseases is the subject of current investigation. This research focus is worthy of pursuit given recent evidence of archaea species identified in subgingival crevices of the oral cavity in 36% of 50 subjects with periodontitis vs none of 31 healthy controls *(22)*. In further support of a role for these archaea species in disease, there was evidence of improvement in clinical periodontitis following eradication therapy.

2.2. Commensal Microflora as an Injurious Agent

Increasingly, it is now recognized that constituents of the normal microbiota can cause damage to susceptible and stressed hosts. For a large variety of murine models of

Table 2
Agents Used as Probiotic Therapy in Humans

Bacteria
Lactobacillus species (acidophilus, rhamnosus, bulgaricus)
Lactococcus lactis
Streptococcus thermophilus
Bifidobacter species (bifidum, breve, infantis)
Escherichia coli, strain Nissle 1917
Fungi
Saccharomyces boulardii

inflammatory bowel disease, mice will only develop inflammation when they are raised in conventional facilities *(23)*. By contrast, animals raised under germ-free conditions do not develop intestinal injury.

We have shown that commensal bacteria, which normally have no effect on gut barrier integrity and function, adhere to the mucosal surface and cause damage when the gut is under stress *(24,25)*. These observations using polarized intestinal epithelia grown in tissue culture have been extended to the in vivo setting using a model of chronic water avoidance stress in rats *(26)*.

There is also the potential that bacterial products, rather than intact viable organisms, induce mucosal inflammation in the genetically susceptible host. For instance, recent evidence suggest that subjects with Crohn's disease have a much greater prevalence of anti-flagellin antibodies in the serum (50%), compared with healthy matched controls (8%). Moreover, adoptive transfer of flagellin-specific CD4[+] T-cells into severe combined immunodeficiency mice induces colitis *(27)*.

Gastrointestinal disorders have been considered here as a paradigm for the identification of microbes as etiological agents in a variety of human disease conditions. These same considerations are also being applied to a variety of disease states related to chronic inflammation, including for example, autoimmune disorders *(28)*.

3. Role of Probiotics

Probiotics is a term that refers to live organisms that are purported to have beneficial effects on health *(29)*. Prebiotics refers to substrates, such as nonabsorbable carbohydrate in the form of fructo-oligosaccharides and inulin, which promote the growth of probiotics *(30)*. Synbiotics is used to describe the approach of combining prebiotics and probiotics with the intention that the combination therapy is better than either approach provided in isolation.

Probiotic agents commonly employed to promote human health or to prevent or treat diseases in humans generally are Gram-positive bacteria resident in the normal colonic microbiota and predominate in the intestinal flora of healthy breast-fed infants (Table 2). The entire genomes of several probiotic bacteria have now been sequenced *(31,32)*.

There is compelling level 1 evidence (that is, results arising from randomized controlled clinical trials) showing that probiotics are effective in treating a variety of diseases affecting the gastrointestinal tract *(33)*. For instance, probiotics have been successfully used to reduce colonization of the stomach by *H. pylori* both in humans *(34)* and in experimental animals *(35)*. Several meta-analyses show that probiotics also reduce the sever-

Table 3
Potential Mechanisms of Action of Probiotics

Enhanced innate immunity
Increased mucin secretion
Enhanced adaptive immunity
Increased sIgA
Reduced NF-κB activation
Altered Th1/Th2/Thr balance
Blocking receptor binding sites
Direct antimicrobial effects
Bacteriocins
Reduced pH

ity and duration of acute virus-induced enteritis in young children *(36,37)*. In addition, probiotics may prevent the onset of infectious diarrhea in a variety of clinical settings, including, for example, antibiotic-associated diarrhea *(38)*, in children attending the day care setting and in the context of travelers' diarrhea.

Gram-negative bacteria have also been employed as a probiotic agent. For instance, *E. coli* strain Nissle 1917 (serotype O6:K5:H1) lacks virulence determinants, such as enterotoxins, encoded by pathogenic *E. coli* strains, but does possess genes encoding bacterial adhesins, including type 1 fimbriae and curli *(39)*. The strain has been used in a variety of experimental studies and clinical trials in humans with apparent efficacy *(40)*. Although there are theoretical concerns about the acquisition of virulence determinants and antimicrobial resistance genes from other organisms in the intestinal flora, such negative outcomes have not been documented.

Nonpathogenic fungi also have been employed as probiotic agents. *Saccharomyces boulardii* has been widely used in the prevention and treatment of antibiotic-induced *Clostridium difficile* pseudomembranous colitis. *S. boulardii* can prevent the damaging effects of enterohemorrhagic *E. coli* O157:H7 infection on polarized epithelia grown in tissue culture *(41)*. There is some debate about whether *S. boulardii* is truly distinct from *S. cerevisiae* used in Baker's yeast *(42)*. Although generally safe, there are reports of fungemia in immunocompromised individuals *(43)*.

Recent interest has focused on the use of a mixture of bacteria as a probiotic cocktail. A mixture of eight different bacteria in concentrations reported as high as 10^{10} live organisms per administration have been employed with success in randomized controlled clinical trials. For instance, in prospective placebo-controlled clinical trials the VSL no. 3 mixture of probiotics (VSL Pharmaceuticals) was used to prevent both the development and the recurrence of pouchitis in the neorectum of subjects with idiopathic ulcerative colitis who have previously undergone sub-total colectomy *(44,45)*.

VSL no. 3 also has been employed to treat experimental colitis in mice deficient in interleukin (IL)-10 *(46)*. The development of colitis in the IL-10 knockout mice also can be prevented by feeding the animals the mixture of probiotics after weaning, before the onset of colitis *(47)*.

There is increasing interest in the mechanism(s) of action underlying the proposed beneficial effects of probiotics (Table 3). Although there is no general consensus *(48)*, current evidence suggests that probiotics exert their effects through multiple mechanisms of action that may well differ amongst different strains and varying species. Some probiotic agents promote innate immune functions; for example, by promoting mucin secretion that prevent pathogen binding to mucosal surfaces (49). Probiotics also can adhere to surface epithelia and compete for available receptor binding sites with pathogenic bacteria. Probiotics also may influence adaptive immunity by promoting production and secretion of polymeric secretory immunoglobulin A and downregulating activation of proinflammatory transcription factors such as nuclear factor-κB (50).

There is increasing interest in the possibility that live organisms might not be absolutely necessary for the observed beneficial effects. For instance, several reports indicate that DNA derived from probiotic bacteria can mimic the effects observed with intact, viable organisms (51,52) Although by definition no longer a probiotic agent, bacterial products, including DNA, offer the potential for use of a product that is cheaper and easier to maintain, store, and transport. Nonviable products also offer, of course, the advantages of avoiding acquired antimicrobial resistance and systemic toxicities. On the other hand, many studies indicate the requirement for viable organisms in order to observe beneficial effects.

Lactococcus lactis has been genetically engineered to produce and secrete biologically active murine forms of IL-10 or trefoil factor. Feeding mice these modified *L. lactis* strains is effective in healing and preventing intestinal injury in murine models of colitis (53,54). Thus, probiotic strains appear to be useful as vehicles for the delivery of anti-inflammatory compounds to sites of injury in the intestinal tract.

It is apparent that probiotic agents may have effects distal to their site of colonization. For instance, both short-term and longer-term prospective randomized studies in infants and toddlers indicate that probiotics can prevent eczematous skin disease in atopic individuals. Verdu and colleagues (55) have shown that *L. paracasei*, but not other probiotic strains, can attenuate muscle hypercontractility in a mouse model of postinfectious irritable bowel syndrome. Intravaginal administration of probiotics has been used to prevent and treat vaginitis and urinary tract infections in women and relevant experimental animal models (56).

Probiotics taken orally colonize the length gastrointestinal tract. However, the colonization is transitory in nature because the organisms are replaced by the resident colonic microflora soon after stopping ingestion. Although this means that probiotics must be taken on a regular (usually daily) basis, it also provides a level of reassurance that the organisms can be removed from the host should untoward or unexpected effects arise.

The previously mentioned discussion has considered the effects of probiotics in humans. There is also great interest in the use of probiotic agents as an alternative to antibiotics in the prevention and treatment of diseases in domesticated animals, fish farming, and the like.

4. Comparative Genomics for the Identification of Virulence Genes

The availability of an increasing number of bacterial genomes now permits analyses undertaken at the whole genome level. For instance, Bansal and Meyer (57) compared the complete genome sequences of 27 bacteria, eight archaea and two eukaryotic species. The study showed a positive relationship between gene content and genome size. Not unexpectedly, smaller genomes were characterized by essential genes. The authors concluded that roughly 2000 genes is the about the minimal size for a free-living organism.

The number of *Fun* (for "function unknown") genes is roughly 20% of open reading frames for most of the prokaryotes sequenced to date *(58)*. Although all such predicted open reading frames may not be transcribed and translated into proteins, it is not yet known whether the hypothetical proteins potentially encoded by at least some of these *Fun* genes are involved either enhancing virulence or in mediating the pathogenesis of disease by pathogenic bacteria.

Eppinger and colleagues *(59)* compared the complete genome sequences available for five organisms from the epsilon subdivision of Proteobacteria, including two strains of *H. pylori, Helicobacter hepaticus, C. jejuni* and the related but nonpathogenic *Wolinella succinogenes*. The study showed that the five bacteria share about one-half of their genes in common. Variations in genes relate to virulence including, for example, genes encoding lipopolysaccharide, flagella, and lipoproteins. Such variability likely accounts for variations in the niche occupied in the infected host (that is, the stomach for *H. pylori* vs the small bowel and large intestine for *H. hepaticus* and *C. jejuni*).

Read and colleagues *(60)* have shown the ability of a genome-based analysis to detect unique polymorphisms between sequenced strains of *Bacillus anthracis*. Such molecular targets could prove helpful, for example, in determining whether outbreaks of infectious disease in the future arise from a common point source. Microbial forensics is a term coined by Cummings and Relman *(61)* to describe the utility of employing comparative genome sequencing to determine the molecular relatedness of bacterial pathogens. Such tools can also be employed, of course, to determine ancestral origin and potential routes of transmission of related strains.

Characterization of microbes in the future is likely to combine assessment of whole genome sequencing with a detailed proteomic analysis *(62)*. Putative virulence genes identified from the genomes sequenced by these analyses must be verified experimentally by the construction of isogenic mutants and testing the constructs using appropriate animal models. Transgenic animals and knockout mice defective in innate immune responses or signal transduction pathways will prove invaluable for such studies (*see* Chapter 4). These complementary analyses are likely to prove fruitful in dissecting mechanisms of virulence and, thereby, provide novel targets for use in the development of strategies that can be employed for interrupting the infectious process.

Acknowledgments

PMS is the recipient of a Canada Research Chair in Gastrointestinal Disease.

References

1. Doolittle, R. F. (1998) Microbial genomes opened up. *Nature* **392,** 339–342.
2. Jenks, P. J. (1998) Sequencing microbial genomes—what will it do for microbiology? *J. Med. Microbiol.* **47,** 375–382.
3. Chan, V. L. (2003) Bacterial genomes and infectious diseases. *Pediatr. Res.* **54,** 1–7.
4. Ussery, D. W. (2004) Genome update: 161 prokaryotic genomes sequenced, and counting. *Microbiology* **150,** 261–263
5. Moran, N. A. (2002) Microbial minimalism: genome reduction in bacterial pathogens. *Cell* **108,** 583–586.
6. Rappe, M. S. and Giovannoni, S. J. (2003) The uncultured microbial majority. *Annu. Rev. Microbiol.* **57,** 369–394.

7. Raoult, D., Birg M. L., La Scola, B., et al. (2000) Cultivation of the bacillus of Whipple's disease. *N. Engl. J. Med.* **342,** 620–625.

8. Bentley, S. D., Malwald, M., Murphy, L. D., et al. (2003) Sequencing and analysis of the genome of the Whipple's disease bacterium *Tropheryma whipplei. Lancet* **361,** 637–644.

9. Maiwald, M., Von Herbay, A., Lepp, P. W., and Relman, D. A. (2000) Organization, structure, and variability of the rRNA operon of the Whipple's disease bacterium (*Tropheryma whipplei*). *J. Bacteriol.* **182,** 3293–3297.

10. Allan, P. (2001) What's the story H pylori? *Lancet* **357,** 694.

11. Peek, R. M. Jr. and Blaser, M. J. (2002) *Helicobacter pylori* and gastrointestinal tract adenocarcinoma. *Nature Rev. Cancer* **2,** 28–37.

12. Houghton, J. M., Stoicov, C., Nomura S., et al. (2004) Gastric cancer originating from bone marrow-derived cells. *Science* **306,** 1568–1571.

13. Lecuit, M., Abachin, E., Martin, A., et al. (2004) Immunoproliferative small intestinal disease associated with *Campylobacter jejuni. N. Engl. J. Med.* **350,** 239–248.

14. Linskens, R. K., Huijsdens, X. W., Savelkoul, P. H. M., Vandenbroucke-Grauls, J. E., and Meuwissen, S. G. M. (2001) The bacterial flora in inflammatory bowel disease: current insight in pathogenesis and the influence of antibiotics and probiotics. *Scand. J. Gastroenterol.* **36(Suppl 234),** 29–40.

15. Selby, W. S. (2004) *Mycobacterium avium* subspecies *paratuberculosis* bacteraemia in patients with inflammatory bowel disease. *Lancet* **364,** 1013–1014.

16. Saunders, K. E., Shen, Z., Dewhirst, F. E., Paster, B. J., Dangler, C. A., and Fox, J. G. (1999) Novel intestinal Helicobacter species isolated from cotton-top tamarins (*Sanguinus oedipus*) with chronic colitis. *J. Clin. Microbiol.* **37,** 146–151.

17. Jiang, H.-Q., Kushnir, N., Thurnheer, M. C., Bos, N. A., and Cebra, J. J. (2002) Monoassociation of SCID mice with *Helicobacter muridarum*, but not four other enterics, provokes IBD upon receipt of T cells. *Gastroenterology* **122,** 1346–1354.

18. Whary, M. T. and Fox, J. G. (2004) Natural and experimental Helicobacter infections. *Comp. Med.* **54,** 128–158.

19. Sestak, K., Merritt, C. K., Borda, J., et al. (2003) Infectious agent and immune response characteristics of chronic enterocolitis in captive rhesus macaques. *Infect. Immun.* **71,** 4079–4086.

20. Campieri, M. and Gionchetti, P. (2001) Bacteria as a cause of ulcerative colitis. *Gut* **48,** 132–135.

21. Kroes, I., Lepp, P. W., and Relman, D. A. (1999) Bacterial diversity within the human subgingival crevice. *Proc. Natl. Acad. Sci. USA* **96,** 14,547–14,552.

22. Lepp, P. W., Brinig, M. M., Ouverney, C. C., Palm, K., Armitage, G. C., and Relman, D. A. (2004) Methanogenic Archaea and human periodontal disease. *Proc. Natl. Acad. Sci. USA* **101,** 6176–6181.

23. Pizarro, T. T., Arseneau, K. O., and Cominelli, F. (2000) Lessons from genetically engineered animal models XI. Novel mouse models to study pathogenic mechanisms of Crohn's disease. *Am. J. Physiol.* **278,** G665–G669.

24. Zareie, M., Singh, P., Irvine, E., Sherman, P., McKay, D., and Perdue, M. (2001) Monocyte/macrophage activation by normal bacteria and bacterial products: implications for epithelial function in Crohn's disease. *Am. J. Pathol.* **158,** 1101–1109.

25. Nazli, A., Yang. P.-C., Jury, J., et al. (2004) Epithelia under metabolic stress perceive commensal bacteria as a threat. *Am. J. Pathol.* **164,** 947–957.

26. Soderholm, J., Yang, P.-C., Ceponis, P., et al. (2002) Chronic stress induces mast cell dependent bacterial adherence and initiates mucosal inflammation in rat intestine. *Gastroenterology* **123,** 1099–1108.

27. Lodes, M. J., Cong, Y., Elson, C. O., et al. (2004) Bacterial flagellin is a dominant antigen in Crohn's disease. *J. Clin. Invest.* **113,** 1296–1306.
28. Waldner, H., Collins, M., and Kuchroo, V. K. (2004) Activation of antigen-presenting cells by microbial products breaks self tolerance and induces autoimmune disease. *J. Clin. Invest.* **113,** 990–997.
29. Agostoni, C., Axelsson, I., Braegger, C., et al. (2004) Probiotic bacteria in dietetic products for infants: a commentary by the ESPGHAN committee on nutrition. *J. Pediatr. Gastroenterol. Nutr.* **38,** 365–374.
30. Agostoni, C., Axelsson, I., Goulet, O., et al. (2004) Prebiotic oligosaccharides in dietetic products for infants: a commentary by the ESPGHAN committee on nutrition. *J. Pediatr. Gastroenterol. Nutr.* **39,** 465–473.
31. Schell, M. A., Karmirantzou, M., Snel, B., et al. (2002) The genome sequence of *Bifidobacterium longum* reflects its adaptation to the human gastrointestinal tract. *Proc. Natl. Acad. Sci. USA* **99,** 14,422–14,227.
32. Boekhorst, J., Siezen, R. J., Zwahlen, M.-C., et al. (2004) The complete genomes of *Lactobacillus plantarum* and *Lactobacillus johnsonii* reveal extensive differences in chromosome organization and gene content. *Microbiology* **150,** 3601–3611.
33. Servin, A. L. (2004) Antagonistic activities of lactobacilli and bifidobacteria against microbial pathogens. *FEMS Microbiol. Rev.* **28,** 40–440.
34. Hamilton-Miller, J. M. (2003) The role of probiotics in the treatment and prevention of *Helicobacter pylori* infection. *Int. J. Antimicrob. Agents* **22,** 360–366.
35. Johnson-Henry, K., Mitchell, D. J., Avitzur, Y., Galindo-Mata, E., Jones, N. L., and Sherman, P. M. (2004) Probiotics reduce bacterial colonization and gastric inflammation in *H. pylori*-infected mice. *Dig. Dis. Sci.* **49,** 1095–1102.
36. Van Niel, C. W., Feudtner, C., Garrison, M. M., and Chritakis, D. A. (2002) Lactobacillus therapy for acute infectious diarrhea in children: a meta-analysis. *Pediatrics* **109,** 678–684.
37. Huang, J. S., Bousvaros, A., Lee, J. W., Diaz, A., and Davidson, E. J. (2002) Efficacy of probiotic use in acute diarrhea in children: a meta-analysis. *Dig. Dis. Sci.* **47,** 2625–2634.
38. Cremonini, F., Di Caro, S., Nista, E. C., et al. (2002) Meta-analysis: the effect of probiotic administration on antibiotic-associated diarrhoea. *Aliment. Pharmacol. Therap.* **16,** 1461–1467.
39. Grozdanov, L., Raasch, C., Schulze, J., et al. (2004) Analysis of the genome structure of the nonpathogenic probiotic *Escherichia coli* strain Nissle 1917. *J. Bacteriol.* **186,** 5432–5441.
40. Kruis, W., Fric, P., Pokrotnieks, J., et al. (2004) Maintaining remission of ulcerative colitis with the probiotic *Escherichia coli* Nissle 1917 is as effective as with standard mesalazine. *Gut* **53,** 1617–1623.
41. Dahan, S., Dalmasso, G., Imbert, V., Peyron, J. F., Rampal, P., and Czerucka, D. (2003) *Saccharomyces boulardii* interferes with enterohemorrhagic *Escherichia coli*-induced signaling pathways in T84 cells. *Infect. Immun.* **71,** 766–733.
42. Stevens, D. A. (2001) Saccharomyces and enteropathogenic *Escherichia coli*. *Infect. Immun.* **69,** 4192.
43. Cassone, M., Serra, P., Mondello, F., et al. (2003) Outbreak of *Saccharomyces cervisiae* subtype *boulardii* fungemia in patients neighboring those treated with a probiotic preparation of the organism. *J. Clin. Microbiol.* **41,** 5340–5343.
44. Gionchetti, P., Rizzello, F., Venturi, A., et al. (2000) Oral bacteriotherapy as maintenance treatment in patients with chronic pouchitis: a double-blind, placebo-controlled trial. *Gastroenterology* **119,** 305–309.
45. Gionchetti, P., Rizello, F., Helwig, U., et al. (2003) Prophylaxis of pouchitis onset with probiotic therapy: a double-blind, placebo-controlled trial. *Gastroenterology* **124,** 1202–1209.

46. Madsen, K. L., Doyle, J. S., Jewell, L. D., Tavernini, M. M., and Fedorak, R. N. (1999) Lactobacillus species prevents colitis in interleukin-10 gene-deficient mice. *Gastroenterology* **116,** 1107–1114.

47. Madsen, K., Cornish, A., Soper, P., et al. (2001) Probiotic bacteria enhance murine and human intestinal epithelial barrier function. *Gastroenterology* **121,** 580–591.

48. Abbott, A. (2004) Gut reaction. *Nature* **427,** 284–286.

49. Mack, D. R., Michail, S., Wei, S., McDougall, L., and Hollingsworth, M. A. (1999) Probiotics inhibit enteropathogenic *Escherichia coli* adherence *in vitro* by inducing intestinal mucin gene expression. *Am. J. Physiol.* **276,** G941–G950.

50. Petrof, E. O., Kojima, K., Ropeleski, M. J., et al. (2004) Probiotics inhibit nuclear factor-kB and induce heat shock proteins in colonic epithelial cells through proteasome inhibition. *Gastroenterology* **127,** 1474–1487.

51. Rachmilewitz, D., Katakura, K., Karmeli, F., et al. (2004) Toll-like receptor 9 signaling mediates the anti-inflammatory effects of probiotics in murine experimental colitis. *Gastroenterology* **126,** 520–528.

52. Jijon, H., Backer, J., Diaz, H., et al. (2004) DNA from probiotic bacteria modulates murine and human epithelial and immune function. *Gastroenterology* **126,** 1358–1373.

53. Steidler, L., Hans, W., Schotte, L., et al. (2000) Treatment of murine colitis by *Lactococcus lactis* secreting interleukin-10. *Science* **289,** 1352–1355.

54. Vandenbroucke, K., Hans, W., Van Huysse, J., et al. (2004) Active delivery of trefoil factors by genetically modified *Lactococcus lactis* prevents and heals acute colitis in mice. *Gastroenterology* **127,** 502–513.

55. Verdue, E. F., Bercik, P., Bergonzelli, G. E., et al. (2004) *Lactobacillus paracasei* normalizes muscle hypercontractility in a murine model of postinfective gut dysfunction. *Gastroenterology* **127,** 826–837.

56. Reid, G., Jass, J., Sebulsky, M. T., and McCormick, J. K. (2003) Potential uses of probiotics in clinical practice. *Clin. Microbiol. Rev.* **16,** 658–672.

57. Bansal, A. K. and Meyer, T. E. (2002) Evolutionary analysis by whole-genome comparisons. *J. Bacteriol.* **184,** 2260–2272.

58. Doolittle, F. R. (2002) Microbial genomes multiply. *Nature* **416,** 697–700.

59. Eppinger, M., Baar, C., Raddatz, G., Huson, D. H., and Schuster, S. C. (2004) Comparative analysis of four Campylobacterales. *Nature Rev. Microbiol.* **2,** 872–885.

60. Read, T. D., Salzberg, S. L., Pop, M., et al. (2002) Comparative genome sequencing for discovery of novel polymorphisms in *Bacillus anthracis*. *Science* **296,** 2028–2033.

61. Cummings, C. A. and Relman, D. A. (2002) Microbial forensics—"cross-examining pathogens." *Science* **296,** 1976–1979

62. Boyce, J. D., Cullen, P. A., and Adler, B. (2004) Genomic-scale analysis of bacterial gene and protein expression in the host. *Emerg. Infect. Dis.* **10,** 1357–1362.

Index